# Introduction to Electrical Power Systems

**Books in the IEEE Press Series on Power Engineering**

*Principles of Electric Machines with Power Electronic Applications,* Second Edition
M.E. El-Hawary

*Pulse Width Modulation for Power Converters: Principles and Practice*
D. Grahame Holmes and Thomas Lipo

*Analysis of Electric Machinery and Drive Systems,* Second Edition
Paul C. Krause, Oleg Wasynczuk, and Scott D. Sudhoff

*Risk Assessment for Power Systems: Models, Methods, and Applications*
Wenyuan Li

*Optimization Principles: Practical Applications to the Operations of Markets of the Electric Power Industry*
Narayan S. Rau

*Electric Economics: Regulation and Deregulation*
Geoffrey Rothwell and Tomas Gomez

*Electric Power Systems: Analysis and Control*
Fabio Saccomanno

*Electrical Insulation for Rotating Machines: Design, Evaluation, Aging, Testing, and Repair*
Greg Stone, Edward A. Boulter, Ian Culbert, and Hussein Dhirani

*Signal Processing of Power Quality Disturbances*
Math H. J. Bollen and Irene Y. H. Gu

*Instantaneous Power Theory and Applications to Power Conditioning*
Hirofumi Akagi, Edson H. Watanabe and Mauricio Aredes

*Maintaining Mission Critical Systems in a 24/7 Environment*
Peter M. Curtis

*Elements of Tidal-Electric Engineering*
Robert H. Clark

*Handbook of Large Turbo-Generator Operation and Maintenance,* Second Edition
Geoff Klempner and Isidor Kerszenbaum

*Introduction to Electrical Power Systems*
Mohamed E. El-Hawary

# Introduction to Electrical Power Systems

**Mohamed E. El-Hawary**

IEEE PRESS SERIES ON POWER ENGINEERING
Mohamed E. El-Hawary, *Series Editor*

IEEE

**IEEE Press**

WILEY

A JOHN WILEY & SONS, INC., PUBLICATION

**IEEE Press**
445 Hoes Lane
Piscataway, NJ 08855





Published by John Wiley & Sons, Inc., Hoboken, New Jersey
Published simultaneously in Canada.



For general information on our other products and services please contact our Customer Care Department within the United States at (800) 762-2974, outside the United States at (317) 572-3993 or fax (317) 572-4002.

Wiley also publishes its books in a variety of electronic formats. Some content that appears in print, however, may not be available in electronic formats. For more information about Wiley products, visit our web site at www.wiley.com.

***Library of Congress Cataloging-in-Publication Data is available.***

ISBN 978-0470-40863-6

10 9 8 7 6 5 4 3 2

# CONTENTS

# PREFACE

This book is written primarily as an introduction to the basics of electrical power systems. It is intended as a general introduction to the area for students in all engineering disciplines, as well as being useful as a reference and self-study guide for those professionals who wish to have a succinct introduction to this important area. The coverage of the book is designed to allow its use in a number of ways including service courses taught to non-electrical majors. The organization and details of the material in this book enables maximum flexibility for the instructor to select topics to include in courses within the modern engineering curriculum.

The book does not assume a level of mathematical sophistication beyond that acquired through completion of undergraduate courses in basic physics and introductory electric circuits. Emphasis is placed on providing an improved appreciation of the operational characteristics of the electrical apparatus discussed, on the basis of linear mathematical models. Almost every key concept is illustrated through the use of in-text examples that are worked out in detail to enforce the reader's understanding. The text coverage includes some usage of MATLAB™ to solve fundamental problems of evaluating basic performance characteristics to assist in understanding power system devices.

The first chapter in this book offers a historical perspective on the t are development of electric power systems. While this topic is not an integral part of conventional coverage in texts and courses in this area, this chapter should provide interesting insights into the influence of these developments on present day civilization. It is through an appreciation of the past developments and achievements that we can understand our present and forge ahead with future advances.

Much of the material in Chapters 2 to 7 is derived from the more detailed expose' found in my earlier book on "Electrical Power System Design and Analysis. The present treatment deals with fundamental topics to be covered in introductory courses in electric power systems. Emphasis is given to practical aspects such as the main performance characteristics of the apparatus discussed and system applications.

In Chapter 2, a review of fundamentals of electric circuits that are necessary for the studies to follow. Chapter 3 discusses the basics of synchronous machines for steady state operations. This is followed by treatments of the transformer in Chapter 4 and power transmission lines in Chapter 5. Unlike the coverage of the more detailed "Electric Power Systems: Design and Analysis," Chapter 6 of the present book deals with Induction Motors both polyphase and single phase machines. In addition, Chapter 7 combines both fault analysis and protection.

The importance of computer control in power system operations is highlighted in Chapter 8 where we discuss the structure and functions involved in a modern energy control center. The chapter includes a brief introduction to functions performed in the electric energy control centre. Naturally some functions that are discussed in detail in "Electric Power Systems: Design and Analysis" such as Power Flow, Stability, optimal operation of power systems, are discussed briefly in this chapter.

Chapter 9 is new to this book, and offers a brief discussion of the Present and Future of Electric Energy Systems. Of specific interest is a discussion of the challenges facing the system designer and operator in view of blackout events that took place during the first decade of the 21st Century.

I have attempted to make this book as self-containing as possible. As a result, the reader will find that many background topics such as the per unit system and three-phase circuits are included in the text's main body as opposed to the recent trend toward including many appendices dealing with these topics. In studying and teaching electrical power systems it has been my experience that a problem solving approach is most effective in exploring this rich area.

A textbook such as this could not have been written without the continuing input of the many students who have gone through many versions of its material as it was developed. I wish to thank the members of the many classes to whom I was privileged to teach this fascinating subject. I wish to acknowledge the able work of Elizabeth Sanford of Dalhousie University in putting this manuscript in a better form than I was able to produce.

As has always been the case, the patience and understanding of my wife, Dr. Ferial El-Hawary, made this project another joy to look forward to completing. It goes without saying that the younger generation of our family deserves a greater share of my appreciation for their continuous understanding.

MOHAMED E. EL-HAWARY

Halifax, Nova Scotia
May 1, 2008

# Chapter 1

# INTRODUCTION

This chapter has three objectives. We first offer a brief perspective on the development of electric power systems. This is not intended to be a detailed historical review, but rather it uses historical landmarks as a background to highlight the features and structure of the modern power systems, which are discussed in Section 1.2. The chapter concludes with an outline of the book.

## *1.1 A BRIEF HISTORY OF ELECTRIC POWER SYSTEMS*

The electric power industry shapes and contributes to the welfare, progress, and technological advances of humanity. The growth of electric energy consumption in the world has been phenomenal. In the United States, for example, electric energy sales have grown to well over 400 times in the period between the turn of the century and the early 1970s. This growth rate was 50 times as much as the growth rate in all other energy forms used during the same period. The installed kW capacity per capita in the U.S. is estimated to be close to 3 kW.

Edison Electric Illuminating Company of New York inaugurated the Pearl Street Station in 1881. The station had a capacity of four 250-hp boilers supplying steam to six engine-dynamo sets. Edison's system used a 110-V dc underground distribution network with copper conductors insulated with a jute wrapping. In 1882, the first water wheel-driven generator was installed in Appleton, Wisconsin. The *low voltage of the circuits* limited the service area of a central station, and consequently, central stations proliferated throughout metropolitan areas.

The invention of the transformer, then known as the "inductorium," made ac systems possible. The first practical ac distribution system in the U.S. was installed by W. Stanley at Great Barrington, Massachusetts, in 1866 for Westinghouse, who acquired the American rights to the transformer from its British inventors Gaulard and Gibbs. Early ac distribution utilized 1000-V overhead lines. The Nikola Tesla invention of the induction motor in 1888 helped replace dc motors and hastened the advance in use of ac systems.

The first American single-phase ac system was installed in Oregon in 1889. Southern California Edison Company established the first three phase 2.3 kV system in 1893.

By 1895, Philadelphia had about twenty electric companies with distribution systems operating at 100-V and 500-V two-wire dc and 220-V three-wire dc, single-phase, two-phase, and three-phase ac, with frequencies of 60, 66, 125, and 133 cycles per second, and feeders at 1000-1200 V and 2000-2400 V.

The subsequent consolidation of electric companies enables the realization of economies of scale in generating facilities, the introduction of equipment standardization, and the utilization of the load diversity between areas. Generating unit sizes of up to 1300 MW are in service, an era that was started by the 1973 Cumberland Station of the Tennessee Valley Authority.

Underground distribution at voltages up to 5 kV was made possible by the development of rubber-base insulated cables and paper-insulated, lead-covered cables in the early 1900s. Since then, higher distribution voltages have been necessitated by load growth that would otherwise overload low-voltage circuits and by the requirement to transmit large blocks of power over great distances. Common distribution voltages presently are in 5-, 15-, 25-, 35-, and 69-kV voltage classes.

The growth in size of power plants and in the higher voltage equipment was accompanied by interconnections of the generating facilities. These interconnections decreased the probability of service interruptions, made the utilization of the most economical units possible, and decreased the total reserve capacity required to meet equipment-forced outages. This was accompanied by use of sophisticated analysis tools such as the network analyzer. Central control of the interconnected systems was introduced for reasons of economy and safety. The advent of the load dispatcher heralded the dawn of power systems engineering, an exciting area that strives to provide the best system to meet the load requirements reliably, safely, and economically, utilizing state-of-the-art computer facilities.

Extra higher voltage (EHV) has become dominant in electric power transmission over great distances. By 1896, an 11-kv three-phase line was transmitting 10 MW from Niagara Falls to Buffalo over a distance of 20 miles. Today, transmission voltages of 230 kV, 287 kV, 345 kV, 500 kV, 735 kV, and 765 kV are commonplacc, with the first 1100-kV line already energized in the early 1990s. The trend is motivated by economy of scale due to the higher transmission capacities possible, more efficient use of right-of-way, lower transmission losses, and reduced environmental impact.

In 1954, the Swedish State Power Board energized the 60-mile, 100-kV dc submarine cable utilizing U. Lamm's Mercury Arc valves at the sending and receiving ends of the world's first high-voltage direct current (HVDC) link connecting the Baltic island of Gotland and the Swedish mainland. Currently, numerous installations with voltages up to 800-kV dc are in operation around the world.

In North America, the majority of electricity generation is produced by investor-owned utilities with a certain portion done by federally and provincially (in Canada) owned entities. In the United States, the Federal Energy Regulatory Commission (FERC) regulates the wholesale pricing of electricity and terms and conditions of service.

The North American transmission system is interconnected into a large power grid known as the North American Power Systems Interconnection. The grid is divided into several pools. The pools consist of several neighboring utilities which operate jointly to schedule generation in a cost-effective manner. A privately regulated organization called the North American Electric Reliability Council (NERC) is responsible for maintaining system standards and reliability. NERC works cooperatively with every provider and distributor of power to ensure reliability. NERC coordinates its efforts with FERC as well as other organizations such as the Edison Electric Institute (EEI). NERC currently has four distinct electrically separated areas. These areas are the Electric Reliability Council of Texas (ERCOT), the Western States Coordination Council (WSCC), the Eastern Interconnect, which includes all the states and provinces of Canada east of the Rocky Mountains (excluding Texas), and Hydro-Quebec. These electrically separate areas exchange with each other but are not synchronized electrically.

The electric power industry in the United States has undergone fundamental changes since the deregulation of the telecommunication, gas, and other industries. The generation business is rapidly becoming market-driven. The power industry was characterized by larger, vertically integrated entities. The advent of open transmission access has resulted in wholesale and retail markets. Utilities may be divided into power generation, transmission, and retail segments. Generating companies (GENCO) sell directly to an independent system operator (ISO). The ISO is responsible for the operation of the grid and matching demand and generation dealing with transmission companies as well (TRANSCO). This scenario is not the only possibility, as the power industry continues to evolve to create a more competitive environment for electricity markets to promote greater efficiency. The industry now faces new challenges and problems associated with the interaction of power system entities in their efforts to make crucial technical decisions while striving to achieve the highest level of human welfare.

## 1.2 THE STRUCTURE OF THE POWER SYSTEM

An interconnected power system is a complex enterprise that may be subdivided into the following major subsystems:

- Generation Subsystem
- Transmission and Subtransmission Subsystem
- Distribution Subsystem
- Utilization Subsystem

### Generation Subsystem

This includes generators and transformers.

***Generators*** – An essential component of power systems is the three-phase ac generator known as synchronous generator or alternator. Synchronous generators have two synchronously rotating fields: One field is produced by the rotor driven at synchronous speed and excited by dc current. The other field is produced in the stator windings by the three-phase armature currents. The dc current for the rotor windings is provided by excitation systems. In the older units, the exciters are dc generators mounted on the same shaft, providing excitation through slip rings. Current systems use ac generators with rotating rectifiers, known as *brushless* excitation systems. The excitation system maintains generator voltage and controls the reactive power flow. Because they lack the commutator, ac generators can generate high power at high voltage, typically 30 kV.

The source of the mechanical power, commonly known as the prime mover, may be hydraulic turbines, steam turbines whose energy comes from the burning of coal, gas and nuclear fuel, gas turbines, or occasionally internal combustion engines burning oil.

Steam turbines operate at relatively high speeds of 3600 or 1800 rpm. The generators to which they are coupled are cylindrical rotor, two-pole for 3600 rpm, or four-pole for 1800 rpm operation. Hydraulic turbines, particularly those operating with a low pressure, operate at low speed. Their generators are usually a salient type rotor with many poles. In a power station, several generators are operated in parallel in the power grid to provide the total power needed. They are connected at a common point called a *bus*.

With concerns for the environment and conservation of fossil fuels, many alternate sources are considered for employing the untapped energy sources of the sun and the earth for generation of power. Some alternate sources used are solar power, geothermal power, wind power, tidal power, and biomass.

***Transformers*** – The transformer transfers power with very high efficiency from one level of voltage to another. The power transferred to the secondary is almost the same as the primary, except for losses in the transformer. Using a step-up transformer will reduce losses in the line, which makes the transmission of power over long distances possible.

Insulation requirements and other practical design problems limit the generated voltage to low values, usually 30 kV. Thus, step-up transformers are used for transmission of power. At the receiving end of the transmission lines step-down transformers are used to reduce the voltage to suitable values for distribution or utilization. The electricity in an electric power system may undergo four or five transformations between generator and consumers.

## Transmission and Subtransmission Subsystem

An overhead transmission network transfers electric power from generating units to the distribution system which ultimately supplies the load. Transmission lines also interconnect neighboring utilities which allow the economic dispatch of power within regions during normal conditions, and the transfer of power between regions during emergencies.

Standard transmission voltages are established in the United States by the American National Standards Institute (ANSI). Transmission voltage lines operating at more than 60 kV are standardized at 69 kV, 115 kV, 138 kV, 161 kV, 230 kV, 345 kV, 500 kV, and 765 kV line-to-line. Transmission voltages above 230 kV are usually referred to as extra-high voltage (EHV).

High voltage transmission lines are terminated in substations, which are called *high-voltage substations*, *receiving substations*, or *primary substations*. The function of some substations is switching circuits in and out of service; they are referred to as switching stations. At the primary substations, the voltage is stepped down to a value more suitable for the next part of the flow toward the load. Very large industrial customers may be served directly from the transmission system.

The portion of the transmission system that connects the high-voltage substations through step-down transformers to the distribution substations is called the subtransmission network. There is no clear distinction between transmission and subtransmission voltage levels. Typically, the subtransmission voltage level ranges from 69 to 138 kV. Some large industrial customers may be served from the subtransmission system. Capacitor banks and reactor banks are usually installed in the substations for maintaining the transmission line voltage.

### Distribution Subsystem

The distribution system connects the distribution substations to the consumers' service-entrance equipment. The primary distribution lines range from 4 to 34.5 kV and supply the load in a well-defined geographical area. Some small industrial customers are served directly by the primary feeders.

The secondary distribution network reduces the voltage for utilization by commercial and residential consumers. Lines and cables not exceeding a few hundred feet in length then deliver power to the individual consumers. The secondary distribution serves most of the customers at levels of 240/120 V, single-phase, three-wire; 208Y/120 V, three-phase, four-wire; or 480Y/277 V, three-phase, four-wire. The power for a typical home is derived from a transformer that reduces the primary feeder voltage to 240/120 V using a three-wire line.

Distribution systems utilize both *overhead* and *underground* conductors. The growth of underground distribution has been extremely rapid and as much as 70 percent of new residential construction in North America is via underground systems.

## Load Subsystems

Power systems loads are divided into industrial, commercial, and residential. Industrial loads are composite loads, and induction motors form a high proportion of these loads. These composite loads are functions of voltage and frequency and form a major part of the system load. Commercial and residential loads consist largely of lighting, heating, and cooking. These loads are independent of frequency and consume negligibly small reactive power.

The load varies throughout the day, and power must be available to consumers on demand. The daily-load curve of a utility is a composite of demands made by various classes of users. The greatest value of load during a 24-hr period is called the peak or *maximum demand.* To assess the usefulness of the generating plant the *load factor* is defined. The load factor is the ratio of average load over a designated period of time to the peak load occurring in that period. Load factors may be given for a day, a month, or a year. The yearly, or annual load factor is the most useful since a year represents a full cycle of time. The daily load factor is

$$\text{Daily L.F.} = \frac{\text{average load}}{\text{peak load}} \tag{1.1}$$

Multiplying the numerator and denominator of (1.1) by a time period of 24 hr, we obtain

$$\text{Daily L.F.} = \frac{\text{average load} \times 24\ \text{hr}}{\text{peak load} \times 24\ \text{hr}} = \frac{\text{energy consumed during 24 hr}}{\text{peak load} \times 24\ \text{hr}} \tag{1.2}$$

The annual load factor is

$$\text{Annual L.F.} = \frac{\text{total annual energy}}{\text{peak load} \times 8760\ \text{hr}} \tag{1.3}$$

Generally there is diversity in the peak load between different classes of loads, which improves the overall system load factor. In order for a power plant to operate economically, it must have a high system load factor. Typical system load factors are in the range of 55 to 70 percent. Load-forecasting at all levels is an important function in the operation, operational planning, and planning of an electric power system. Other devices and systems are required for the satisfactory operation and protection of a power system. Some of the protective devices directly connected to the circuits are called *switchgear*. They include instrument transformers, circuit breakers, disconnect switches, fuses and lightning arresters. These devices are necessary to deenergize either for normal operation or on the occurrence of faults. The associated control equipment and protective relays are placed on *switchboards* in *control houses*.

For reliable and economical operation of the power system it is necessary to monitor the entire system in a so-called control center. The modern control center is called the *energy control center* (ECC). Energy control centers are equipped with on-line computers performing all signal processing through the remote acquisition system. Computers work in a hierarchical structure to properly coordinate different functional requirements in normal as well as emergency conditions. Every energy control center contains control consoles which consist of a visual display unit (VDU), keyboard, and light pen. Computers may give alarms as advance warnings to the operators (dispatchers) when deviation from the normal state occurs. The dispatcher makes decisions and executes them with the aid of a computer. Simulation tools and software packages are implemented for efficient operation and reliable control of the system. In addition, SCADA, an acronym for "supervisory control and data acquisition," systems are auxiliaries to the energy control center.

## 1.3 OUTLINE OF THE TEXT

Chapter 2 lays the foundations for the development in the rest of the book. The intention of the discussion offered here is to provide a brief review of fundamentals including power concepts, three-phase systems, principles of electromagnetism, and electromechanical energy conversion. Chapter 3 treats the synchronous machine from an operational modeling point of view. Emphasis here is on performance characteristics of importance to the electric power specialist. Chapter 4 provides a comprehensive treatment of transformers. This is followed, in Chapter 5, by a brief coverage of induction motors including the fractional horsepower category.

Chapter 6 is concerned with transmission lines starting from parameter evaluation for different circuit and conductor configurations. Various transmission line performance modeling approaches are covered.

Faults on electric energy systems are considered in Chapter 7. Here we start with the transient phenomenon of a symmetrical short circuit, followed by a treatment of unbalanced and balanced faults. Realizing the crucial part that system protection plays in maintaining service integrity is the basis for the remainder of this chapter. Here an introduction to this important area is given.

Chapter 8 is concerned with the Energy Control Center, its structure, and role in the operation of a modern power system. We outline the objectives and aims of many of the decision support functions adopted in these significant "smarts" of the power system. In chapter 9, we focus our attention on a number of technological challenges that face power system planners and operators. The major North East blackout of 8/14 in 2003 is discussed along with possible mitigation measures. Wherever relevant, we introduce MATLAB™ scripts that allow the student to automate many of the computational details. This feature is deemed important for this textbook's coverage.

# Chapter 2

# BASICS OF ELECTRIC ENERGY SYSTEM THEORY

## 2.1 INTRODUCTION

This chapter lays the groundwork for the study of electric energy systems. We develop some basic tools involving fundamental concepts, definitions, and procedures. The chapter can be considered as simply a review of topics utilized throughout this work. We start by introducing the principal electrical quantities.

## 2.2 CONCEPTS OF POWER IN ALTERNATING CURRENT SYSTEMS

Generally, the electric power systems specialist is more concerned with electric power in the circuit rather than the currents. The value of instantaneous power flowing into an element is the product of voltage across and current through it. It seems, then, reasonable to exchange the current for power without losing any information. In treating sinusoidal steady-state behavior of circuits, some further definitions are necessary. To illustrate the concepts, we use a cosine representation of the waveforms.

Consider the impedance element $Z = Z\angle\phi$. For a sinusoidal voltage, $\upsilon(t)$ given by

$$\upsilon(t) = V_m \cos \omega t$$

The instantaneous current in the circuit is

$$i(t) = I_m \cos(\omega t - \phi)$$

Here

$$I_m = V_m / |Z|$$

The instantaneous power into the element is given by

$$p(t) = \upsilon(t) i(t) = V_m I_m [\cos(\omega t)\cos(\omega t - \phi)]$$

This reduces to

$$p(t) = \frac{V_m I_m}{2}[\cos\phi + \cos(2\omega t - \phi)]$$

Since the average of cos ($2\omega t$ - $\phi$) is zero, through 1 cycle, this term therefore contributes nothing to the average of $p$, and the average power $p_{\text{av}}$ is given by

$$p_{\text{av}} = \frac{V_m I_m}{2} \cos\phi \qquad \textbf{(2.1)}$$

Using the effective (rms) values of voltage and current and substituting $V_m = \sqrt{2}(V_{\text{rms}})$, and $I_m = \sqrt{2}(I_{\text{rms}})$, we get

$$p_{\text{av}} = V_{\text{rms}} I_{\text{rms}} \cos\phi \qquad \textbf{(2.2)}$$

The power entering any network is the product of the effective values of terminal voltage and current and the cosine of the phase angle $\phi$, which is, called the *power factor* (PF). This applies to sinusoidal voltages and currents only. When reactance and resistance are present, a component of the current in the circuit is engaged in conveying the energy that is periodically stored in and discharged from the reactance. This stored energy, being shuttled to and from the magnetic field of an inductance or the electric field of a capacitance, adds to the current in the circuit but does not add to the average power.

The value of the average power consumed in a circuit is called *active power*, and the power that supplies the stored energy in reactive elements is called *reactive power*. Active power is $P$, and the reactive power, designated $Q$, are thus*

$$P = VI \cos\phi \qquad \textbf{(2.3)}$$

$$Q = VI \sin\phi \qquad \textbf{(2.4)}$$

In both equations, $V$ and $I$ are rms values of terminal voltage and current, and $\phi$ is the phase angle by which the current lags the voltage.

To emphasize that the $Q$ represents the nonactive power, it is measured in reactive voltampere units (var).

---

* If we write the instantaneous power as

$$p(t) = V_{\text{rms}} I_{\text{rms}} [\cos\phi(1 + \cos 2\omega t)] + V_{\text{rms}} I_{\text{rms}} \sin\phi \sin 2\omega t$$

then it is seen that

$$p(t) = P(1 + \cos 2\omega t) + Q \sin 2\omega t$$

Thus $P$ and $Q$ are the average power and the amplitude of the pulsating power, respectively.

Figure 2.1 shows the time variation of the various variables discussed.

Assume that $V$, $V\cos\phi$, and $V\sin\phi$, all shown in Figure 2.2, are each multiplied by $I$, the rms values of current. When the components of voltage $V\cos\phi$ and $V\sin\phi$ are multiplied by current, they become $P$ and $Q$, respectively. Similarly, if $I$, $I\cos\phi$, and $I\sin\phi$ are each multiplied by $V$, they become $VI$, $P$, and $Q$, respectively. This defines a power triangle.

We now define a quantity called the *complex* or *apparent power*, designated $S$, of which $P$ and $Q$ are components. By definition,

$$\begin{aligned} S &= P + jQ \\ &= VI(\cos\phi + j\sin\phi) \end{aligned}$$

Using Euler's identity, we thus have

$$S = VIe^{j\phi}$$

or

$$S = VI\angle\phi$$

It is clear that an equivalent definition of complex or apparent power is

$$S = VI^* \tag{2.5}$$

We can write the complex power in two alternative forms by using the relationships

$$V = ZI \qquad \text{and} \qquad I = YV$$

This leads to

$$S = ZII^* = Z|I|^2 \tag{2.6}$$

or

$$S = VY^*V^* = Y^*|V|^2 \tag{2.7}$$

Consider the series circuit shown in Figure 2.3. The applied voltage to the overall arrangement is equal to the sum of the voltage drops:

$$V = I(Z_1 + Z_2 + \ldots + Z_n)$$

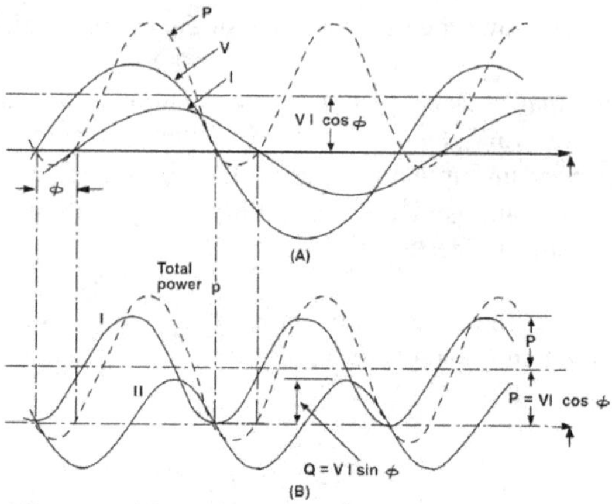


**Figure 2.1** Voltage, Current, and Power in a Single-Phase Circuit.

**Figure 2.2** Phasor Diagrams Leading to Power Triangles.

Multiplying both sides of this relation by $I^*$ results in

$$S = \sum_{i=1}^{n} S_i \tag{2.8}$$

with the individual element's complex power.

$$S_i = |I|^2 Z_i \tag{2.9}$$

Equation (2.8) is known as the summation rule for complex powers. The rule also applies to parallel circuits.

The phasor diagram shown in Figure 2.2 can be converted into complex power diagrams by simply following the definitions relating complex power to voltage and current.

Consider an inductive circuit in which the current lags the voltage by the angle $\phi$. The conjugate of the current will be in the first quadrant in the complex plane as shown in Figure 2.4(a). Multiplying the phasors by **V**, we obtain the complex power diagram shown in Figure 2.4(b). Inspection of the diagram as well as the previous development leads to a relation for the power factor of the circuit:

$$\cos\phi = \frac{P}{|S|}$$

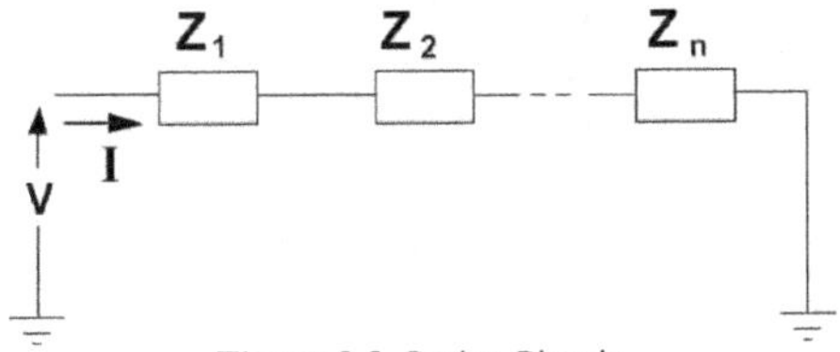


**Figure 2.3** Series Circuit.

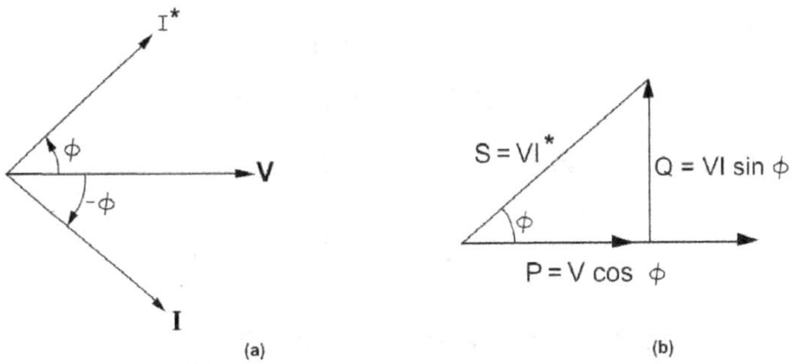


**Figure 2.4** Complex Power Diagram

### Example 2.1

Consider the circuit composed of a series *R-L* branch in parallel with capacitance with the following parameters:

$$R = 0.5 \text{ ohms}$$
$$X_L = 0.8 \text{ ohms}$$
$$B_c = 0.6 \text{ siemens}$$

Assume that:

$$V = 200\angle 0 \text{ V}$$

Calculate the input current and the active, reactive, and apparent power into the circuit.

*Solution*

The current into the *R-L* branch is given by

$$I_Z = \frac{200}{0.5 + j0.8} = 212\angle -57.99^\circ \text{ A}$$

The power factor (PF) of the *R-L* branch is

$$\begin{aligned} \text{PF}_Z &= \cos\phi_Z = \cos 57.99^\circ \\ &= 0.53 \end{aligned}$$

The current into the capacitance is

$$I_c = j(0.6)(200) = 120\angle 90^\circ \text{ A}$$

The input current $I_t$ is

$$\begin{aligned} I_t &= I_c + I_Z \\ &= 212\angle -57.99^\circ + 120\angle 90^\circ \\ &= 127.28\angle -28.01^\circ \end{aligned}$$

The power factor (PF) of the overall circuit is

$$\text{PF}_t = \cos\phi_t = \cos 28.01^\circ = 0.88$$

Note that the magnitude of $I_t$ is less than that of $I_Z$, and that $\cos\phi_t$ is higher than $\cos\phi_Z$. This is the effect of the capacitor, and its action is called *power factor correction* in power system terminology.

The apparent power into the circuit is

$$\begin{aligned} S_t &= VI_t^* \\ &= (200\angle 0)(127.28)\angle 28.01^\circ \\ &= 25{,}456.00\angle 28.01^\circ \text{ VA} \end{aligned}$$

In rectangular coordinates we get

$$S_t = 22{,}471.92 + j11{,}955.04$$

Thus, the active and reactive powers are

$$\begin{aligned} P_t &= 22{,}471.92 \text{ W} \\ Q_t &= 11{,}955.04 \text{ var} \end{aligned}$$

The major portion of all electric power presently used in generation, transmission, and distribution uses balanced three-phase systems. Three-phase operation makes more efficient use of generator copper and iron. Power flow in

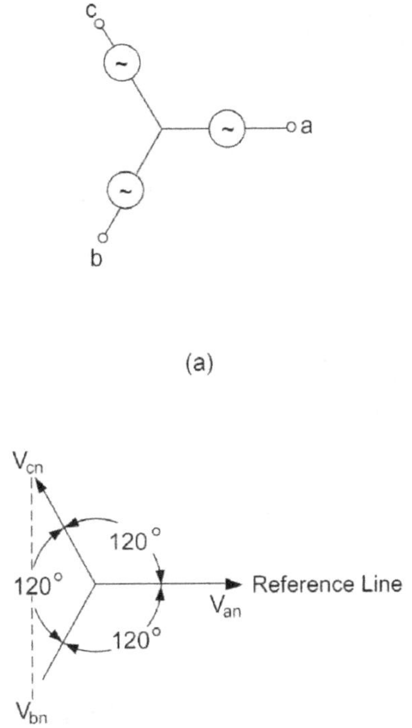


**Figure 2.5** A Y-Connected Three-Phase System and the Corresponding Phasor Diagram.

single-phase circuits was shown in the previous section to be pulsating. This drawback is not present in a three-phase system. Also, three-phase motors start more conveniently and, having constant torque, run more satisfactorily than single-phase motors. However, the complications of additional phases are not compensated for by the slight increase of operating efficiency when polyphase systems other than three-phase are used.

A balanced three-phase voltage system consists of three single-phase voltages with the same magnitude and frequency but time-displaced from one another by 120°. Figure 2.5(a) shows a schematic representation where the three single-phase voltage sources appear in a Y connection; a Δ configuration is also possible. A phasor diagram showing each of the phase voltages is also given in Figure 2.5(b).

## Phase Sequence

As the phasors rotate at the angular frequency $\omega$ with respect to the reference line in the counterclockwise (positive) direction, the positive maximum value first occurs for phase *a* and then in succession for phases *b* and *c*. Stated in a different way, to an observer in the phasor space, the voltage of phase *a* arrives first followed by that of *b* and then that of *c*. The three-phase voltage of Figure 2.5 is then said to have the phase sequence *abc* (*order* or *phase*

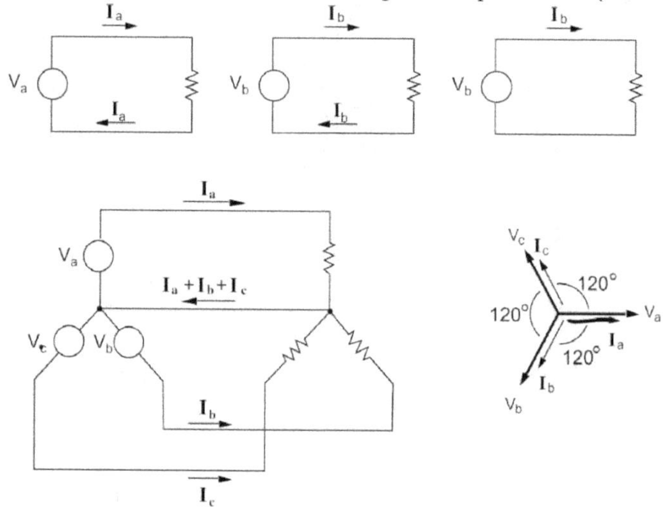


**Figure 2.6** A Three-Phase System.

*sequence* or *rotation* are all synonymous terms). This is important for applications such as three-phase induction motors, where the phase sequence determines whether the motor turns clockwise or counterclockwise.

With very few exceptions, synchronous generators (commonly referred to as *alternators*) are three-phase machines. To produce a set of three voltages phase-displaced by 120 electrical degrees in time, it follows that we must use a minimum of three coils phase-displaced 120 electrical degrees in space.

It is convenient to consider representing each coil as a separate generator. An immediate extension of the single-phase circuits discussed above would be to carry the power from the three generators along six wires. However, instead of having a return wire from each load to each generator, a single wire is used for the return of all three. The current in the return wire will be $I_a + I_b + I_c$; and for a balanced load, these will cancel out. If the load is unbalanced, the value of the return current will still be small compared to either $I_a$, $I_b$, or $I_c$. Thus the return wire could be made smaller than the other three. This connection is known as a four-wire three-phase system. It is desirable for safety and system protection to have a connection from the electrical system to the ground (equipotential surface at zero potential). A logical point for grounding is the generator neutral point.

## Current and Voltage Relations

Balanced three-phase systems can be studied using techniques developed for single-phase circuits. The arrangement of the three single-phase voltages into a $Y$ or a $\Delta$ configuration requires some modification in dealing with the overall system.

### *Y Connection*

With reference to Figure 2.7, the common terminal $n$ is called the *neutral* or *star* ($Y$) *point*. The voltages appearing between any two of the line terminals $a$, $b$, and $c$ have different relationships in magnitude and phase to the voltages appearing between any one line terminal and the neutral point $n$. The set of voltages $V_{ab}$, $V_{bc}$, and $V_{ca}$ are called the *line voltages*, and the set of voltages $V_{an}$, $V_{bn}$, and $V_{cn}$ are referred to as the *phase voltages*. Analysis of phasor diagrams provides the required relationships.

The effective values of the phase voltages are shown in Figure 2.7 as $V_{an}$, $V_{bn}$, and $V_{cn}$. Each has the same magnitude, and each is displaced 120° from the other two phasors.

Observe that the voltage existing from $a$ to $b$ is equal to the voltage from $a$ to $n$ (i.e., $V_{an}$) plus the voltage from $n$ to $b$.

For a balanced system, each phase voltage has the same magnitude, and we define

$$|V_{an}| = |V_{bn}| = |V_{cn}| = V_p \tag{2.10}$$

where $V_p$ denotes the effective magnitude of the phase voltage.

We can show that

$$\begin{aligned} V_{ab} &= V_p(1 - 1\angle -120^\circ) \\ &= \sqrt{3}V_p\angle 30^\circ \end{aligned} \tag{2.11}$$

Similarly, we obtain

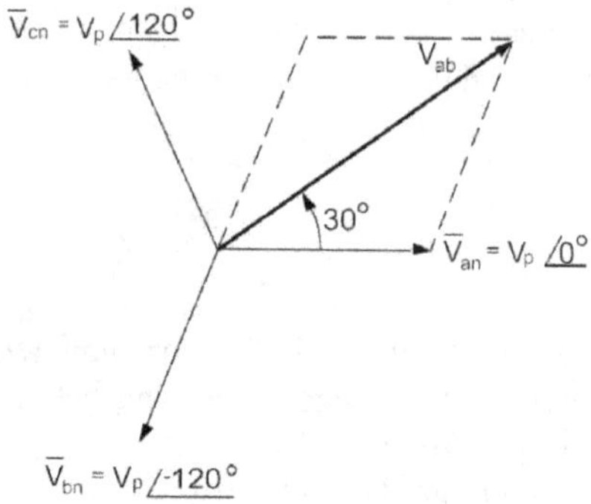


**Figure 2.7** Illustrating the Phase and Magnitude Relations Between the Phase and Line Voltage of a Y Connection.

$$V_{bc} = \sqrt{3}V_p\angle -90^\circ \tag{2.12}$$

$$V_{ca} = \sqrt{3}V_p\angle 150^\circ \tag{2.13}$$

The line voltages constitute a balanced three-phase voltage system whose magnitudes are $\sqrt{3}$ times the phase voltages. Thus, we write

$$V_L = \sqrt{3}V_p \tag{2.14}$$

A current flowing out of a line terminal $a$ (or $b$ or $c$) is the same as that flowing through the phase source voltage appearing between terminals $n$ and $a$ (or $n$ and $b$, or $n$ and $c$). We can thus conclude that for a $Y$-connected three-phase source, the line current equals the phase current. Thus,

$$I_L = I_p \tag{2.15}$$

Here $I_L$ denotes the effective value of the line current and $I_p$ denotes the effective value for the phase current.

### $\Delta$ Connection

Consider the case when the three single-phase sources are rearranged to form a three-phase Δ connection as shown in Figure 2.8. The line and phase voltages have the same magnitude:

$$|V_L| = |V_P| \tag{2.16}$$

The phase and line currents, however, are not identical, and the relationship

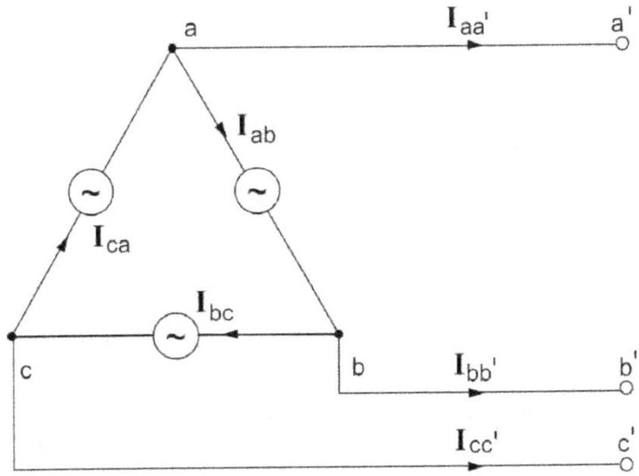


**Figure 2.8** A Δ-Connected Three-Phase Source.

between them can be obtained using Kirchhoff's current law at one of the line terminals.

In a manner similar to that adopted for the *Y*-connected source, let us consider the phasor diagram shown in Figure 2.9. Assume the phase currents to be

$$I_{ab} = I_p \angle 0$$
$$I_{bc} = I_p \angle -120^\circ$$
$$I_{ca} = I_p \angle 120^\circ$$

The current that flows in the line joining $a$ to $a'$ is denoted $I_{aa'}$ and is given by

$$I_{aa'} = I_{ca} - I_{ab}$$

As a result, we have

$$I_{aa'} = \sqrt{3} I_p \angle 150^\circ$$

Similarly,

$$I_{bb'} = \sqrt{3} I_p \angle 30^\circ$$
$$I_{cc'} = \sqrt{3} I_p \angle -90^\circ$$

Note that a set of balanced three phase currents yields a corresponding set of balanced line currents that are $\sqrt{3}$ times the phase values:

$$I_L = \sqrt{3} I_p \tag{2.17}$$

where $I_L$ denotes the magnitude of any of the three line currents.

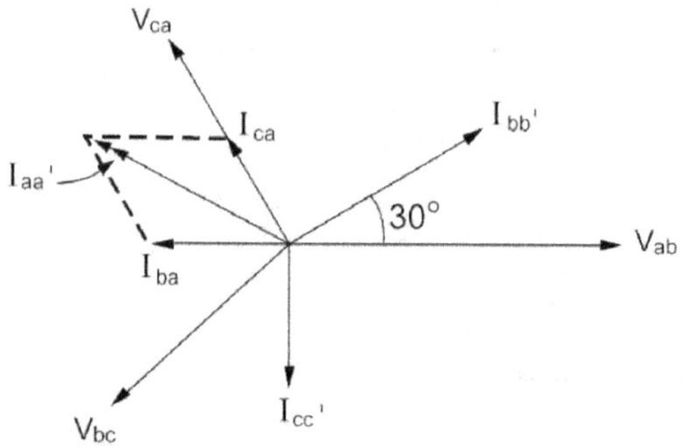


**Figure 2.9** Illustrating Relation Between Phase and Line Currents in a Δ Connection.

## Power Relationships

Assume that the three-phase generator is supplying a balanced load with the three sinusoidal phase voltages

$$v_a(t) = \sqrt{2}V_p \sin \omega t$$
$$v_b(t) = \sqrt{2}V_p \sin(\omega t - 120^\circ)$$
$$v_c(t) = \sqrt{2}V_p \sin(\omega t + 120^\circ)$$

With the currents given by

$$i_a(t) = \sqrt{2}I_p \sin(\omega t - \phi)$$
$$i_b(t) = \sqrt{2}I_p \sin(\omega t - 120^\circ - \phi)$$
$$i_c(t) = \sqrt{2}I_p \sin(\omega t + 120^\circ - \phi)$$

where $\phi$ is the phase angle between the current and voltage in each phase. The total power flowing into the load is

$$p_{3\phi}(t) = v_a(t)i_a(t) + v_b(t)i_b(t) + v_c(t)i_c(t)$$

This turns out to be

$$p_{3\phi}(t) = V_p I_p\{3\cos\phi - [\cos(2\omega t - \phi) + \cos(2\omega t - 240 - \phi) + \cos(2\omega t + 240 - \phi)]\}$$

Note that the last three terms in the above equation are the reactive power terms and they add up to zero. Thus we obtain

$$p_{3\phi}(t) = 3V_p I_p \cos\phi \tag{2.18}$$

The relationship between the line and phase voltages in a *Y*-connected system is

$$|V_L| = \sqrt{3}|V|$$

The power equation thus reads in terms of line quantities:

$$p_{3\phi} = \sqrt{3}|V_L||I_L|\cos\phi \tag{2.19}$$

The value of the total instantaneous power is constant, having a magnitude of three times the real power per phase. We may be tempted to assume that the reactive power is of no importance in a three-phase system since the *Q* terms cancel out. However, this situation is analogous to the summation of balanced three-phase currents and voltages that also cancel out. Although the sum cancels out, these quantities are still very much in evidence in each phase. We thus extend the concept of complex or apparent power (*S*) to three-phase systems by defining

$$S_{3\phi} = 3V_p I_p^* \tag{2.20}$$

where the active power and reactive power are obtained from

$$S_{3\phi} = P_{3\phi} + jQ_{3\phi}$$

as

$$P_{3\phi} = 3|V_p||I_p|\cos\phi \tag{2.21}$$

$$Q_{3\phi} = 3|V_p||I_p|\sin\phi \tag{2.22}$$

and

$$P_{3\phi} = \sqrt{3}|V_L||I_L|\cos\phi \tag{2.23}$$

$$Q_{3\phi} = \sqrt{3}|V_L||I_L|\sin\phi \tag{2.24}$$

In specifying rated values for power system apparatus and equipment such as generators, transformers, circuit breakers, etc., we use the magnitude of the apparent power $S_{3\phi}$ as well as line voltage for specification values. In specifying three-phase motor loads, we use the horsepower output rating and voltage.

**Example 2.2**

A *Y*-connected, balanced three-phase load consisting of three impedances of $20\angle 30^\circ$ ohms each as shown in Figure 2.10 is supplied with the balanced line-to-neutral voltages:

$$V_{an} = 220\angle 0 \text{ V}$$
$$V_{bn} = 220\angle 240^\circ \text{ V}$$
$$V_{cn} = 220\angle 120^\circ \text{ V}$$

A. Calculate the phase currents in each line.
B. Calculate the line-to-line phasor voltages.
C. Calculate the total active and reactive power supplied to the load.

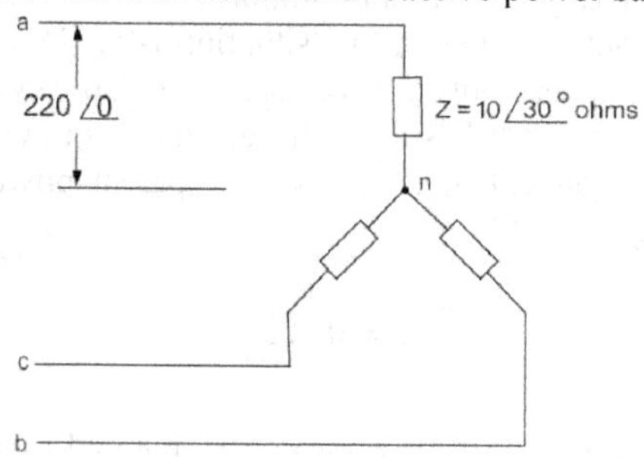


**Figure 2.10** Load Connection for Example 2.2.

*Solution*

A. The phase currents are obtained as

$$I_{an} = \frac{220}{20\angle 30} = 11\angle -30^\circ \text{ A}$$
$$I_{bn} = \frac{220\angle 240}{20\angle 30} = 11\angle 210^\circ \text{ A}$$
$$I_{cn} = \frac{220\angle 120}{20\angle 30} = 11\angle 90^\circ \text{ A}$$

B. The line-to-line voltages are obtained as

$$\begin{aligned} V_{ab} &= V_{an} - V_{bn} \\ &= 220\angle 0 - 220\angle 240^\circ \\ &= 220\sqrt{3}\angle 30^\circ \\ V_{bc} &= 220\sqrt{3}\angle 30 - 120 = 220\sqrt{3}\angle -90^\circ \\ V_{ca} &= 220\sqrt{3}\angle -210^\circ \end{aligned}$$

C. The apparent power into phase $a$ is given by

$$\begin{aligned} S_a &= V_{an} I_{an}^* \\ &= (220)(11)\angle 30^\circ \\ &= 2420\angle 30^\circ \text{ VA} \end{aligned}$$

The total apparent power is three times the phase value:

$$\begin{aligned} S_t &= 2420 \times 3\angle 30^\circ = 7260.00\angle 30^\circ \text{ VA} \\ &= 6287.35 + j3630.00 \end{aligned}$$

Thus

$$\begin{aligned} P_t &= 6287.35 \text{ W} \\ Q_t &= 3630.00 \text{ var} \end{aligned}$$

**Example 2.3**

Repeat Example 2.2 as if the same three impedances were connected in a Δ connection.

*Solution*

From Example 2.2 we have

$$\begin{aligned} V_{ab} &= 220\sqrt{3}\angle 30^\circ \\ V_{bc} &= 220\sqrt{3}\angle -90^\circ \\ V_{ca} &= 220\sqrt{3}\angle -210^\circ \end{aligned}$$

The currents in each of the impedances are

$$\begin{aligned} I_{ab} &= \frac{220\sqrt{3}\angle 30^\circ}{20\angle 30} = 11\sqrt{3}\angle 0^\circ \\ I_{bc} &= 11\sqrt{3}\angle -120^\circ \\ I_{ca} &= 11\sqrt{3}\angle 120^\circ \end{aligned}$$

The line currents are obtained with reference to Figure 2.11 as

$$
\begin{aligned}
I_a &= I_{ab} - I_{ca} \\
&= 11\sqrt{3}\angle 0 - 11\sqrt{3}\angle -120^\circ \\
&= 33\angle 30^\circ \\
I_b &= I_{bc} - I_{ab} \\
&= 33\angle -90^\circ \\
I_c &= I_{ca} - I_{bc} \\
&= 33\angle -210^\circ
\end{aligned}
$$

The apparent power in the impedance between $a$ and $b$ is

$$
\begin{aligned}
S_{ab} &= V_{ab} I_{ab}^* \\
&= (220\sqrt{3}\angle 30^\circ)(22\sqrt{3}\angle 0) \\
&= 7260\angle 30^\circ
\end{aligned}
$$

The total three-phase power is then

$$
\begin{aligned}
S_t &= 21780\angle 30^\circ \\
&= 18{,}862.02 + j10890.00
\end{aligned}
$$

As a result,

$$
\begin{aligned}
P_t &= 37724.04 \text{ W} \\
Q_t &= 21780.00 \text{ var}
\end{aligned}
$$

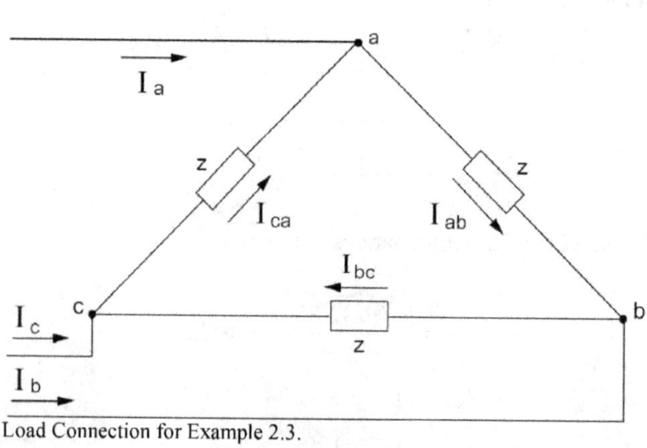


**Figure 2.11** Load Connection for Example 2.3.

## 2.4 THE PER UNIT SYSTEM

The per unit (p.u.) value representation of electrical variables in power system problems is favored in electric power systems. The numerical per unit value of any quantity is its ratio to a chosen base quantity of the same dimension. Thus a per unit quantity is a *normalized* quantity with respect to the chosen base value. The per unit value of a quantity is thus defined as

$$\text{p.u. value} = \frac{\text{Actual value}}{\text{Reference or base value of the same dimension}} \tag{2.25}$$

Five quantities are involved in the calculations. These are the current *I*, the voltage *V*, the complex power *S*, the impedance *Z*, and the phase angles. The angles are dimensionless; the other four quantities are completely described by knowledge of only two of them. An arbitrary choice of two base quantities will fix the other base quantities. Let $|I_b|$ and $|V_b|$ represent the base current and base voltage expressed in kiloamperes and kilovolts, respectively. The product of the two yields the base complex power in megavoltamperes (MVA)

$$|S_b| = |V_b||I_b| \text{MVA} \tag{2.26}$$

The base impedance will also be given by

$$|Z_b| = \frac{|V_b|}{|I_b|} = \frac{|V_b|^2}{|S_b|} \text{ohms} \tag{2.27}$$

The base admittance will naturally be the inverse of the base impedance. Thus,

$$\begin{aligned} |Y_b| &= \frac{1}{|Z_b|} = \frac{|I_b|}{|V_b|} \\ &= \frac{|S_b|}{|V_b|^2} \text{siemens} \end{aligned} \tag{2.28}$$

The nominal voltage of lines and equipment is almost always known as well as the apparent (complex) power in megavoltamperes, so these two quantities are usually chosen for base value calculation. The same megavoltampere base is used in all parts of a given system. Once base voltage is chosen; all other base voltages must then be related to the one chosen by the turns ratios of the connecting transformers.

From the definition of per unit impedance, we can express the ohmic impedance $Z_\Omega$ in the per unit value $Z_{\text{p.u.}}$ as

$$Z_{\text{p.u.}} = \frac{Z_{\Omega}|S_b|}{|V_b|^2} \text{ p.u.} \tag{2.29}$$

As for admittances, we have

$$Y_{\text{p.u.}} \stackrel{\Delta}{=} \frac{1}{Z_{\text{p.u.}}} = \frac{|V_b|^2}{Z_{\Omega}|S_b|} = Y_S \frac{|V_b|^2}{|S_b|} \text{ p.u.} \tag{2.30}$$

Note that $Z_{\text{p.u.}}$ can be interpreted as the ratio of the voltage drop across $Z$ with base current injected to the base voltage.

**Example 2.4**

Consider a transmission line with $Z = 3.346 + j77.299\Omega$. Assume that

$$S_b = 100 \text{ MVA}$$
$$V_b = 735 \text{ kV}$$

We thus have

$$Z_{\text{p.u.}} = Z_{\Omega} \cdot \frac{S_b}{|V_b|^2} = Z_{\Omega} \cdot \frac{1000}{(735)^2}$$
$$= 1.851 \times 10^{-4} (Z_{\Omega})$$

For $R = 3.346$ ohms we obtain

$$R_{\text{p.u.}} = (3.346)(1.851 \times 10^{-4}) = 6.19372 \times 10^{-4}$$

For $X = 77.299$ ohms, we obtain

$$X_{\text{p.u.}} = (77.299)(1.851 \times 10^{-4}) = 1.43 \times 10^{-2}$$

For the admittance we have

$$Y_{\text{p.u.}} = Y_S \cdot \frac{|V_b|^2}{S_b}$$
$$= Y_S \frac{(735)^2}{100}$$
$$= 5.4 \times 10^3 (Y_S)$$

For $Y = 1.106065 \times 10^{-3}$ siemens, we obtain

$$Y_{\text{p.u.}} = \left(5.40225\times10^{3}\right)\left(1.106065\times10^{-3}\right)$$
$$= 5.97524$$

## Base Conversions

Given an impedance in per unit on a given base $S_{b_0}$ and $V_{b_0}$, it is sometimes required to obtain the per unit value referred to a new base set $S_{b_n}$ and $V_{b_n}$. The conversion expression is obtained as:

$$Z_{\text{p.u.}_n} = Z_{\text{p.u.}_0} \frac{\left|S_{b_n}\right|}{\left|S_{b_0}\right|} \cdot \frac{\left|V_{b_0}\right|^2}{\left|V_{b_n}\right|^2} \tag{2.31}$$

which is the required conversion formula. The admittance case simply follows the inverse rule. Thus,

$$Y_{\text{p.u.}_n} = Y_{\text{p.u.}_0} \frac{\left|S_{b_0}\right|}{\left|S_{b_n}\right|} \cdot \frac{\left|V_{b_n}\right|^2}{\left|V_{b_0}\right|^2} \tag{2.32}$$

### Example 2.5

Convert the impedance and admittance values of Example 2.4 to the new base of 200 MVA and 345 kV.

*Solution*

We have

$$Z_{\text{p.u.}_0} = 6.19372\times10^{-4} + j1.430867\times10^{-2}$$

for a 100-MVA, 735-kV base. With a new base of 200 MVA and 345 kV, we have, using the impedance conversion formula,

$$Z_{\text{p.u.}_n} = Z_{\text{p.u.}_0}\left(\frac{200}{100}\right)\cdot\left(\frac{735}{345}\right)^2$$
$$= 9.0775 Z_{\text{p.u.}_0}$$

Thus,

$$Z_{\text{p.u.}_n} = 5.6224\times10^{-3} + j1.2989\times10^{-1} \text{ p.u.}$$

For the admittance we have

$$Y_{\text{p.u.}_n} = Y_{\text{p.u.}_0}\left(\frac{100}{200}\right)\cdot\left(\frac{345}{735}\right)^2$$
$$= 0.11016 Y_{\text{p.u.}_0}$$

Thus,

$$Y_{\text{p.u.}_n} = (5.97524)(0.11016)$$
$$= 0.65825 \text{ p.u.}$$

## 2.5 ELECTROMAGNETISM AND ELECTROMECHANICAL ENERGY CONVERSION

An electromechanical energy conversion device transfers energy between an input side and an output side, as shown in Figure 2.12. Within an electric motor, the input is electrical energy drawn from the supply and the output is the mechanical energy supplied to the load, which may be a pump, fan, hoist, or any other mechanical load. An electric generator converts mechanical energy supplied by a prime mover to electrical form at the output side. The operation of electromechanical energy conversion devices is based on fundamental principles resulting from experimental work.

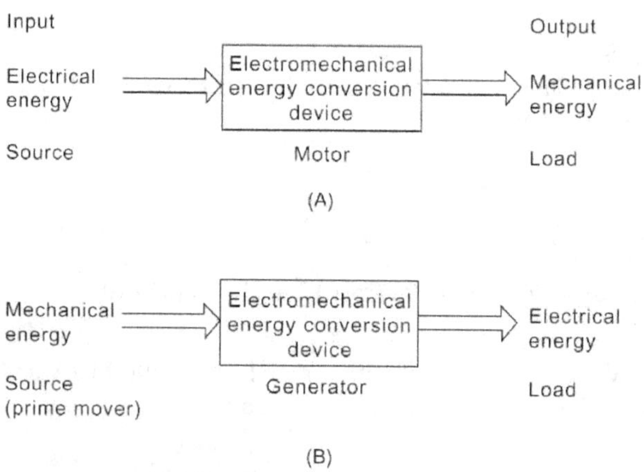


**Figure 2.12** Functional block diagram of electromechanical energy conversion devices as (A) motor, and (B) generator.

The presence of stationary electric charges is associated with electric fields. On the other hand, a magnetic field is associated with moving charges and as a result we conclude that electric currents are sources of magnetic fields. A magnetic field is identified by a vector **B** called the magnetic flux density. The unit of **B** is the tesla (T). The magnetic flux $\Phi = \mathbf{B}.\mathbf{A}$. The unit of magnetic flux $\Phi$ is the weber (Wb).

## The Lorentz Force Law

A charged particle $q$, moving at a velocity **V** in a magnetic field of flux density **B**, is found experimentally to experience a force whose magnitude is proportional to the product of the magnitude of the charge $q$, its velocity, and the flux density **B** and to the sine of the angle between the vectors **V** and **B** and is given by a vector in the direction of the cross product **V** × **B**. Thus we write

$$\mathbf{F} = q\mathbf{V} \times \mathbf{B} \qquad (2.33)$$

Equation (2.33) is known as the Lorentz force equation. The direction of the force is orthogonal to the plane of **V** and **B** and follows the right-hand rule. An interpretation of Eq. (2.33) is given in Figure 2.13.

The tesla can then be defined as the magnetic flux density that exists when a charge $q$ of 1 coulomb, moving normal to the field at a velocity of 1 m/s, experiences a force of 1 Newton.

A distribution of charge experiences a differential force $d\mathbf{F}$ on each moving incremental charge element $dq$ given by

$$d\mathbf{F} = dq\,(\mathbf{V} \times \mathbf{B})$$

Moving charges over a line constitute a line current and thus:

$$d\mathbf{F} = (I \times \mathbf{B})\,\mathbf{dl} \qquad (2.34)$$

Equation (2.34) simply states that a current element $I\,\mathbf{dl}$ in a magnetic field **B** will experience a force $d\mathbf{F}$ given by the cross product of $I\,\mathbf{dl}$ and **B**. A pictorial presentation of Eq. (2.34) is given in Figure 2.14.

The current element $I\,\mathbf{dl}$ cannot exist by itself and must be a part of a complete circuit. The force on an entire loop can be obtained by integrating the current element

$$\mathbf{F} = \oint I\,\mathbf{dl} \times \mathbf{B} \qquad (2.35)$$

Equations (2.34) and (2.35) are fundamental in the analysis and design of electric motors, as will be seen later.

The Biot-Savart law is based on Ampère's work showing that electric currents exert forces on each other and that a magnet could be replaced by an equivalent current.

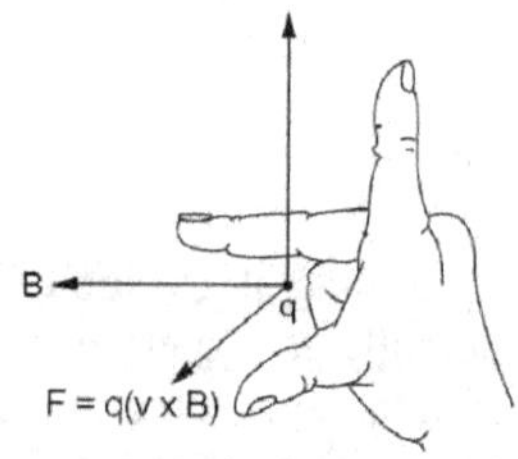


**Figure 2.13** Lorentz force law.

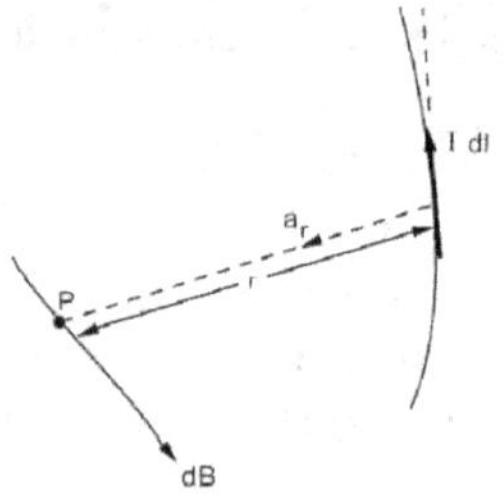


**Figure 2.14** Interpreting the Biot-Savart law.

Consider a long straight wire carrying a current $I$ as shown in Figure 2.15. Application of the Biot-Savart law allows us to find the total field at $P$ as:

$$\mathrm{B} = \frac{\mu_0 I}{2\pi R} \tag{2.36}$$

The constant $\mu_0$ is called the permeability of free space and in SI units is given by

$$\mu_0 = 4\pi \times 10^{-7}$$

The magnetic field is in the form of concentric circles about the wire,

with a magnitude that increases in proportion to the current $I$ and decreases as the distance from the wire is increased.

The Biot-Savart law provides a relation between current and the resulting magnetic flux density **B**. An alternative to this relation is Ampère's circuital law, which says that the line integral of **B** about any closed path in free space is equal to the current enclosed by that path times $\mu_0$.

$$\oint_c \mathbf{B} \cdot \mathbf{dl} = \begin{cases} \mu_0 I & \text{path } c \text{ encloses } I \\ 0 & \text{path } c \text{ does not enclose } I \end{cases} \tag{2.37}$$

Note that the path $c$ can be arbitrarily shaped closed loop about the net current $I$.

## 2.6 PERMEABILITY AND MAGNETIC FIELD INTENSITY

To extend magnetic field laws to materials that exhibit a linear variation of **B** with *I*, all expressions are valid provided that $\mu_0$ is replaced by the permeability corresponding to the material considered. From a **B**-*I* – variation point of view we divide materials into two classes:

1. Nonmagnetic material such as all dielectrics and metals with permeability equal to $\mu_0$ for all practical purposes.
2. Magnetic material such as ferromagnetic material (the iron group), where a given current produces a much larger **B** field than in free space. The permeability in this case is much higher than that of free space and varies with current in a nonlinear manner over a wide range. Ferromagnetic material can be further categorized into two classes:

   a) Soft ferromagnetic material for which a linearization of the **B**-*I* variation in a region is possible. The source of **B** in the case of soft ferromagnetic material can be modeled as due to the current *I*.
   b) Hard ferromagnetic material for which it is difficult to give a meaning to the term *permeability*. Material in this group is suitable for permanent magnets.

For hard ferromagnetic material, the source of **B** is a combined effect of current *I* and material magnetization *M*, which originates entirely in the medium. To separate the two sources of the magnetic **B** field, the concept of magnetic field intensity **H** is introduced.

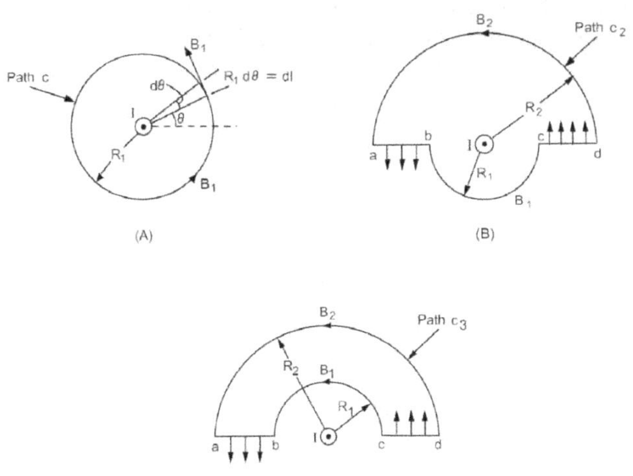


**Figure 2.15** Illustrating Ampère's circuital law: (A) path $c_1$ is a circle enclosing current *I*, (B) path $c_2$ is not a circle but encloses current *I*, and (C) path $c_3$ does not enclose current *I*.

## Magnetic Field Intensity

The magnetic field intensity (or strength) denoted by **H** is a vector defined by the relation

$$\mathbf{B} = \mu\mathbf{H} \tag{2.38}$$

For isotropic media (having the same properties in all directions), $\mu$ is a scalar and thus **B** and **H** are in the same direction. On the basis of Eq. (2.38), we can write the statement of Ampère's circuital law as

$$\oint \mathbf{H} \cdot \mathbf{dl} = \begin{cases} I & \text{path } c \text{ encloses } I \\ 0 & \text{path } c \text{ does not enclose } I \end{cases} \tag{2.39}$$

The expression in Eq. (2.39) is independent of the medium and relates the magnetic field intensity **H** to the current causing it, $I$.

In general, permeability $\mu$ is not a constant but is dependent on **H** and, strictly speaking, one should state this dependence in the form

$$\mu = \mu(\mathbf{H}) \tag{2.40}$$

For nonmagnetic material, $\mu$ is constant at a value equal to $\mu_0 = 4\pi \times 10^{-7}$ for all practical purposes. The **B-H** characteristic of nonmagnetic materials is shown in Figure 2.16

The **B-H** characteristics of soft ferromagnetic material, often called the magnetization curve, follow the typical pattern displayed in Figure 2.17. In accordance with Eq. (2.38,) the permeability of the material is the ratio of **B** to **H** and is a function of **H**, as indicated by Eq. (2.40).

$$\mu = \frac{\mathbf{B}}{\mathbf{H}} \tag{2.41}$$

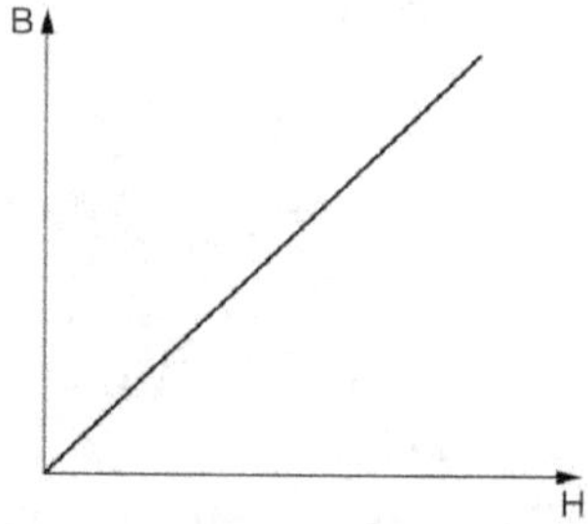


**Figure 2.16** B-H characteristic for nonmagnetic material.

The permeability at low values of **H** is called the initial permeability and is much lower than the permeability at higher values of **H**. The maximum value of $\mu$ occurs at the knee of the **B-H** characteristic. The permeability of soft ferromagnetic material $\mu$ is much larger than $\mu_0$ and we define the relative permeability $\mu_r$ as:

$$\mu_r = \frac{\mu}{\mu_0} \tag{2.42}$$

A typical variation of $\mu_r$ with **H** for a ferromagnetic material is shown in Figure 2.18.

For practical electromechanical energy conversion devices, a linear approximation to the magnetization curve provides satisfactory answers in the normal region of operation. The main idea is to fit a straight line passing through the origin of the **B-H** curve that best fits the data points which is drawn and taken to represent the characteristics of the material considered. Within the acceptable range of **H** values, one may then use the following relation to model the ferromagnetic material:

$$\mathbf{B} = \mu_0 \mu_r \mathbf{H} \tag{2.43}$$

It should be noted that $\mu_r$ is in the order of thousands for magnetic materials used in electromechanical energy conversion devices (2000 to 80,000, typically). Properties of magnetic materials are discussed further in the following sections. Presently, we assume that $\mu_r$ is constant.

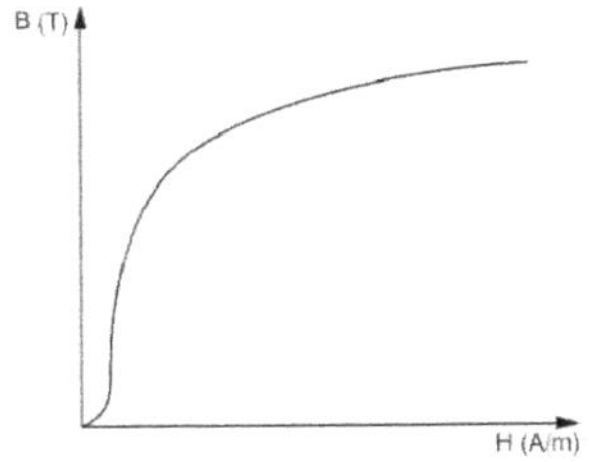


**Figure 2.17** B-H characteristic for a typical ferromagnetic material.

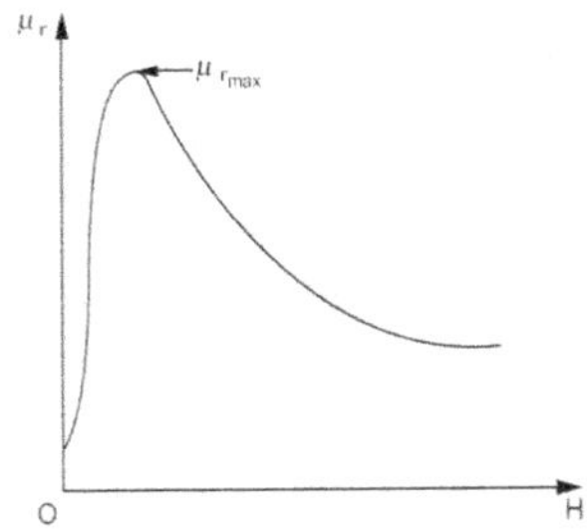


**Figure 2.18** Typical variation of $\mu_r$ with H for a ferromagnetic material.

## 2.7 FLUX LINKAGES, INDUCED VOLTAGES, INDUCTANCE, AND ENERGY

Faraday's experiments established that a change in a magnetic field causes an induced voltage, which is expressed by Faraday's law of electromagnetic induction.

Consider a toroidal coil with $N$ turns through which a current $i$ flows producing a total flux$\Phi$. Each turn encloses or links the total flux and we also note that the total flux links each of the $N$ turns. Following Henry, we introduce the concept of flux linkages $\lambda$ as the product of the number of turns $N$ and the flux $\Phi$ linking each turn.

$$\lambda = N\Phi \tag{2.44}$$

The flux linkages $\lambda$ can be related to the current $i$ in the coil by the definition of inductance $L$ through the following relation conceived by Joseph Henry as an analog to momentum

$$\lambda = Li \tag{2.45}$$

Inductance is the passive circuit element that is related to the geometry and material properties of the structure. From this point of view, inductance is the ratio of total flux linkages to the current, which the flux links. The inductance $L$ is related to the reluctance $\Re$ of the magnetic structure of a single-loop structure.

$$L = \frac{N^2}{\Re} \tag{2.46}$$

In the case of a toroid with a linear **B-H** curve, we have

$$L = \frac{N^2 A}{l}\mu_0\mu_r \tag{2.47}$$

There is no single definition of inductance that is useful in all cases for which the medium is not linear. The unit of inductance is the henry or weber-turns per ampere.

In terms of flux linkages, Faraday's law is stated as

$$e = \frac{d\lambda}{dt} = N\frac{d\Phi}{dt} \tag{2.48}$$

The electromotive force (EMF), or induced voltage, is thus equal to the rate of change of flux linkages in the structure. We also write:

$$e = \frac{d}{dt} Li \tag{2.49}$$

In electromechanical energy conversion devices, the reluctance varies with time and thus $L$ also varies with time. In this case,

$$e = L\frac{di}{dt} + i\frac{dL}{dt} \tag{2.50}$$

Note that if $L$ is constant, we get the familiar equation for modeling an inductor in elementary circuit analysis.

Power and energy relationships in a magnetic circuit are important in evaluating performance of electromechanical energy conversion devices. We treat some basic relationships, starting with the fundamental definition of power *p(t)* given by

$$p(t) = e(t)i(t) \tag{2.51}$$

The power into a component (the coil in the case of toroid) is given as the product of the voltage across its terminals *e(t)* and the current through *i(t)*. Using Faraday's law, see Eq. (2.50), we can write

$$p(t) = i(t)\frac{d\lambda}{dt} \tag{2.52}$$

The units of power are watts (or joules per second).

Recall the basic relation stating that power *p(t)* is the rate of change of energy *W(t)*:

$$p(t) = \frac{dW}{dt} \tag{2.53}$$

We can show that

$$\mathrm{dW} = (\mathrm{lA})\mathrm{H\ dB} \tag{2.54}$$

Consider the case of a magnetic structure that experiences a change in state between the time instants $t_1$ and $t_2$. Then, change in energy into the system is denoted by $\Delta W$ and is given by

$$\Delta \mathrm{W} = \mathrm{W}(\mathrm{t}_2) - \mathrm{W}(\mathrm{t}_1) \tag{2.55}$$

We can show that

$$\Delta W = lA\int_{B_1}^{B_2} \mathrm{H}d\mathrm{B} \tag{2.56}$$

It is clear that the energy per unit volume expended between $t_1$ and $t_2$ is the area between the **B-H** curve and the B axis between $B_1$ and $B_2$.

It is important to realize that the energy relations obtained thus far do not require linear characteristics. For a linear structure, we can develop these relations further and show that

$$\Delta W = \frac{1}{2L}(\lambda_2^2 - \lambda_1^2) \tag{2.57}$$

or

$$\Delta W = \tfrac{1}{2} L(i_2^2 - i_1^2) \tag{2.58}$$

The energy expressions obtained provide us with measures of energy stored in the magnetic field treated.

## 2.8 HYSTERESIS LOOP

Ferromagnetic materials are characterized by nonlinear and multi-valued **B-H** characteristics, which a hysteresis characteristic. To illustrate this phenomenon, we use the sequence of portraits of Figure 2.19 showing the evolution of a hysteresis loop for a toroid with virgin ferromagnetic core. Assume that the MMF (and hence **H**) is a slowly varying sinusoidal waveform with period $T$ as shown in the lower portion graphs of Figure 2.19. We will discuss the evolution of the **B-H** hysteresis loop in the following intervals.

*Interval I*: Between $t = 0$ and $T/4$, the magnetic field intensity **H** is positive and increasing. The flux density increases along the initial curve (*oa*) up to the saturation value $\mathbf{B}_s$. Increasing **H** beyond the saturation level does not result in an increase in **B**.

*Interval II*: Between $t = T/4$ and $T/2$, the magnetic field intensity is positive but decreasing. The flux density **B** is seen to decrease along the segment *ab*. Note that *ab* is above *oa* and thus for the same value of **H**, we get a different value of **B**. This is true at *b*, where there is a value for $\mathbf{B} = \mathbf{B}_r$ different from zero even though **H** is zero at that point in time $t = T/2$. The value of $\mathbf{B}_r$ is referred to as the residual field, remanence, or retentivity. If we leave the coil unenergized, the core will still be magnetized.

*Interval III*: Between $t = T/2$ and $3T/4$, the magnetic field intensity **H** is reversed and increases in magnitude. **B** decreases to zero at point *c*. The value of **H**, at which magnetization is zero, is called the coercive force $\mathbf{H}_c$. Further decrease in **H** results in reversal of **B** up to point *d*, corresponding to $t = 3T/4$.

*Interval IV*: Between $t = 3T/4$ and $T$, the value of **H** is negative but increasing. The flux density **B** is negative and increases from $d$ to $e$. Residual field is observed at $e$ with $\mathbf{H} = 0$.
*Interval V*: Between $t = T$ and $5T/4$, **H** is increased from 0, and the flux density is negative but increasing up to $f$, where the material is demagnetized. Beyond $f$, we find that **B** increases up to $a$ again.

A typical hysteresis loop is shown in Figure 2.20. On the same graph, the **B-H** characteristic for nonmagnetic material is superimposed to show the relative magnitudes involved. As shown in Figure 2.21, there is a steady-state loop for each maximum value of the ac magnetic field intensity cycle. The dashed curve connecting the tips of the loops in the figure is the dc magnetization curve for the material. Table 2.1 lists some typical values for $\mathbf{H}_c$, $\mathbf{B}_r$, and $\mathbf{B}_s$ for common magnetic materials.

The energy supplied by the source per unit volume of the magnetic structure is given by

$$d\widetilde{W} = \mathrm{H}d\mathrm{B}$$

and

$$\Delta\widetilde{W} = \int_{\mathbf{B}_1}^{\mathbf{B}} \mathrm{H}d\mathrm{B}$$

The energy supplied by the source in moving from a to b in the graph of Figure 2.22(A) is negative since **H** is positive but **B** is decreasing. If we continue on from $b$ to $d$ through $c$, the energy is positive as **H** is negative but **B** is decreasing.

The second half of the loop is treated in Figure 2.22(B) and is self-explanatory. Superimposing both halves of the loop, we obtain Figure 2.22(C), which clearly shows that the net energy per unit volume supplied by the source is the area enclosed by the hysteresis loop. This energy is expended in the magnetization-demagnetization process and is dissipated as heat loss. Note that the loop is described in one cycle and as a result, the hysteresis loss per second is equal to the product of the loop area and the frequency $f$ of the waveform applied. The area of the loop depends on the maximum flux density and we assert that the power dissipated through hysteresis $P_h$ is given by

$$P_h = k_h f(\mathrm{B}_m)^n$$

Here $k_h$ is a constant, $f$ is the frequency, and $\mathrm{B}_m$ is the maximum value of the flux density. The exponent $n$ is determined from experimental results and ranges between 1.5 and 2.5.

## 2.9 EDDY CURRENT AND CORE LOSSES

When the core is subject to a time-varying magnetic field (sinusoidal input was assumed), energy is extracted from the source in the form of hysteresis losses. There is another loss mechanism that arises in connection with the application of time-varying magnetic field, called eddy-current loss. A rigorous analysis of the eddy-current phenomenon is a complex process but the basic model can be explained in simple terms on the basis of Faraday's law.

The change in flux induces voltages in the core material resulting in currents circulating in the core. The induced currents tend to establish a flux that opposes the original change imposed by the source. The induced currents, which are essentially the eddy currents, will result in power loss due to heating of the core material. To minimize eddy current losses, the magnetic core is made of stackings of sheet steel laminations, ideally separated by highly resistive material. It is clear that this effectively results in the actual area of the

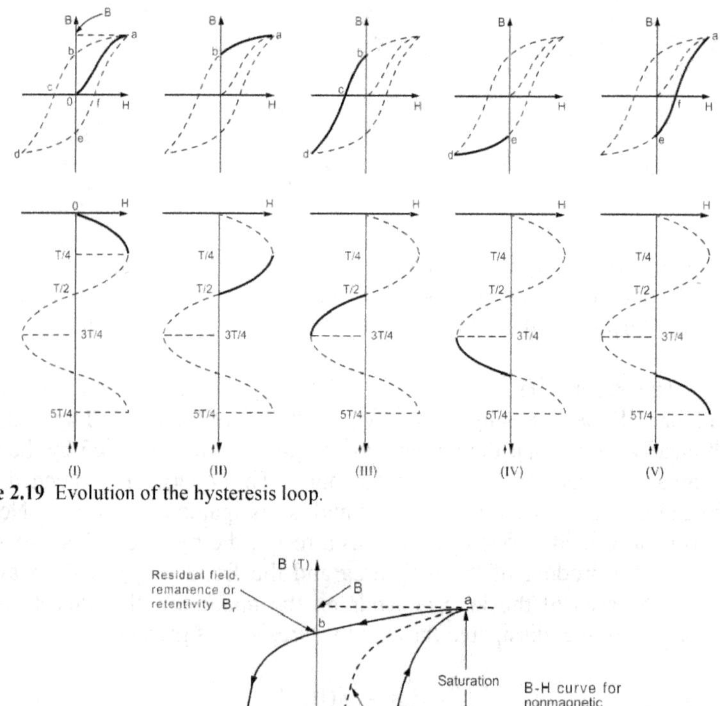


**Figure 2.19** Evolution of the hysteresis loop.

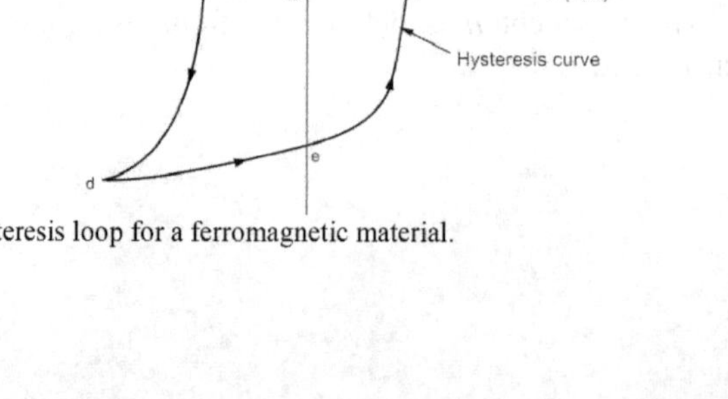


**Figure 2.20** Hysteresis loop for a ferromagnetic material.

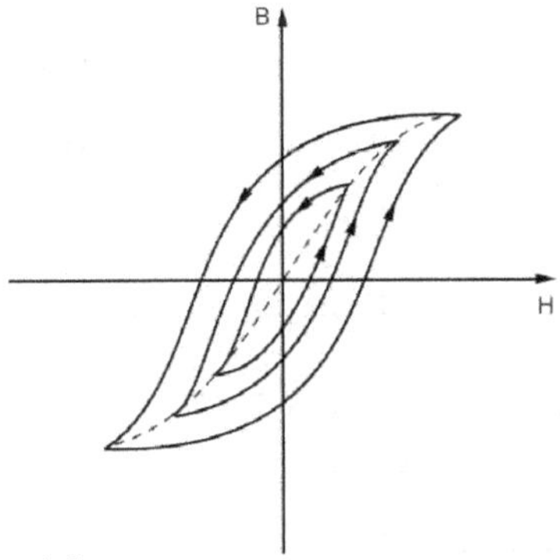


**Figure 2.21** Family of hysteresis loops.

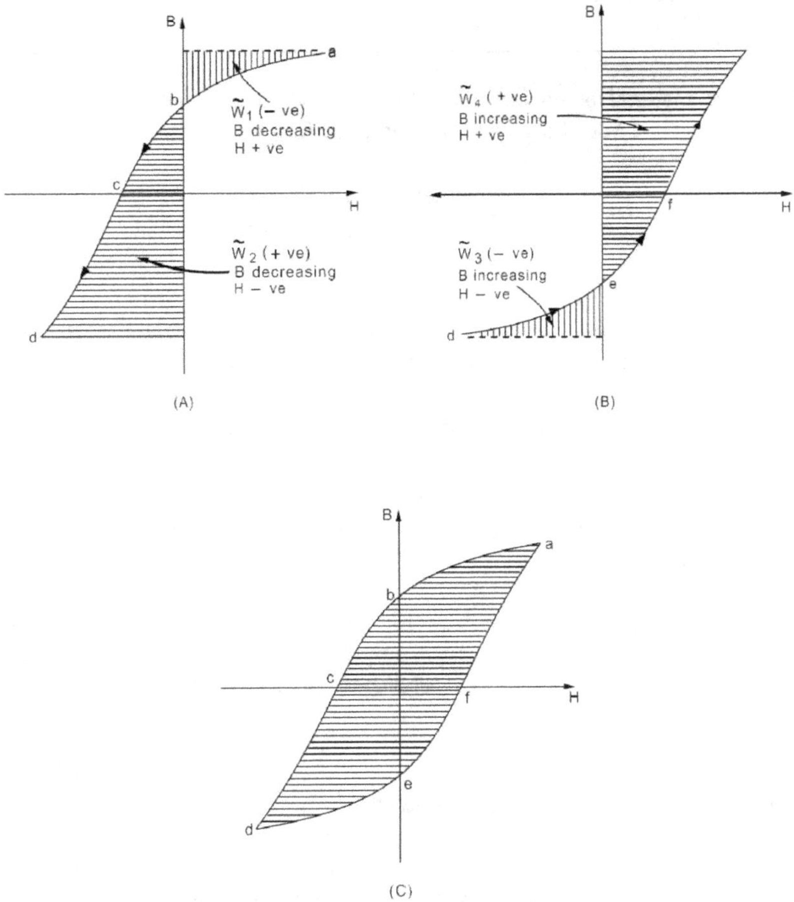


**Figure 2.22** Illustrating the concept of energy loss in the hysteresis process.

**Table 2.1**

Properties of Magnetic Materials and Magnetic Alloys

| *Material (Composition)* | *Initial Relative Permeability,* $\mu_i/\mu_0$ | *Maximum Relative Permeability,* $\mu_{max}/\mu_0$ | *Coercive Force* $H_r$ (A/m) | *Residual Field* $B_r$ (Wb/m$^2$) | *Saturation Field* $B_s$ (Wb/m$^2$) |
|---|---|---|---|---|---|
| Commercial iron (0.2 imp) | 250 | 9,000 | ≅ 80 | 0.77 | 2.15 |
| Silicon-iron (4 Si) | 1,500 | 7,000 | 20 | 0.5 | 1.95 |
| Silicon-iron (3 Si) | 7,500 | 55,000 | 8 | 0.95 | 2.00 |
| Mu metal (5 Cu, 2 Cr, 77 Ni) | 20,000 | 100,000 | 4 | 0.23 | 0.65 |
| 78 Permalloy (78.5 Ni) | 8,000 | 100,000 | 4 | 0.6 | 1.08 |
| Supermalloy (79 Ni, 5 Mo) | 100,000 | 1,000,000 | 0.16 | 0.5 | 0.79 |

magnetic material being less than the gross area presented by the stack. To account for this, a stacking factor is employed for practical circuit calculations.

$$\text{Stacking factor} = \frac{\text{actual magnetic cross-sectional area}}{\text{gross cross-sectional area}}$$

Typically, lamination thickness ranges from 0.01 mm to 0.35 m with associated stacking factors ranging between 0.5 to 0.95. The eddy-current power loss per unit volume can be expressed by the empirical formula

$$P_c = K_e(f\mathrm{B}_m t_1)^2 \qquad \mathrm{W/m^3}$$

The eddy-current power loss per unit volume varies with the square of frequency $f$, maximum flux density $B_m$, and the lamination thickness $t_1$. $K_e$ is a proportionality constant.

The term *core loss* is used to denote the combination of eddy-current and hysteresis power losses in the material. In practice, manufacturer-supplied data are used to estimate the core loss $P_c$ for given frequencies and flux densities for a particular type of material.

## 2.10 ENERGY FLOW APPROACH

From an energy flow point of view consider an electromechanical energy conversion device operating as a motor. We develop a model of the process that is practical and easy to follow and therefore take a macroscopic approach based on the principle of energy conservation. The situation is illustrated using the diagrams of Figure 2.23. We assume that an incremental change in electric energy supply $dW_e$ has taken place. This energy flow into the device can be visualized as being made up of three components, as shown in

Figure 2.23(A). Part of the energy will be imparted to the magnetic field of the device and will result in an increase in the energy stored in the field, denoted by $dW_f$. A second component of energy will be expended as heat losses $dW_{\text{loss}}$. The third and most important component is that output energy be made available to the load ($dW_{\text{mech}}$). The heat losses are due to ohmic ($I^2R$) losses in the stator (stationary member) and rotor (rotating member); iron or field losses through eddy current and hysteresis; and mechanical losses in the form of friction and windage.

In part (B) of Figure 2.23, the energy flow is shown in a form that is closer to reality by visualizing Ampère's bonne homme making a trip through the machine. Starting in the stator, ohmic losses will be encountered, followed by field losses and a change in the energy stored in the magnetic field. Having crossed the air gap, our friend will witness ohmic losses in the rotor windings taking place, and in passing to the shaft, bearing frictional losses are also encountered. Finally, a mechanical energy output is available to the load. It is emphasized that the phenomena dealt with are distributed in nature and what we are doing is simply developing an understanding in the form of mathematical expressions called models. The trip by our Amperean friend can never take place in real life but is a helpful means of visualizing the process.

We write the energy balance equation as:

$$dW_e = dW_{\text{fld}} + dW_{\text{loss}} + dW_{\text{mech}} \tag{2.59}$$

To simplify matters assume that losses are negligible.

The electric power input $P_e(t)$ to the device is given in terms of the terminal voltage $e(t)$ and current $i(t)$, and using it, we write Faraday's law:

$$P_e(t)dt = i(t)d\lambda \tag{2.60}$$

We recognize the left-hand side of the equation as being the increment in electric energy $dW_e$, and we therefore write

$$dW_e = id\lambda \tag{2.61}$$

Assuming a lossless device, we can therefore write an energy balance equation which is a modification of Eq. (2.61).

$$dW_e = dW_{\text{fld}} + dW_{\text{mech}} \tag{2.62}$$

The increment in mechanical output energy can be expressed in the case of a translational (linear motion) increment $dx$ and the associated force exerted by the field $F_{\text{fld}}$ as

$$dW_{\text{mech}} = F_{\text{fld}}dx \tag{2.63}$$

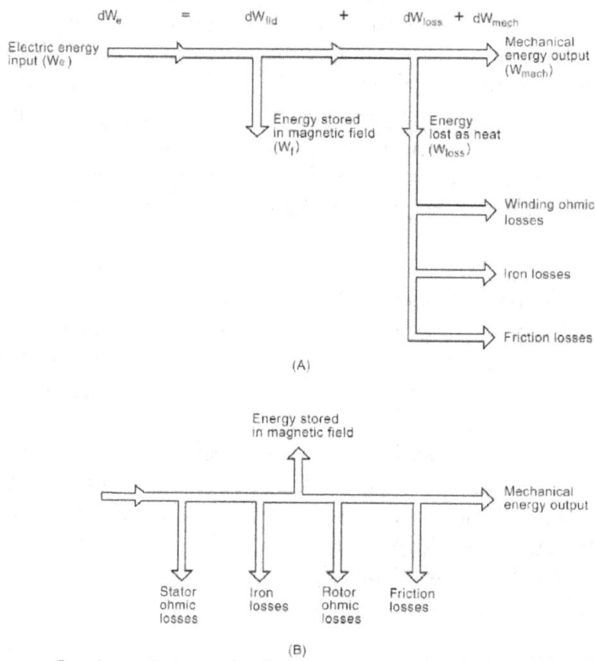


**Figure 2.23** Energy flow in an electromechanical energy conversion device: (A) with losses segregated, and (B) more realistic representation.

In the case of rotary motion, the force is replaced by torque $T_{\text{fld}}$ and the linear increment $dx$ is replaced by the angular increment $d\theta$.

$$dW_{\text{mech}} = T_{\text{fld}} d\theta \tag{2.64}$$

As a result, we have for the case of linear motion,

$$dW_{\text{fld}} = id\lambda - F_{\text{fld}} dx \tag{2.65}$$

And for rotary motion,

$$dW_{\text{fld}} = id\lambda - T_{\text{fld}} d\theta \tag{2.66}$$

The preceding results say that the net change in the field energy is obtained through knowledge of the incremental electric energy input ($i\ d\lambda$) and the mechanical increment of work done.

The field energy is a function of two states of the system. The first is the displacement variable $x$ (or $\theta$ for rotary motion), and the second is either the flux linkages $\lambda$ or the current $i$. This follows since knowledge of $\lambda$ completely specifies $i$ through the $\lambda$-$i$ characteristic. Let us first take dependence of $W_f$ on $\lambda$ and $x$, and write

$$dW_{\text{fld}}(\lambda, x) = \frac{\partial W_f}{\partial \lambda} d\lambda + \frac{\partial W_f}{\partial x} dx \tag{2.67}$$

The incremental increase in field energy $W_f$ is made up of two components. The first is the product of $d\lambda$ and a (gain factor) coefficient equal to the partial derivative of $W_f$ with respect to $\lambda$ ($x$ is held constant); the second component is equal to the product of $dx$ and the partial derivative of $W_f$ with respect to $x$ ($\lambda$ is held constant). This is a consequence of Taylor's series for a function of two variables. We conclude that

$$i = \frac{\partial W_f(\lambda, x)}{\partial \lambda} \tag{2.68}$$

$$F_{\text{fld}} = -\frac{\partial W_f(\lambda, x)}{\partial x} \tag{2.69}$$

This result states that if the energy stored in the field is known as a function of $\lambda$ and $x$, then the electric force developed can be obtained by the partial differentiation shown in Eq. (2.69).

For rotary motion, we replace $x$ by $\theta$ in the foregoing development to arrive at

$$T_{\text{fld}} = \frac{-\partial W_f(\lambda, x)}{\partial \theta} \tag{2.70}$$

Of course, $W_f$ as a function of $\lambda$ and $\theta$ must be available to obtain the developed torque. Our next task, therefore, is to determine the variations of the field energy with $\lambda$ and $x$ for linear motion and that with $\lambda$ and $\theta$ for rotary motion.

## Field Energy

To find the field force we need an expression for the field energy $W_f(\lambda_p, x_p)$ at a given state $\lambda_p$ and $x_p$. This can be obtained by integrating the relation of Eq. (2.70) to obtain

$$W_f(\lambda_p, x_p) = \int_0^{\lambda_p} i(\lambda_p, x_p) d\lambda \tag{2.71}$$

If the $\lambda$–$i$ characteristic is linear in $i$ then

$$W_f(\lambda_p, x_p) = \frac{\lambda_p^2}{2L} \tag{2.72}$$

Note that $L$ can be a function of $x$.

### Coil Voltage

Using Faraday's law, we have

$$e(t) = \frac{d\lambda}{dt} = \frac{d}{dt}(Li)$$

Thus, since $L$ is time dependent, we have

$$e(t) = L\frac{di}{dt} + i\frac{dL}{dt}$$

However,

$$\frac{dL}{dt} = \frac{dL}{dx}\left(\frac{dx}{dt}\right) = \upsilon\frac{dL}{dx}$$

As a result, we assert that the coil voltage is given by

$$e(t) = L\frac{di}{dt} + i\upsilon\frac{dL}{dx} \tag{2.73}$$

where $\upsilon = dx/dt$.

## 2.11 MULTIPLY EXCITED SYSTEMS

Some rotating electromechanical energy conversion devices have more than one exciting winding and are referred to as multiply excited systems. The torque produced can be obtained by a simple extension of the techniques discussed earlier. Consider a system with three windings as shown in Figure 2.24.

The differential electric energy input is

$$dW_e = i_1 d\lambda_1 + i_2 d\lambda_2 + i_3 d\lambda_3 \tag{2.74}$$

The mechanical energy increment is given by

$$dW_{\text{mech}} = T_{\text{fld}} d\theta$$

Thus, the field energy increment is obtained as

$$\begin{aligned} dW_{\text{fld}} &= dW_e - dW_{\text{mech}} \\ &= i_1 d\lambda_1 + i_2 d\lambda_2 + i_3 d\lambda_3 - T_{\text{fld}} d\theta \end{aligned} \tag{2.75}$$

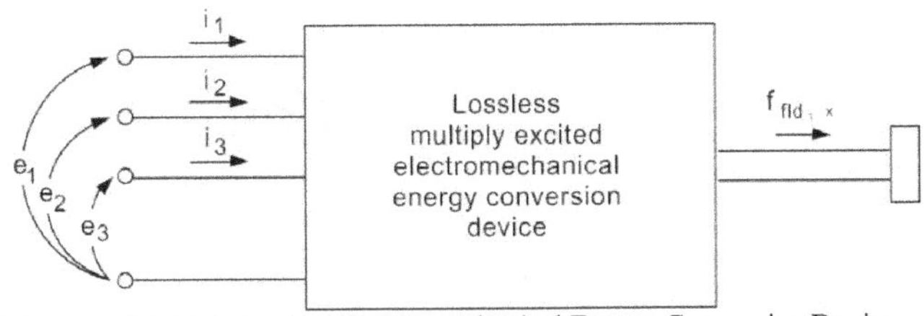


**Figure 2.24** Lossless Multiply Excited Electromechanical Energy Conversion Device.

If we express $W_{\text{fld}}$ in terms of $\lambda_1$, $\lambda_2$, $\lambda_3$, and $\theta$, we have

$$dW_f = \frac{\partial W_f}{\partial \lambda_1} d\lambda_1 + \frac{\partial W_f}{\partial \lambda_2} d\lambda_2 + \frac{\partial W_f}{\partial \lambda_3} d\lambda_3 + \frac{\partial W_f}{\partial \theta} d\theta \tag{2.76}$$

Comparing Eqs. (2.75) and (2.76), we conclude:

$$T_{\text{fld}} = -\frac{\partial W_f(\lambda_1, \lambda_2, \lambda_3, \theta)}{\partial \theta} \tag{2.77}$$

$$i_k = \frac{\partial W_f(\lambda_1, \lambda_2, \lambda_3, \theta)}{\partial \lambda_k} \tag{2.78}$$

The field energy at a state corresponding to point $P$, where $\lambda_1 = \lambda_{1p}$, $\lambda_2 = \lambda_{2p}$, $\lambda_3 = \lambda_{3p}$, and $\theta = \theta_p$ is obtained as:

$$W_f(\lambda_{1p}, \lambda_{2p}, \lambda_{3p}, \theta_p) = \sum_{i=1}^{3} \sum_{j=1}^{3} \frac{1}{2} \lambda_{ip} \Gamma_{ij} \lambda_{jp} \tag{2.79}$$

where

$$i_1 = \Gamma_{11}\lambda_1 + \Gamma_{12}\lambda_2 + \Gamma_{13}\lambda_3 \tag{2.80}$$

$$i_2 = \Gamma_{12}\lambda_1 + \Gamma_{22}\lambda_2 + \Gamma_{23}\lambda_3 \tag{2.81}$$

$$i_3 = \Gamma_{13}\lambda_1 + \Gamma_{23}\lambda_2 + \Gamma_{33}\lambda_3 \tag{2.82}$$

The matrix $\mathbf{\Gamma}$ is the inverse of the inductance matrix $\mathbf{L}$:

$$\Gamma = L^{-1} \tag{2.83}$$

## 2.12 DOUBLY EXCITED SYSTEMS

Rotating electric machines are characterized by more than one exciting winding. In the system shown in Figure 2.25, a coil on the stator is fed by an

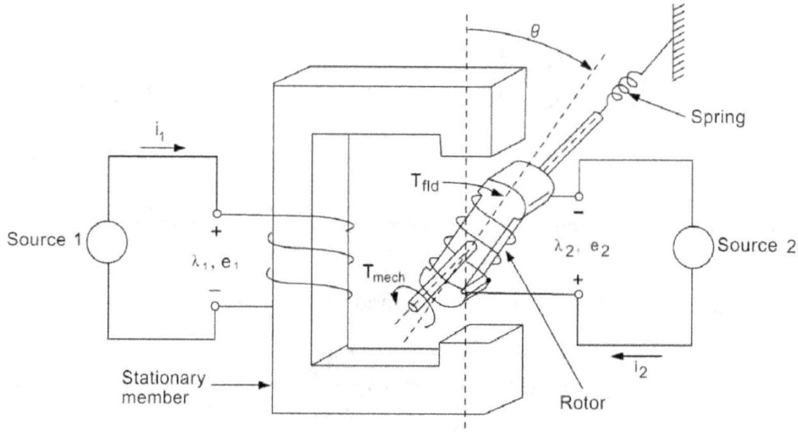


**Figure 2.25 Doubly Excited Electromechanical Energy Conversion Device.**

electric energy source 1 and a second coil is mounted on the rotor and fed by source 2. For this doubly excited system, we write the relation between flux linkages and currents as

$$\lambda_1 = L_{11}(\theta)i_1 + M(\theta)i_2 \tag{2.84}$$

$$\lambda_2 = M(\theta)i_1 + L_{22}(\theta)i_2 \tag{2.85}$$

The self-inductances $L_{11}$ and $L_{22}$ and the mutual inductance $M$ are given as functions of $\theta$ as follows:

$$L_{11}(\theta) = L_1 + \Delta L_1 \cos 2\theta \tag{2.86}$$

where

$$L_1 = \tfrac{1}{2}(L_{max} + L_{min}) \tag{2.87}$$

$$\Delta L_1 = \frac{1}{2}(L_{max} - L_{min}) \tag{2.88}$$

$$M(\theta) = M_0 \cos\theta \tag{2.89}$$

$$L_{22}(\theta) = L_2 + \Delta L_2 \cos 2\theta \tag{2.90}$$

In many practical applications, $\Delta L_2$ is considerably less than $L_2$ and we may conclude that $L_{22}$ is independent of the rotor position.

$$T_{fld} = -[(i_1^2 \Delta L_1 + i_2^2 \Delta L_2)\sin 2\theta + i_1 i_2 M_0 \sin\theta)] \tag{2.91}$$

Let us define

$$T_R = i_1^2 \Delta L_1 + i_2^2 \Delta L_2 \tag{2.92}$$

$$T_M = i_1 i_2 M_0 \tag{2.93}$$

Thus the torque developed by the field is written as

$$T_{\text{fld}} = -(T_M \sin\theta + T_R \sin 2\theta) \tag{2.94}$$

Note that for a round rotor, the reluctance of the air gap is constant and hence the self-inductances $L_{11}$ and $L_{22}$ are constant, with the result that $\Delta L_1 = \Delta L_2 = 0$. We therefore see that for a round rotor $T_R = 0$, and in this case

$$T_{\text{fld}} = -T_M \sin\theta \tag{2.95}$$

For an unsymmetrical rotor the torque is made up of a reluctance torque $T_R \sin 2\theta$ and the primary torque $T_M \sin\theta$.

## 2.13 SALIENT-POLE MACHINES

The majority of electromechanical energy conversion devices used in present-day applications are rotating electric machines category with symmetrical stator structure. From a broad geometric configuration point of view, such machines can be classified as being either of the salient-pole type, as this class is a simple extension of the discussion of the preceding section, or round-rotor.

In a salient-pole machine, one member (the rotor in our discussion) has protruding or salient poles and thus the air gap between stator and rotor is not uniform, as shown in Figure 2.26. It is clear that results of Section 2.12 are applicable here and we simply modify these results to conform to common machine terminology, shown in Figure 2.26. Subscript 1 is replaced by $s$ to represent stator quantities and subscript 2 is replaced by $r$ to represent rotor quantities. Thus we rewrite Eq. (2.84) as

$$\lambda_s = (L_s + \Delta L_s \cos 2\theta) i_s + (M_0 \cos\theta) i_r \tag{2.96}$$

Similarly, Eq. (2.85) is rewritten as

$$\lambda_r = (M_0 \cos\theta) i_s + L_r i_r \tag{2.97}$$

Note that we assume that $L_{22}$ is independent of $\theta$ and is represented by $L_r$. Thus, $\Delta L_2 = 0$ under this assumption. The developed torque given by Eq. (2.91) is written as

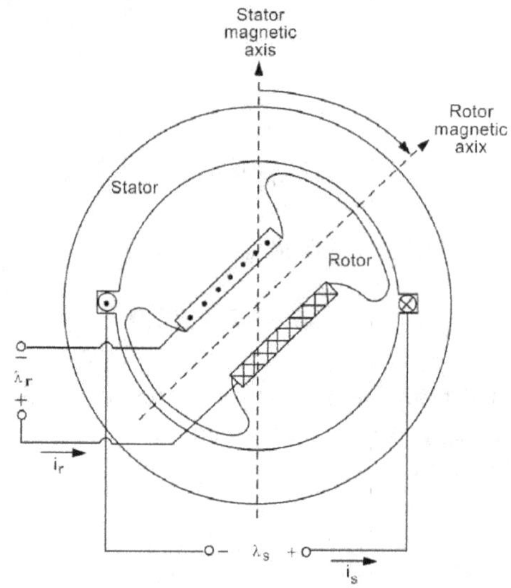


**Figure 2.26** Two-Pole Single-Phase Salient-Pole Machine With Saliency On The Rotor.

$$T_{\text{fld}} = -i_s i_r M_0 \sin\theta - i_s^2 \Delta L_s \sin 2\theta \tag{2.98}$$

We define the primary or main torque $T_1$ by

$$T_1 = -i_s i_r M_0 \sin\theta \tag{2.99}$$

We also define the reluctance torque $T_2$ by

$$T_2 = -i_s^2 \Delta L_s \sin 2\theta \tag{2.100}$$

Thus we have

$$T_{\text{fld}} = T_1 + T_2 \tag{2.101}$$

Let us assume that the source currents are sinusoidal.

$$i_s(t) = I_s \sin\omega_s t \tag{2.102}$$

$$i_r(t) = I_r \sin\omega_r t \tag{2.103}$$

Assume also that the rotor is rotating at an angular speed $\omega_m$ and hence,

$$\theta(t) = \omega_m t + \theta_0 \tag{2.104}$$

We examine the nature of the instantaneous torque developed under these conditions.

The primary or main torque $T_1$, expressed by Eq. (2.99), reduces to the following form under the assumptions of Eqs. (2.102) to (2.104):

$$T_1 = -\frac{I_s I_r M_0}{4}\{[(\omega_m + \omega_s - \omega_r)t + \theta_0] + \sin[(\omega_m - \omega_s + \omega_r)t + \theta_0]$$
$$-\sin[(\omega_m + \omega_s + \omega_r)t + \theta_0] - \sin[(\omega_m - \omega_s - \omega_r)t + \theta_0]]\} \tag{2.105}$$

An important characteristic of an electric machine is the average torque developed. Examining Eq. (2.105), we note that $T_1$ is made of four sinusoidal components each of zero average value if the coefficient of $t$ is different from zero. It thus follows that as a condition for non-zero average of $T_1$, we must satisfy one of the following:

$$\omega_m = \pm\omega_s \pm \omega_r \tag{2.106}$$

For example, when

$$\omega_m = -\omega_s + \omega_r$$

then

$$T_{1_{\text{av}}} = -\frac{\mathrm{I}_s \mathrm{I}_r \mathrm{M}_0}{4}\sin\theta_0$$

and when

$$\omega_m = \omega_s + \omega_r$$

then

$$T_{1_{\text{av}}} = -\frac{I_s I_r M_0}{4}\sin\theta_0$$

The reluctance torque $T_2$ of Eq. (2.100) can be written using Eqs. (2.102) to (2.104) as

$$T_2 = -\frac{I_s^2 \Delta L_s}{4}\{2\sin(2\omega_m t + 2\theta_0) - \sin[2(\omega_m + \omega_s)t + 2\theta_0]$$
$$-\sin[2(\omega_m - \omega_s)t + 2\theta_0]\} \tag{2.107}$$

The reluctance torque will have an average value for

$$\omega_m = \pm\omega_s \tag{2.108}$$

When either of the two conditions is satisfied,

$$T_{2_{\text{av}}} = \frac{I_s^2 \Delta L_2}{4} \sin 2\theta_0$$

## 2.14 ROUND OR SMOOTH AIR-GAP MACHINES

A round-rotor machine is a special case of the salient-pole machine where the air gap between the stator and rotor is (relatively) uniform. The term *smooth air gap* is an idealization of the situation illustrated in Figure 2.27. It is clear that for the case of a smooth air-gap machine the term $\Delta L_s$ is zero, as the reluctance does not vary with the angular displacement $\theta$. Therefore, for the machine of Figure 2.27, we have

$$\lambda_s = L_s i_s + M_0 \cos\theta i_r \tag{2.109}$$

$$\lambda_r = M_0 \cos\theta i_s + L_r i_r \tag{2.110}$$

Under the assumptions of Eqs. (2.102) to (2.104), we obtain

$$T_{\text{fld}} = T_1 \tag{2.111}$$

where $T_1$ is as defined in Eq. (2.105).

We have concluded that for an average value of $T_1$ to exist, one of the conditions of Eq. (2.106) must be satisfied:

$$\omega_m = \pm\omega_s \pm \omega_r \tag{2.112}$$

We have seen that for

$$\omega_m = -\omega_s + \omega_r \tag{2.113}$$

then

$$T_{\text{av}} = -\frac{I_s I_r M_0}{4} \sin\theta \tag{2.114}$$

Now substituting Eq. (2.113) in to Eq. (2.105), we get

$$T_{fld} = -\frac{I_s I_r M_0}{4} \{\sin\theta_0 + \sin[2(\omega_r - \omega_s)t + \theta_0] - \sin(2\omega_r t + \theta_0) - \sin(-2\omega_s t + \theta_0)\} \tag{2.115}$$

The first term is a constant, whereas the other three terms are still sinusoidal

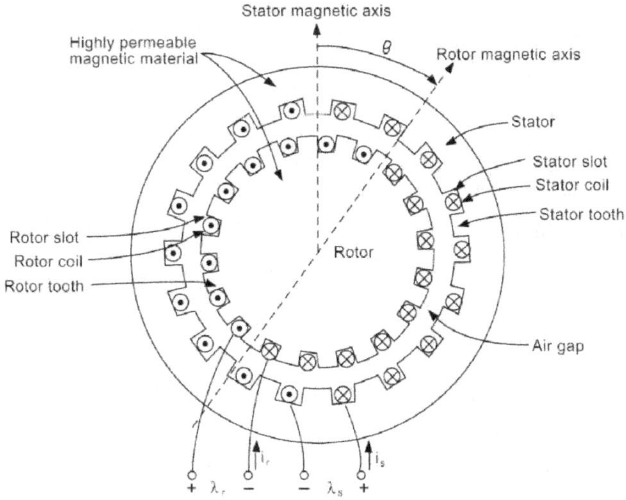


**Figure 2.27** Smooth Air-Gap Machine.

time functions and each represents an alternating torque. Although these terms are of zero average value, they can cause speed pulsations and vibrations that may be harmful to the machine's operation and life. The alternating torques can be eliminated by adding additional windings to the stator and rotor, as discussed presently.

## Two-Phase Machines

Consider the machine of Figure 2.28, where each of the distributed windings is represented by a single coil. It is clear that this is an extension of the machine of Figure 2.27 by adding one additional stator winding ($bs$) and one additional rotor winding ($br$) with the relative orientation shown in Figure 2.28(B). Our analysis of this machine requires first setting up the inductances required. This can be best done using vector terminology. We can write for this four-winding system:

$$\begin{bmatrix} \lambda_{as} \\ \lambda_{ar} \\ \lambda_{bs} \\ \lambda_{br} \end{bmatrix} = \left[\begin{array}{cc|cc} L_s & M_0\cos\theta & 0 & M_0\sin\theta \\ M_0\cos\theta & L_r & -M_0\sin\theta & 0 \\ \hline 0 & M_0\sin\theta & L_s & M_0\cos\theta \\ -M_0\sin\theta & 0 & M_0\cos\theta & L_r \end{array}\right] \begin{bmatrix} i_{as} \\ i_{ar} \\ i_{bs} \\ i_{br} \end{bmatrix} \tag{2.116}$$

The field energy is the same as given by Eq. (2.108). The torque is obtained in the usual manner. Let us now assume that the terminal currents are given by the balanced, two-phase current sources

$$i_{as} = I_s \cos\omega_s t \tag{2.117}$$

$$i_{bs} = I_s \sin \omega_s t \tag{2.118}$$

$$i_{ar} = I_r \cos \omega_r t \tag{2.119}$$

$$i_{br} = I_r \sin \omega_r t \tag{2.120}$$

We also assume that

$$\theta(t) = \omega_m t + \theta_0 \tag{2.121}$$

The torque is given by

$$T_{\text{fld}} = M_0[(i_{ar}i_{bs} - i_{br}i_{as})\cos\theta - (i_{ar}i_{as} + i_{br}i_{bs})\sin\theta] \tag{2.122}$$

Substituting Eqs. (2.117) through (2.121) into (2.122), we obtain (after some manipulations)

$$T_{\text{fld}} = M_0 I_s I_r \sin[(\omega_m - \omega_s - \omega_r)t + \theta_0] \tag{2.123}$$

The condition for nonzero average torque is given by

$$\omega_m = \omega_s - \omega_r \tag{2.124}$$

For this condition, we have

$$T_{\text{fld}} = M_0 I_s I_r \sin\theta_0 \tag{2.125}$$

The instantaneous torque in this case is constant in spite of the excitation being sinusoidal.

## 2.15 MACHINE-TYPE CLASSIFICATION

The results of the preceding discussion provides a basis for defining conventional machine types.

### Synchronous Machines

The two-phase machine of Figure 2.29 is excited with direct current applied to the rotor ($\omega_r = 0$) and balanced two-phase currents of frequency $\omega_s$ applied to the stator. With $\omega_r = 0$ we get

$$\omega_m = \omega_s \tag{2.126}$$

Thus, the rotor of the machine should be running at the single value defined by

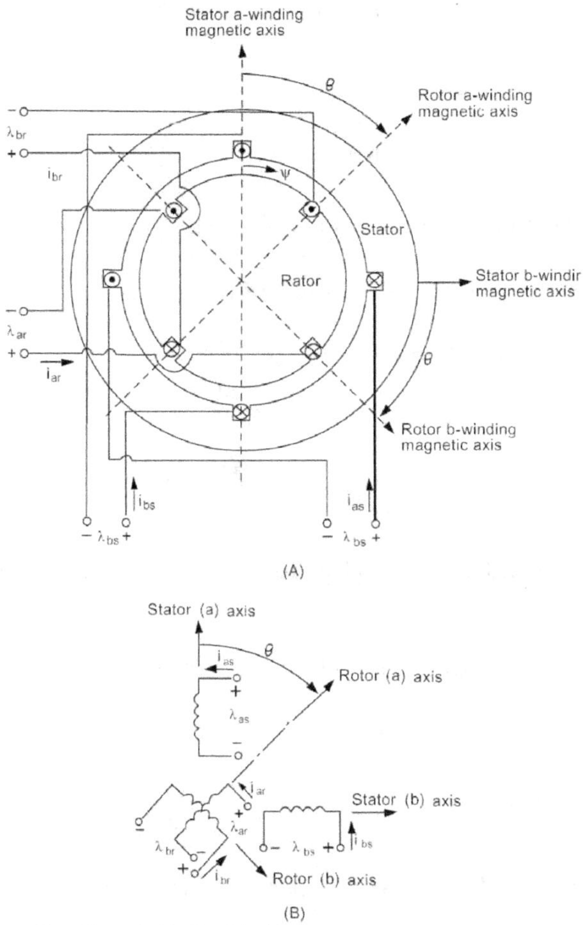

**Figure 2.28** Two-Phase Smooth Air-Gap Machine.

the stator sources to produce a torque with nonzero average value. This mode of operation yields a synchronous machine which is so named because it can convert average power only at one mechanical speed – the synchronous speed, $\omega_s$. The synchronous machine is the main source of electric energy in modern power systems acting as a generator.

### *Induction Machines*

Single-frequency alternating currents are fed into the stator circuits and the rotor circuits are all short circuited in a conventional induction machine. The machine in Figure 2.28 is used again for the analysis. Equations (2.117) and (2.118) still apply and are repeated here:

$$i_{a_s} = I_s \cos \omega_s t \tag{2.127}$$

$$i_{b_s} = I_s \sin \omega_s t \tag{2.128}$$

With the rotor circuits short-circuited,

$$v_{ar} = v_{br} = 0 \tag{2.129}$$

and the rotor is running according to Eq. (2.121):

$$\theta(t) = \omega_m t + \theta_0 \tag{2.130}$$

Conditions (2.129) are written as

$$v_{ar} = R_r i_{ar} + \frac{d\lambda_{ar}}{dt} = 0 \tag{2.131}$$

$$v_{br} = R_r i_{br} + \frac{d\lambda_{br}}{dt} = 0 \tag{2.132}$$

Here we assume that each rotor phase has a resistance of $R_r \Omega$. We have, by Eq. (2.116),

$$\lambda_{ar} = M_0 \cos\theta i_{as} + L_r i_{ar} + M_0 \sin\theta i_{bs} \tag{2.133}$$

$$\lambda_{br} = -M_0 \sin\theta i_{as} + M_0 \cos\theta i_{bs} + L_r i_{br} \tag{2.134}$$

As a result, we have

$$0 = R_r i_{ar} + L_r \frac{di_{ar}}{dt} + M_0 I_s \frac{d}{dt}[\cos\omega_s t \cos(\omega_m t + \theta_0) + \sin\omega_s t \sin(\omega_m t + \theta_0) \tag{2.135}$$

and

$$0 = R_r i_{br} + L_r \frac{di_{br}}{dt} + M_0 I_s \frac{d}{dt}[-\cos\omega_s t \sin(\omega_m t + \theta_0) + \sin\omega_s t \cos(\omega_m t + \theta \tag{2.136}$$

A few manipulations provide us with

$$M_0 I_s(\omega_s - \omega_m)\sin[(\omega_s - \omega_m)t - \theta_0] = L_r \frac{di_{ar}}{dt} + R_r i_{ar} \tag{2.137}$$

$$-M_0 I_s(\omega_s - \omega_m)\cos[(\omega_s - \omega_m)t - \theta_0] = L_r \frac{di_{br}}{dt} + R_r i_{br} \tag{2.138}$$

The right-hand sides are identical linear first-order differential operators. The

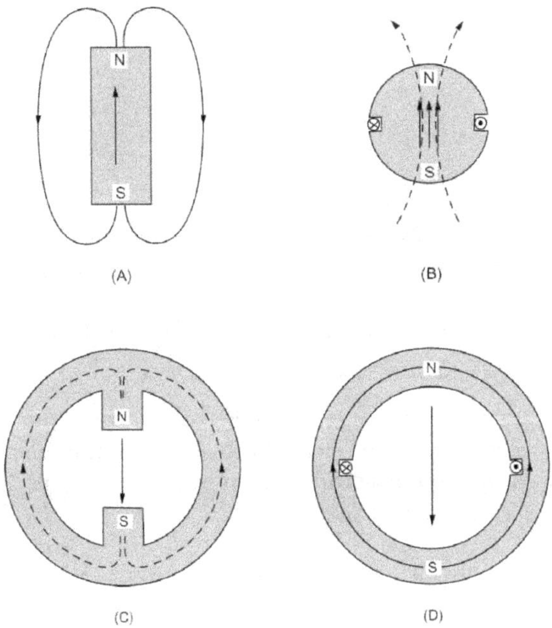


**Figure 2.29** Two-Pole Configurations.

left sides are sinusoidal voltages of equal magnitude but 90° apart in phase. The rotor currents will have a frequency of $(\omega_s - \omega_m)$, which satisfies condition (2.124), and thus an average power and an average torque will be produced by the induction machine. We emphasize the fact that currents induced in the rotor have a frequency of $(\omega_s - \omega_m)$ and that average torque can be produced.

## 2.16 P-POLE MACHINES

The configuration of the magnetic field resulting from coil placement in the magnetic structure determines the number of poles in an electric machine. An important point to consider is the convention adopted for assigning polarities in schematic diagrams, which is discussed presently. Consider the bar magnet of Figure 2.29(A). The magnetic flux lines are shown as closed loops oriented from the South Pole to the North Pole within the magnetic material. Figure 2.29(B) shows a two-pole rotor with a single coil with current flowing in the direction indicated by the dot and cross convention. According to the right-hand rule, the flux lines are directed upward inside the rotor material, and as a result we assert that the south pole of the electromagnet is on the bottom part and that the North Pole is at the top, as shown.

The situation with a two-pole stator is explained in Figure 2.29(C) and 2.29(D). First consider 2.30(C), showing a permanent magnet shaped as shown. According to our convention, the flux lines are oriented away from the South Pole toward the North Pole within the magnetic material (not in air gaps). For

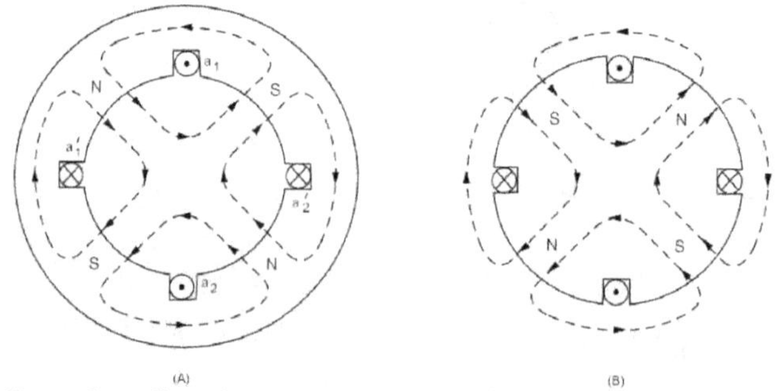


**Figure 2.30** Four-pole configurations: (A) stator arrangement, and (B) rotor arrangement.

2.29(D), we have an electromagnet resulting from the insertion of a single coil in slots on the periphery of the stator as shown. The flux lines are oriented in accordance with the right-hand rule and we conclude that the North and South Pole orientations are as shown in the figure.

Consider now the situation illustrated in Figure 2.30, where two coils are connected in series and placed on the periphery of the stator in part (a) and on the rotor in part (b). An extension of the prior arguments concerning a two-pole machine results from the combination of the stator and rotor of Figure 2.31 and is shown in Figure 2.31 to illustrate the orientation of the magnetic axes of rotor and stator.

It is clear that any arbitrary even number of poles can be achieved by placing the coils of a given phase in symmetry around the periphery of stator and rotor of a given machine. The number of poles is simply the number encountered in one round trip around the periphery of the air gap. It is necessary for successful operation of the machine to have the same number of poles on the stator and rotor.

Consider the four-pole, single-phase machine of Figure 2.31. Because of the symmetries involved, the mutual inductance can be seen to be

$$M(\theta) = M_0 \cos 2\theta \tag{2.139}$$

Compared with Eq. (2.89) for a two-pole machine, we can immediately assert that for a $P$-pole machine,

$$M(\theta) = M_0 \cos \frac{P\theta}{2} \tag{2.140}$$

where $P$ is the number of poles.

We note here that our treatment of the electric machines was focused on two-pole configurations. It is clear that extending our analytic results to a $P$-pole machine can easily be done by replacing the mechanical angle $\theta$ in a

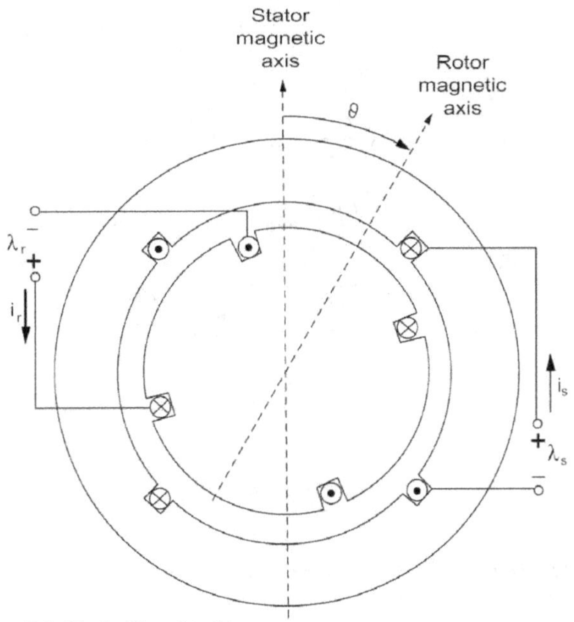


**Figure 2.31** Four-Pole Single-Phase Machine.

relation developed for a two-pole machine by the angle $P\theta/2$ to arrive at the corresponding relation for a $P$-pole machine. As an example, Eqs. (2.109) and (2.110) for a $P$-pole machine are written as

$$\lambda_s = L_s i_s + \left( M_0 \cos\frac{P\theta}{2} \right) i_r \qquad \textbf{(2.141)}$$

$$\lambda_r = \left( M_0 \cos\frac{P\theta}{2} \right) i_s + L_r i_r \qquad \textbf{(2.142)}$$

Similarly, the torque expression in Eq. (2.99) becomes

$$T_1 = -i_s i_r M_0 \sin\frac{P\theta}{2} \qquad \textbf{(2.143)}$$

Note that $\theta$ in the expressions above is in mechanical degrees.

The torque $T_1$ under the sinusoidal excitation conditions (2.102) and (2.103) given by Eq. (2.105) is rewritten for a $P$-pole machine as

$$\begin{aligned} T_1 = -\frac{I_s I_r M_0}{4} \Bigg\{ & \sin\left[\left(\frac{P\omega_m}{2} + \omega_s - \omega_r\right)t + \frac{P\theta_0}{2}\right] \\ & + \sin\left[\left(\frac{P\omega_m}{2} - \omega_2 + \omega_r\right)t + \frac{P\theta_0}{2}\right] \\ & - \sin\left[\left(\frac{P\omega_m}{2} + \omega_s + \omega_r\right)t + \frac{P\theta_0}{2}\right] \\ & - \sin\left[\left(\frac{P\omega_m}{2} - \omega_s - \omega_r\right)t + \frac{P\theta_0}{2}\right] \Bigg\} \end{aligned} \quad \textbf{(2.144)}$$

The conditions for average torque production of Eq. (2.106) are written for a $P$-pole machine as

$$\omega_m = \frac{2}{P}(\pm\omega_s \pm \omega_r) \quad \textbf{(2.145)}$$

Thus, for given electrical frequencies the mechanical speed is reduced as the number of poles is increased.

A time saving and intuitively appealing concept in dealing with $P$-pole machines is that of electrical degrees. Let us define the angle $\theta_e$ corresponding to take a mechanical angle $\theta$, in a $P$-pole machine the

$$\theta_e = \frac{P}{2}\theta \quad \textbf{(2.146)}$$

With this definition we see that all statements, including $\theta$ for a two-pole machine apply to any $P$-pole machine with $\theta$ taken as an electrical angle.

Consider the first condition of Eq. (2.145) with $\omega_r = 0$ corresponding to synchronous machine operation:

$$\omega_m = \frac{2}{P}\omega_s \quad \textbf{(2.147)}$$

The stator angular speed $\omega_s$ is related to frequency $f_s$ in hertz by

$$\omega_s = 2\pi f_s \quad \textbf{(2.148)}$$

The mechanical angular speed $\omega_m$ is related to the mechanical speed $n$ in revolutions per minute by

$$\omega_m = \frac{2\pi n}{60} \quad \textbf{(2.149)}$$

Combining Eq. (2.147) with Eq. (2.149), we obtain

$$f_s = \frac{Pn}{120} \tag{2.150}$$

This is an important relation in the analysis of rotating electrical machines.

## 2.17 POWER SYSTEM REPRESENTATION

A major portion of the modern power system utilizes three-phase ac circuits and devices. A detailed representation of each of the three phases in the system is cumbersome and can also obscure information about the system. A balanced three-phase system is solved as a single-phase circuit made of one line and the neutral return; thus a simpler representation would involve retaining one line to represent the three phases and omitting the neutral. Standard symbols are used to indicate the various components. A transmission line is represented by a single line between two ends. The simplified diagram is called the *single-line diagram.*

The one-line diagram summarizes the relevant information about the system for the particular problem studied. For example, relays and circuit breakers are not important when dealing with a normal state problem. However, when fault conditions are considered, the location of relays and circuit breakers is important and is thus included in the single-line diagram.

The International Electro technical Commission (IEC), the American National Standards Institute (ANSI), and the Institute of Electrical and Electronics Engineers (IEEE) have published a set of standard symbols for electrical diagrams. A basic symbol for a rotating machine is a circle. Figure 2.32(A) shows rotating machine symbols. If the winding connection is desired, the connection symbols may be shown in the basic circle using the representations given in Figure 2.32(B). The symbols commonly used for transformer representation are given in Figure 2.33(A). The two-circle symbol is the symbol to be used on schematics for equipment having international usage according to IEC. Figure 2.33(B) shows symbols for a number of single-phase transformers, and Figure 2.34 shows both single-line symbols and three-line symbols for three-phase transformers.

## PROBLEMS

### Problem 2.1

In the circuit shown in Figure 2.35, the source phasor voltage is $V = 30\angle 15^\circ$. Determine the phasor currents $I_2$ and $I_3$ and the impedance $Z_2$. Assume that $I_1$ is equal to five $A$. Calculate the apparent power produced by the source and the individual apparent powers consumed by the 1-ohm resistor, the impedance $Z_2$, and the resistance $R_3$. Show that conservation of power holds true.

### Problem 2.2

A three-phase transmission link is rated 100 kVA at 2300 V. When operating at rated load, the total resistive and reactive voltage drops in the link are, respectively, 2.4 and 3.6 % of the rated voltage. Determine the input power and power factor when the link delivers 60 kW at 0.8 PF lagging at 2300 V.

### Problem 2.3

A 60-hp, three-phase, 440-V induction motor operates at 0.8 PF lagging.

a) Find the active, reactive, and apparent power consumed per phase.
b) Suppose the motor is supplied from a 440-V source through a feeder whose impedance is $0.5 + j0.3$ ohm per phase.

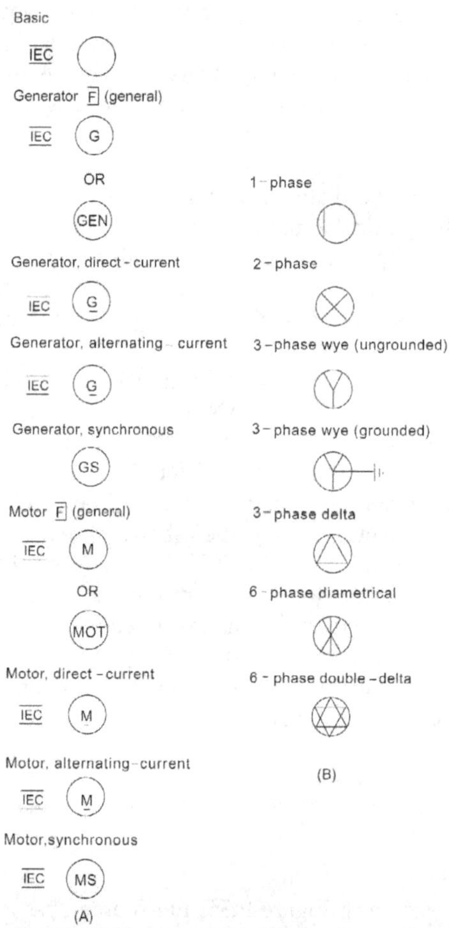


**Figure 2.32** Symbols for Rotating Machines (A) and Their Winding Connections (B).

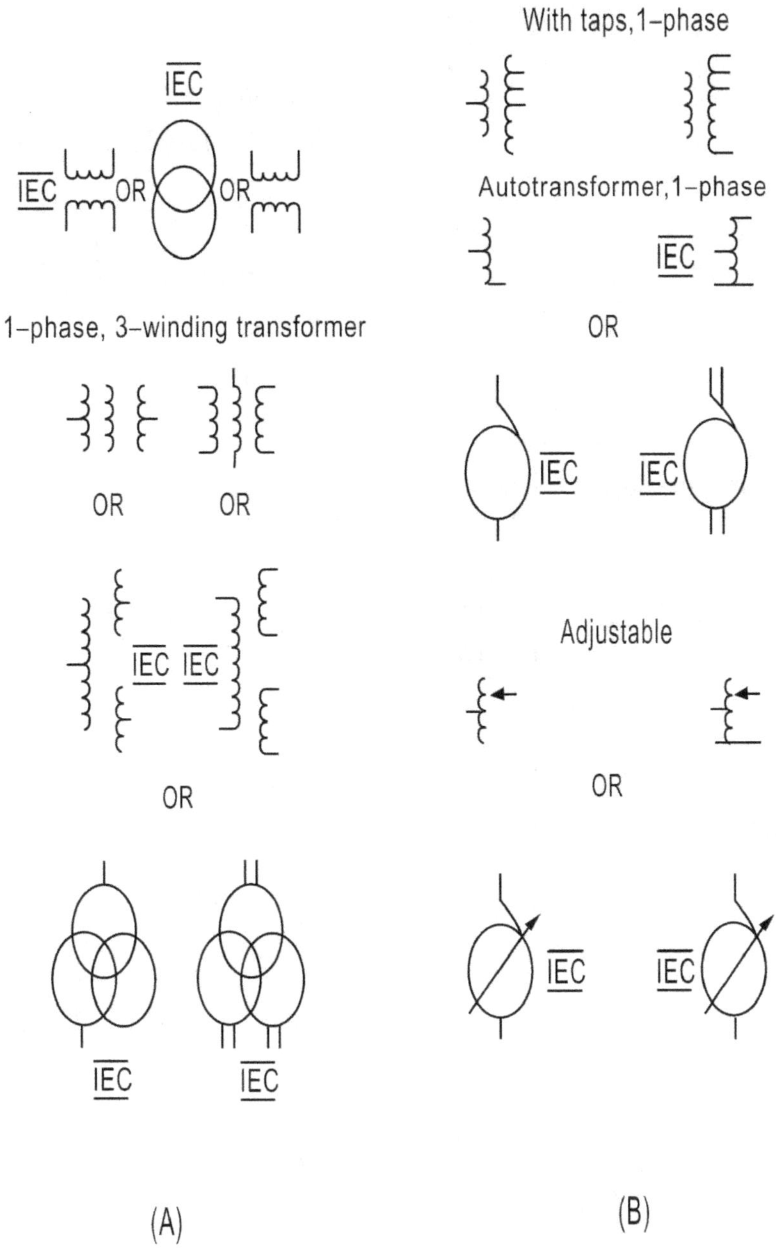


**Figure 2.33** (A) Transformer Symbols, and (B) Symbols for Single-Phase Transformers.

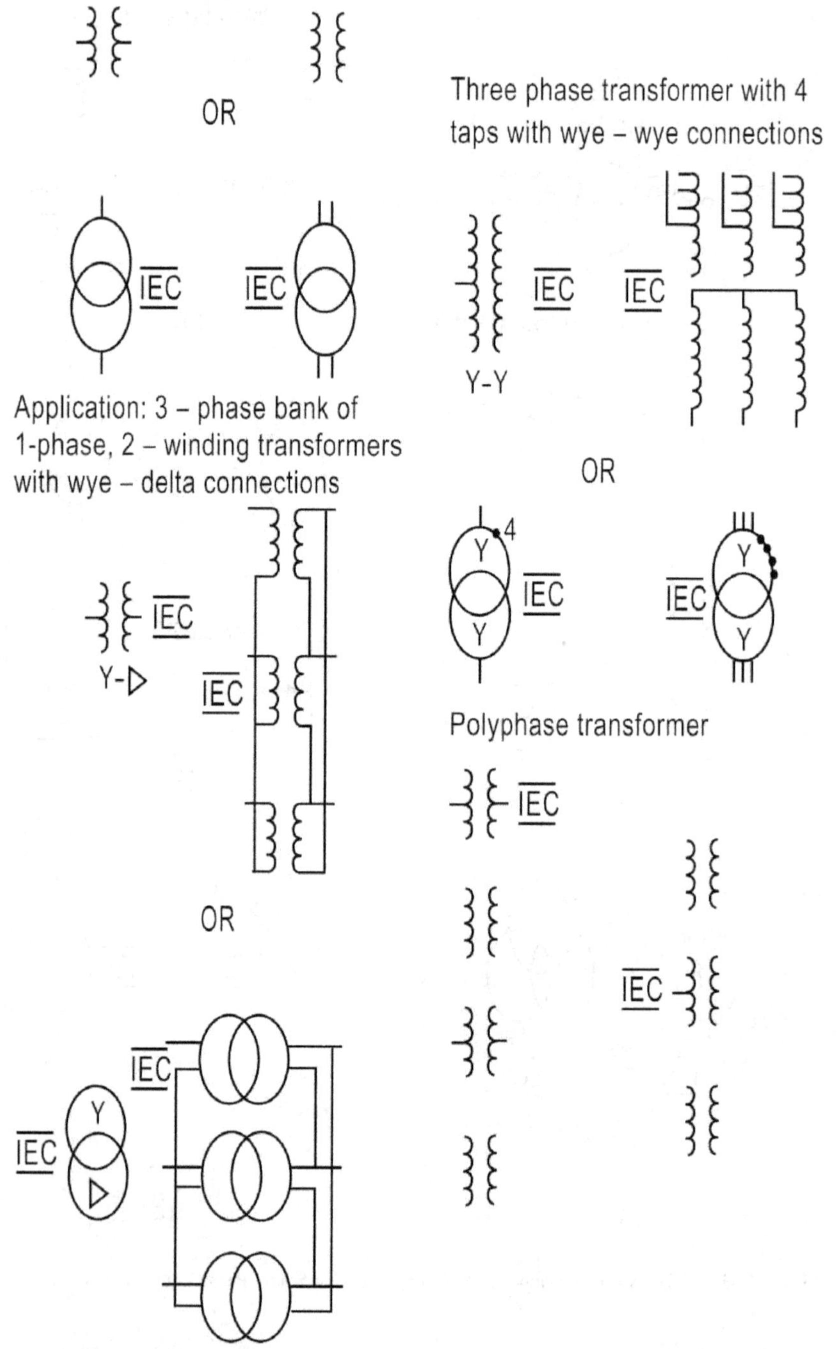


**Figure 2.34** Symbols for Three-Phase Transformers.

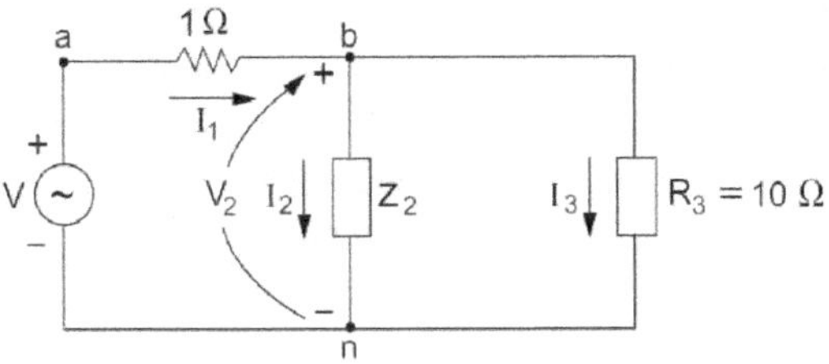


**Figure 2.35** Circuit for Problem 2.1.

c) Calculate the voltage at the motor side, the source power factor, and the efficiency of transmission.

**Problem 2.4**
Repeat Problem 2.3 if the motor's efficiency is 85%.

**Problem 2.5**
Repeat Problem 2.4 if the PF is 0.7 lagging.

**Problem 2.6**
Consider a 100 kW load operating at a lagging power factor of 0.7. A capacitor is connected in parallel with the load to raise the source power factor to 0.9 p.f. lagging. Find the reactive power supplied by the capacitor assuming that the voltage remains constant.

**Problem 2.7**
A balanced Y-connected 3 phase source with voltage $V_{ab} = 240\angle 0^\circ$ V is connected to a balanced Δ load with $Z_\Delta = 30\angle 35^\circ\ \Omega$. Find the currents in each of the load phases and hence obtain the current through each phase of the source.

**Problem 2.8**
Assume that the load of Problem 2.7 is connected to the source using a line whose impedance is $Z_L = 1\angle 80^\circ\ \Omega$ for each phase. Calculate the line currents, the Δ-load currents, and the voltages at the load terminals.

**Problem 2.9**
A balanced, three-phase 240-V source supplies a balanced three-phase load. If the line current $I_A$ is measured to be 5 A, and is in phase with the line-to-line voltage $V_{BC}$, find the per phase load impedance if the load is (a) Y-connected, and (b) Δ-connected.

**Problem 2.10**
Two balanced Y-connected loads, one drawing 20 kW at 0.8 p.f. lagging and the other 30 kW at 0.9 p.f. leading, are connected in parallel and supplied by a balanced three-phase Y-connected, 480-V source. Determine the impedance per phase of each load and the source currents.

**Problem 2.11**

A load of 30 MW at 0.8 p.f. lagging is served by two lines from two generating sources. Source 1 supplies 15 MW at 0.8 p.f. lagging with a terminal voltage of 4600 V line-to-line. The line impedance is $(1.4 + j1.6)\ \Omega$ per phase between source 1 and the load, and $(0.8 + j1)\ \Omega$ per phase between source 2 and the load. Find

a) The voltage at the load terminals
b) The voltage at the terminals of source 2, and
c) The active and reactive power supplied by source 2.

**Problem 2.12**

The impedance of a three-phase line is $0.3 + j2.4$ per phase. The line feeds two balanced three-phase loads connected in parallel. The first load takes 600 kVA at 0.7 p.f. lagging. The second takes 150 kW at unity power factor. The line to line voltage at the load end of the line is 3810.5 V. Find

a) The magnitude of the line voltage at the source end of the line.
b) The total active and reactive power loss in the line.
c) The active and reactive power supplied at the sending end of the line.

**Problem 2.13**

Three loads are connected in parallel across a 12.47 kV three-phase supply. The first is a resistive 60 kW load, the second is a motor (inductive) load of 60 kW and 660 kvar, and the third is a capacitive load drawing 240 kW at 0.8 p.f. Find the total apparent power, power factor, and supply current.

**Problem 2.14**

A Y-connected capacitor bank is connected in parallel with the loads of Problem 2.13. Find

a) The total kvar and capacitance per phase in $\mu$F to improve the overall power factor to 0.8 lagging.
b) The corresponding line current.

**Problem 2.15**

Assume that 30 V and 5 A are chosen as base voltage and current for the circuit of Problem 2.1.

a) Find the corresponding base impedance and VA.
b) Find the phasor currents $I_2$ and $I_3$ in per unit.
c) Determine the source apparent power in per unit.

**Problem 2.16**

Consider the transmission link of Problem 2.2 and choose 100 kVA and 2300 V as base kVA and voltage. Determine the input power in per unit under the conditions of Problem 2.2.

**Problem 2.17**

Assume for the motor of Problem 2.3 that 50 kW and 440 V are taken as base values. Find the voltage in per unit at the motor side.

**Problem 2.18**

Assume that the base voltage is 4600 V in the system of Problem 2.11, and that 50 MVA is the corresponding apparent power base. Repeat Problem 2.11 using per unit values.

**Problem 2.19**

Repeat Problem 2.12 using per unit values assuming that 1000 kVA is base apparent power and 3 Ω is the base impedance.

**Problem 2.20**

The following information is available about a 40 MVA 20-kV/400 kV, single-phase transformer:

$$Z_1 = 0.9 + j1.8\ \Omega$$
$$Z_2 = 128 + j288\ \Omega$$

Using the transformer rating as base, determine the per unit impedance of the transformer from the ohmic value referred to the low voltage side. Find the per unit impedance using the ohmic value referred to the high voltage side.

**Problem 2.21**

Consider a toroidal coil with relative permeability of 1500 with a circular cross section whose radius is 0.025 cm. The outside radius of the toroid is 0.2 cm. Find the inductance of the coil assuming that $N = 10$ turns.

**Problem 2.22**

The eddy-current and hysteresis losses in a transformer are 450 and 550 W, respectively, when operating from a 60-Hz supply with an increase of 10% in flux density. Find the change in core losses.

**Problem 2.23**

The relationship between current, displacement, and flux linkages in a conservative electromechanical device is given by

$$i = \lambda[0.7\lambda + 2.0(x-1)^2]$$

Find expressions for the stored energy and the magnetic field force in terms of λ and $x$. Find the force for $x = 0.9$.

**Problem 2.24**

Repeat Problem 2.23 for the relationship

$$i = \lambda^3 + \lambda(0.2\lambda + 0.9x)$$

**Problem 2.25**

A plunger-type solenoid is characterized by the relation

$$\lambda = \frac{8i}{1 + 10^4 x/2.54}$$

Find the force exerted by the field for $x = 2.54 \times 10^{-3}$ and $i = 12$ A.

**Problem 2.26**

The inductance of a coil used with a plunger-type electromechanical device is given by

$$L = \frac{1.75 \times 10^{-5}}{x}$$

where $x$ is the plunger displacement. Assume that the current in the coil is given by

$$i(t) = 8 \sin \omega t$$

where $\omega = 2\pi(60)$. Find the force exerted by the field for $x = 10^{-2}$ m. Assume that $x$ is fixed and find the necessary voltage applied to the coil terminals given that its resistance is 1 Ω.

**Problem 2.27**

A rotating electromechanical conversion device has a stator and rotor, each with a single coil. The inductances of the device are

$$L_{11} = 0.5 \text{ H} \qquad L_{22} = 2.5 \text{ H}$$
$$L_{12} = 1.25 \cos \theta \text{ H}$$

Where the subscript 1 refers to stator and the subscript 2 refers to rotor. The angle $\theta$ is the rotor angular displacement from the stator coil axis. Express the torque as a function of currents $i_1$, $i_2$ and $\theta$ and compute the torque for $i_1 = 3$ A and $i_2 = 1$ A.

**Problem 2.28**

Assume for the device of Problem 2.27 that

$$L_{11} = 0.3 + 0.2 \cos 2\theta$$

All other parameters are unchanged. Find the torque in terms of $\theta$ for $i_1 = 2.5$ A and (a) $i_2 = 0$; (b) $i_2 = 1.5$ A.

**Problem 2.29**

Assume for the device of Problem 2.27 that the stator and rotor coils are connected in series, with the current being

$$i(t) = I_m \sin \omega t$$

Find the instantaneous torque and its average value over one cycle of the supply current in terms of $I_m$ and $\omega$.

**Problem 2.30**

A rotating electromechanical energy conversion device has the following inductances in terms of $\theta$ in radians (angle between rotor and stator axes):

$$L_{11} = 0.8\theta$$
$$L_{22} = -0.25 + 1.8\theta$$
$$L_{12} = -0.75 + 1.4\theta$$

Find the torque developed for the following excitations.

a) $i_1 = 15$ A, $i_2 = 0$.
b) $i_1 = 0$ A, $i_2 = 15$ A.
c) $i_1 = 15$ A, $i_2 = 15$ A.
d) $i_1 = 15$ A, $i_2 = -15$ A.

**Problem 2.31**

For the machine of Problem 2.27, assume that the rotor coil terminals are shorted ($e_2 = 0$) and that the stator current is given by

$$i_1(t) = I \sin \omega t$$

Find the torque developed as a function of $I$, $\theta$, and time.

**Problem 2.32**

For the device of Problem 2.31, the rotor coil terminals are connected to a 10-Ω resistor. Find the rotor current in the steady state and the torque developed.

# Chapter 3

# POWER GENERATION AND THE SYNCHRONOUS MACHINE

## 3.1 INTRODUCTION

The backbone of any electric power system is a number of generating stations operating in synchronism. At each station there may be several synchronous generators operating in parallel. Synchronous machines represent the largest single-unit electric machine in production. Generators with power ratings of several hundred to over a thousand megavoltamperes (MVA) are fairly common in many utility systems. A synchronous machine provides a reliable and efficient means for energy conversion.

The operation of a synchronous generator is (like all other electromechanical energy conversion devices) based on Faraday's law of electromagnetic induction. The term synchronous refers to the fact that this type of machine operates at constant speed and frequency under steady-state conditions. Synchronous machines are equally capable of operating as motors, in which case the electric energy supplied at the armature terminals of the unit is converted into mechanical form.

## 3.2 THE SYNCHRONOUS MACHINE: PRELIMINARIES

The armature winding of a synchronous machine is on the stator, and the field winding is on the rotor as shown in Figure 3.1. The field is excited by the direct current that is conducted through carbon brushes bearing on slip (or collector) rings. The dc source is called the exciter and is often mounted on the same shaft as the synchronous machine. Various excitation systems with ac exciters and solid-state rectifiers are used with large turbine generators. The main advantages of these systems include the elimination of cooling and maintenance problems associated with slip rings, commutators, and brushes. The pole faces are shaped such that the radial distribution of the air-gap flux density B is approximately sinusoidal as shown in Figure 3.2.

The armature winding includes many coils. One coil is shown in Figure 3.1 and has two coil sides ($a$ and $-a$) placed in diametrically opposite slots on the inner periphery of the stator with conductors parallel to the shaft of the machine. The rotor is turned at a constant speed by a power mover connected to its shaft. As a result, the flux waveform sweeps by the coil sides $a$ and $-a$. The induced voltage in the coil is a sinusoidal time function. For each revolution of the two poles, the coil voltage passes through a complete cycle of values. The frequency of the voltage in cycles per second (hertz) is the same as the rotor speed in revolutions per second. Thus, a two-pole synchronous machine must revolve at 3600 r/min to produce a 60-Hz voltage.

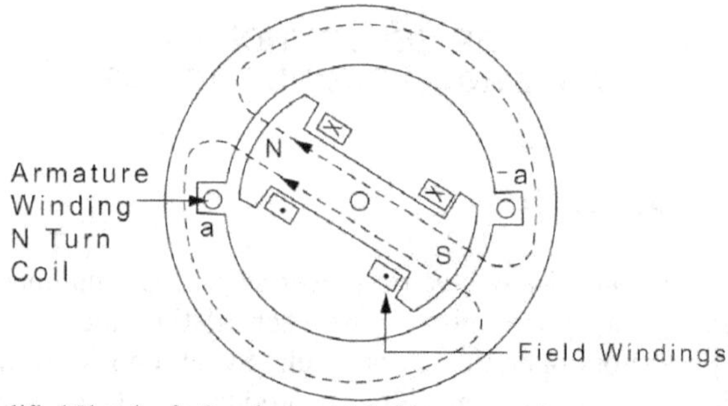


**Figure 3.1** Simplified Sketch of a Synchronous Machine.

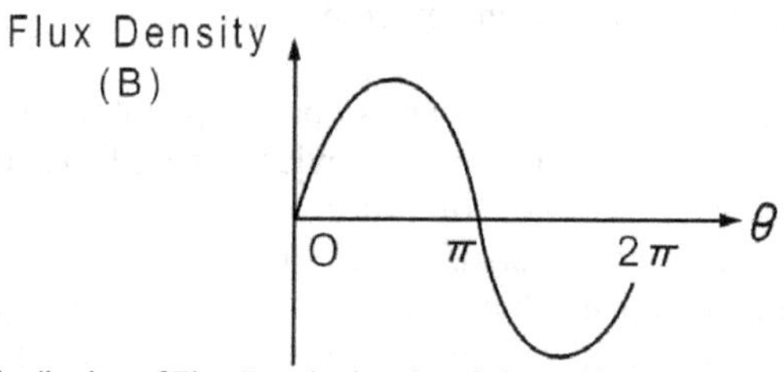


**Figure 3.2** Space Distribution of Flux Density in a Synchronous Generator.

### *P*-Pole Machines

Many synchronous machines have more than two poles. A P-pole machine is one with P poles. As an example, we consider an elementary, single-phase, four-pole generator shown in Figure 3.3. There are two complete cycles in the flux distribution around the periphery as shown in Figure 3.4. The armature winding in this case consists of two coils ($a_1$, $-a_1$, and $a_2$, $-a_2$) connected in series. The generated voltage goes through two complete cycles per revolution of the rotor, and thus the frequency $f$ in hertz is twice the speed in revolutions per second. In general, the coil voltage of a machine with $P$-poles passes through a complete cycle every time a pair of poles sweeps by, or $P/2$ times for each revolution. The frequency $f$ is therefore given by

$$f = \frac{P}{2}\left(\frac{n}{60}\right) \tag{3.1}$$

where $n$ is the shaft speed in revolutions per minute (r/min).

In treating $P$-pole synchronous machines, it is more convenient to express angles in electrical degrees rather than in the more familiar mechanical units. Here we concentrate on a single pair of poles and recognize that the conditions associated with any other pair are simply repetitions of those of the pair under consideration. A full cycle of generated voltage will be described when the rotor of a four-pole machine has turned 180 mechanical degrees. This

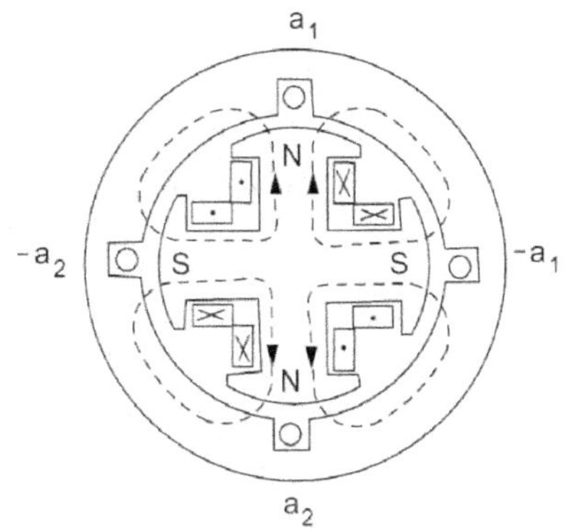


**Figure 3.3** Four-Pole Synchronous Machine.

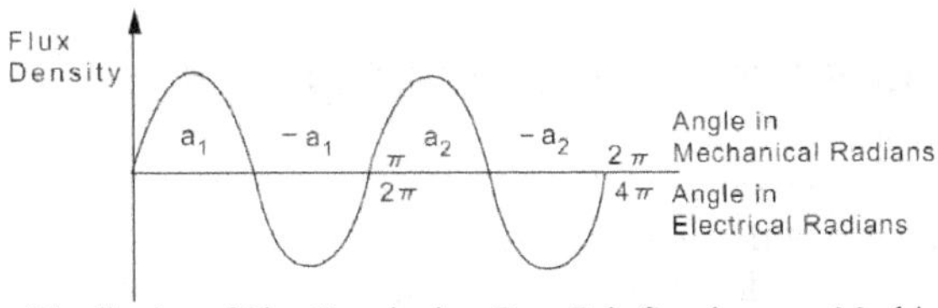


**Figure 3.4** Space Distribution of Flux Density in a Four-Pole Synchronous Machine.

cycle represents 360 electrical degrees in the voltage wave. Extending this argument to a $P$-pole machine leads to the conclusion that

$$\theta_e = \left(\frac{P}{2}\right)\theta_m$$

where $\theta_e$ and $\theta_m$ are angles in electrical and mechanical degrees, respectively.

## Cylindrical vs. Salient-Pole Construction

Machines like the ones illustrated in Figures 3.1 and 3.3 have rotors with distinct or *salient* poles. There is another type of rotor, which is shown in Figure 3.5. The machine with such a rotor is called a *cylindrical rotor* or *nonsalient*-pole machine. The choice between the two designs (salient or nonsalient) for a specific application depends on the prime mover. For hydroelectric generation, a salient-pole construction is employed, because hydraulic turbines run at relatively low speeds, and a large number of poles is required to produce the desired frequency as indicated by Eq. (3.1). Steam and gas turbines perform better at relatively high speeds, and two- or four-pole cylindrical rotor turboalternators are used to avoid the use of protruding parts on the rotor.

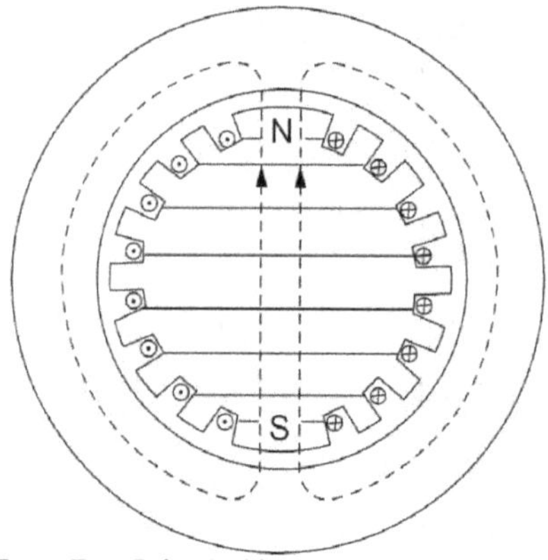


**Figure 3.5** A Cylindrical Rotor Two-Pole Machine.

## 3.3 SYNCHRONOUS MACHINE FIELDS

An understanding of the nature of the magnetic field produced by a polyphase winding is necessary for the analysis of polyphase ac machines. We will consider a two-pole, three-phase machine. The windings of the individual phases are displaced spatially from each other by 120 electrical degrees. The magnetomotive forces developed in the air gap due to currents in the windings will also be displaced from each other by 120 electrical degrees in space. Assuming sinusoidal, balanced three-phase operation, the phase currents are displaced by 120 electrical degrees in time.

Assume that $I_m$ is the maximum value of the current, and the time origin is arbitrarily taken as the instant when the phase *a* current is a positive maximum. The phase sequence is assumed to be *abc* which means that any event such as the maximum of current will take place on phase *a* followed by phase *b* and then phase *c*.

The magnetomotive force (MMF) of each phase is proportional to the corresponding current, and hence, the peak MMF is given by

$$F_{\max} = KI_m$$

where *K* is a constant of proportionality that depends on the winding distribution and the number of series turns in the winding per phase. We thus have

$$A_{a(p)} = F_{\max} \cos \omega t \quad (3.2)$$

$$A_{b(p)} = F_{\max} \cos(\omega t - 120^\circ) \quad (3.3)$$

$$A_{c(p)} = F_{\max} \cos(\omega t - 240^\circ) \quad (3.4)$$

where $A_{a(p)}$ is the amplitude of the MMF component wave at time $t$.

At time $t$, all three phases contribute to the air-gap MMF at a point $P$ (whose spatial angle is $\theta$). The resultant MMF is then given by

$$A_p = A_{a(p)} \cos\theta + A_{b(p)} \cos(\theta - 120^\circ) + A_{c(p)} \cos(\theta - 240^\circ) \quad \textbf{(3.5)}$$

This reduces to

$$A_p = \tfrac{3}{2}[F_{\max} \cos(\theta - \omega t)] \quad \textbf{(3.6)}$$

The wave represented in Eq. (3.6) depends on the spatial position of the point $P$ $\theta$ as well as time. The angle $\omega t$ provides rotation of the entire wave around the air gap at the constant angular velocity $\omega$. At time $t_1$, the wave is a sinusoid with its positive peak displaced $\omega t_1$ from the point $P$ (at $\theta$); at a later instant ($t_2$) the wave has its positive peak displaced $\omega t_2$ from the same point. We thus see that a polyphase winding excited by balanced polyphase currents produces the same effect as a permanent magnet rotating within the stator.

The MMF wave created by the three-phase armature current in a synchronous machine is commonly called *armature-reaction* MMF. It is a wave that rotates at synchronous speed and is directly opposite to phase $a$ at the instant when phase $a$ has its maximum current ($t = 0$). Due to the rotation, the dc field winding produces a sinusoid $F$ on an axis which is 90° ahead of the axis of phase $a$ in accordance with Faraday's law.

The resultant magnetic field at any point in the machine is the sum of the two contributions from the field and armature reaction. Figure 3.6 shows a sketch of the armature and field windings of a cylindrical rotor generator. The space MMF produced by the field winding is shown by the sinusoid $F$. This is shown for the specific instant when the electromotive force (EMF) of phase $a$ due to excitation is at its maximum value. The time rate of change of flux linkages with phase $a$ is a maximum under these conditions, and thus the axis of the field is 90° ahead of phase $a$. The armature-reaction wave is shown as the sinusoid $A$ in the figure. This is drawn opposite phase $a$ because at this instant both $I_a$ and the EMF of the field $E_f$ (also called excitation voltage) have their maximum value. The resultant magnetic field in the machine is denoted $R$ and is obtained by graphically adding the $F$ and $A$ waves.

Sinusoids can conveniently be handled using phasor methods. We can thus perform the addition of the $A$ and $F$ waves using phasor notation. Figure 3.7 shows a space phasor diagram where the fluxes $\phi_f$ (due to the field), $\phi_{ar}$ (due to armature reaction), and $\phi_r$ (the resultant flux) are represented. It is clear that under the assumption of a uniform air gap and no saturation, these are proportional to the MMF waves $F$, $A$, and $R$, respectively. The figure is drawn for the case when the armature current is in phase with the excitation voltage.

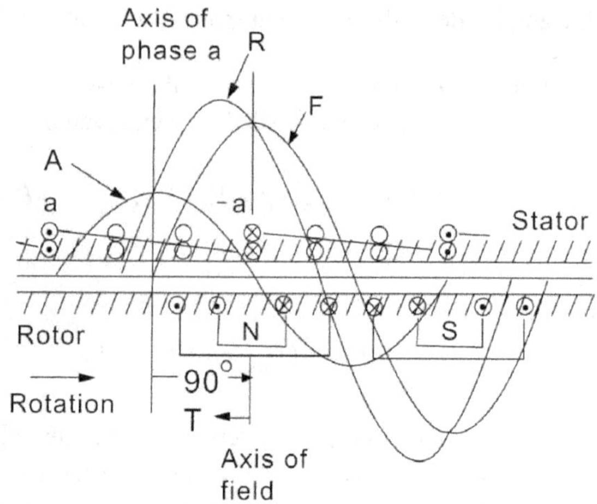


**Figure 3.6** Spatial MMF Waves in a Cylindrical Rotor Synchronous Generator.

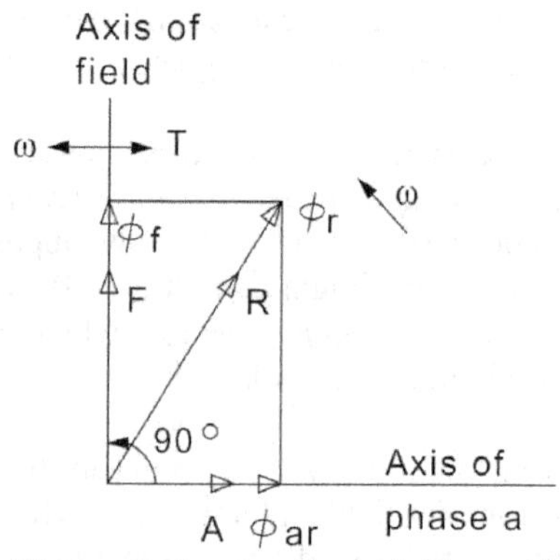


**Figure 3.7** A Space Phasor Diagram for Armature Current in Phase with Excitation Voltage.

## 3.4 A SIMPLE EQUIVALENT CIRCUIT

The simplest model of a synchronous machine with cylindrical rotor can be obtained if the effect of the armature-reaction flux is represented by an inductive reactance. The basis for this is shown in Figure 3.8, where the phasor diagram of component fluxes and corresponding voltages is given. The field flux $\phi_f$ is added to the armature-reaction flux $\phi_{ar}$ to yield the resultant air-gap flux $\phi_r$. The armature-reaction flux $\phi_{ar}$ is in phase with the armature current $I_a$. The excitation voltage $E_f$ is generated by the field flux, and $E_f$ lags $\phi_f$ by 90°. Similarly, $E_{ar}$ and $E_r$ are generated by $\phi_{ar}$ and $\phi_r$ respectively, with each of the voltages lagging the flux causing it by 90°.

Introduce the constant of proportionality $x_\phi$ to relate the rms values of $E_{ar}$ and $I_a$, to write

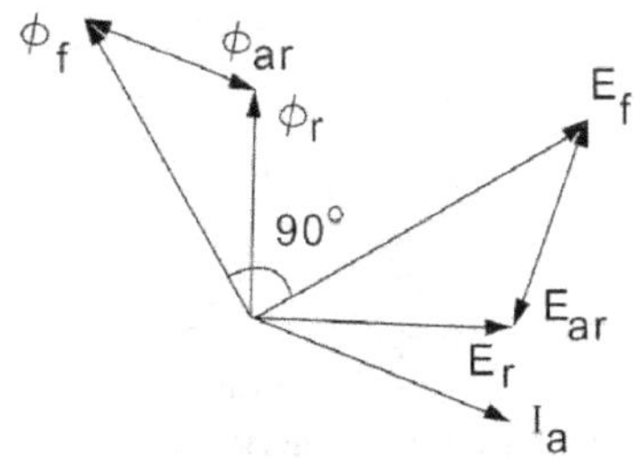


**Figure 3.8** Phasor Diagram for Fluxes and Resulting Voltages in a Synchronous Machine.

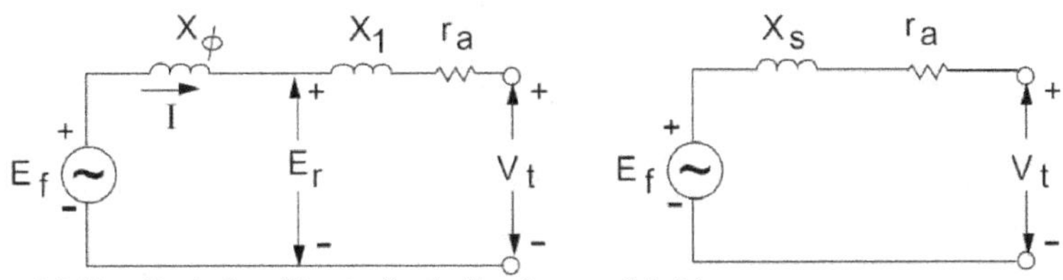


**Figure 3.9** Two Equivalent Circuits for the Synchronous Machine.

$$E_{ar} = -jx_\phi I_a \tag{3.7}$$

where the $-j$ represents the 90° lagging effect. We therefore have

$$E_r = E_f - jx_\phi I_a \tag{3.8}$$

An equivalent circuit based on Eq. (3.8) is given in Figure 3.9. We thus conclude that the inductive reactance $x_\phi$ accounts for the armature-reaction effects. This reactance is known as the *magnetizing reactance* of the machine.

The terminal voltage of the machine denoted by $V_t$ is the difference between the air-gap voltage $E_r$ and the voltage drops in the armature resistance $r_a$, and the leakage-reactance $x_l$. Here $x_l$ accounts for the effects of leakage flux as well as space harmonic field effects not accounted for by $x_\phi$. A simple impedance commonly known as the *synchronous impedance* $Z_s$ is obtained by combining $x_\phi$, $x_l$, and $r_a$ according to

$$Z_s = r_a + jX_s \tag{3.9}$$

The synchronous reactance $X_s$ is given by

$$X_s = x_l + x_\phi \tag{3.10}$$

The model obtained here applies to an unsaturated cylindrical rotor machine supplying balanced polyphase currents to its load. The term unsaturated is used whenever the value of flux in the machine is in the linear region (unsaturated) of the magnetization characteristic of the core material. The voltage relationship is now given by

$$E_f = V_t + I_a Z_s \quad \textbf{(3.11)}$$

**Example 3.1**

A 10 MVA, 13.8 kV, 60 Hz, two-pole, Y-connected, three-phase alternator has an armature winding resistance of 0.07 ohms per phase and a leakage reactance of 1.9 ohms per phase. The armature reaction EMF for the machine is related to the armature current by

$$E_{ar} = -j19.91 I_a$$

Assume that the generated EMF is related to the field current by

$$E_f = 60 I_f$$

A. Compute the field current required to establish rated voltage across the terminals of a load when rated armature current is delivered at 0.8 PF lagging.
B. Compute the field current needed to provide rated terminal voltage to a load that draws 100 per cent of rated current at 0.85 PF lagging.

*Solution*

The rated current is given by

$$I_a = \frac{10 \times 10^6}{\sqrt{3} \times 13800} = 418.37 \text{ A}$$

The phase value of terminal voltage is

$$V_t = \frac{13,800}{\sqrt{3}} = 7967.43 \text{ V}$$

With reference to the equivalent circuit of Figure 3.9, we have

A.

$$\begin{aligned} E_r &= V_t + I_a Z_a \\ &= 7967.43 + +(418.37\angle -\cos^{-1} 0.8)(0.07 + j1.9) \\ &= 8490.35\angle 4.18^\circ \end{aligned}$$

$$E_{ar} = -j(19.91)(418.37\angle -\cos^{-1} 0.8) = -8329.75\angle 53.13^\circ$$

The required field excitation voltage $E_f$ is therefore,

$$\begin{aligned} E_f &= E_r - E_{ar} \\ &= 8490.35\angle 4.18^\circ + 8329.75\angle 53.13^\circ \\ &= 15308.61\angle 28.4^\circ \text{ V} \end{aligned}$$

Consequently, using the given field voltage versus current relation,

$$I_f = \frac{E_f}{60} = 255.14 \text{ A}$$

B. With conditions given, we have

$$\begin{aligned} I_a &= (418.37)(1\angle -\cos^{-1} 0.85) = 418.37\angle -31.79^\circ \\ E_r &= 7967.43 + (418.37\angle -31.79^\circ)(0.07 + j1.9) \\ &= 8436.94\angle 4.49^\circ \text{ V} \\ E_{ar} &= -j(19.91)(418.37\angle -31.79^\circ) \\ &= -8329.74\angle 58.21^\circ \\ E_f &= E_r - E_{ar} \\ &= 8436.94\angle 4.48^\circ + 8329.74\angle 58.21^\circ \\ &= 14{,}957.72\angle 31.16^\circ \text{ V} \end{aligned}$$

We therefore calculate the required field current as

$$I_f = \frac{14{,}957.72}{60} = 249.30 \text{ A}$$

## 3.5 PRINCIPAL STEADY-STATE CHARACTERISTICS

Consider a synchronous generator delivering power to a constant power factor load at a constant frequency. A *compounding curve* shows the variation of the field current required to maintain rated terminal voltage with the load. Typical compounding curves for various power factors are shown in Figure 3.10. The computation of points on the curve follows easily from applying Eq. (3.11). Figure 3.11 shows phasor diagram representations for three different power factors.

**Example 3.2**

A 1,250-kVA, three-phase, Y-connected, 4,160-V (line-to-line), ten-pole, 60-Hz generator has an armature resistance of 0.126 ohms per phase and a synchronous reactance of 3 ohms per phase. Find the full load generated voltage per phase at a power factor of 0.8 lagging.

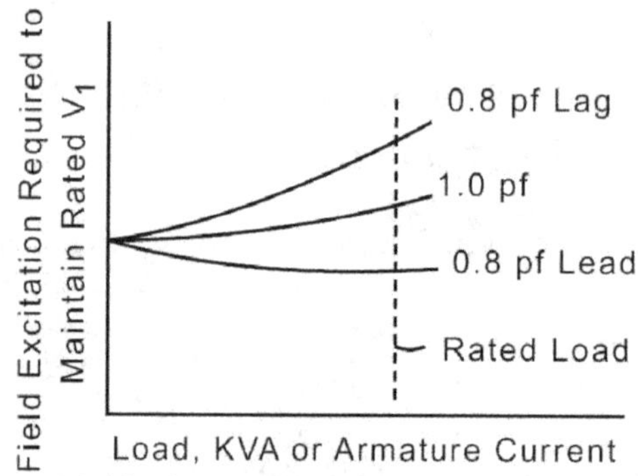


**Figure 3.10** Synchronous-Machine Compounding Curves.

*Solution*

The magnitude of full load current is obtained as

$$I_a = \frac{1{,}250 \times 10^3}{\sqrt{3} \times 4{,}160} = 173.48 \text{ A}$$

The terminal voltage per phase is taken as reference

$$V_t = \frac{4{,}160}{\sqrt{3}} = 2{,}401.77\angle 0 \text{ V}$$

The synchronous impedance is obtained as

$$\begin{aligned} Z_s &= r_a + jX_s \\ &= 0.126 + j3 \\ &= 3.0026\angle 87.59^\circ \text{ ohms per phase} \end{aligned}$$

The generated voltage per phase is obtained for a power factor of 0.8 lagging using Eq. (3.11) as:

$$\begin{aligned} I_a &= 173.48\angle -36.87^\circ \text{ A} \\ E_f &= 2{,}401.77 + \left(173.48\angle -36.87^\circ\right)\left(3.0026\angle 87.59^\circ\right) \\ &= 2{,}761.137\angle 8.397^\circ \text{ V} \end{aligned}$$

A significant characteristic of the synchronous machine is the reactive-capability curves. These give the maximum allowed reactive power loadings corresponding to various active power loadings for rated voltage operation. Armature heating constraints govern the machine for power factors from rated to unity. Field heating represents the constraints for lower power factors. Figure 3.12 shows a typical set of curves for a large turbine generator.

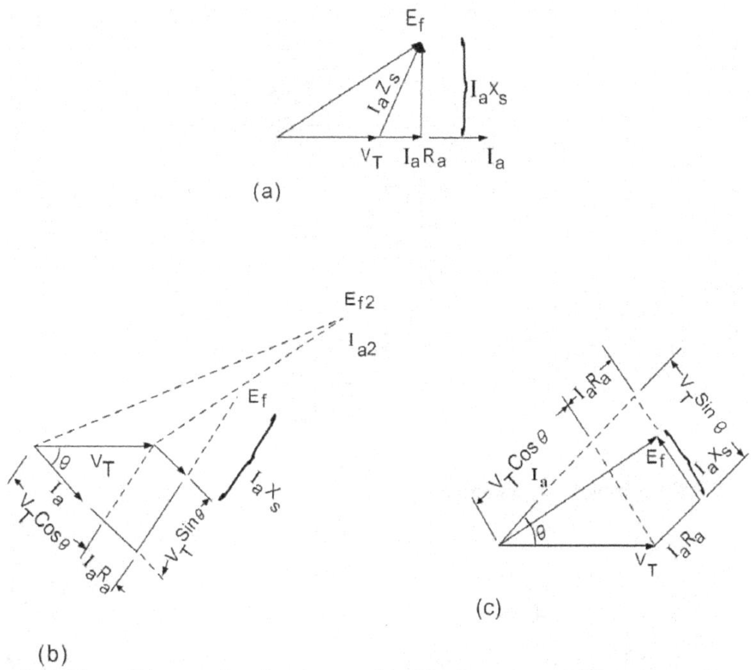


**Figure 3.11** Phasor Diagrams for a Synchronous Machine Operating at Different Power Factors are: (a) Unity PF Loads, (b) Lagging PF Loads, and (c) Leading PF Loads.

## 3.6 POWER-ANGLE CHARACTERISTICS AND THE INFINITE BUS CONCEPT

Consider the simple circuit shown in Figure 3.13. The impedance $Z$ connects the sending end, whose voltage is $E$ and receiving end, with voltage $V$. Let us assume that in polar form we have

$$E = E\angle\delta$$
$$V = V\angle 0$$
$$Z = Z\angle\psi$$

We therefore conclude that the current $I$ is given by

$$I = \frac{E - V}{Z}$$

The complex power $S_1$ at the sending end is given by

$$S_1^* = E^* I$$

Similarly, the complex power $S_2$ at the receiving end is

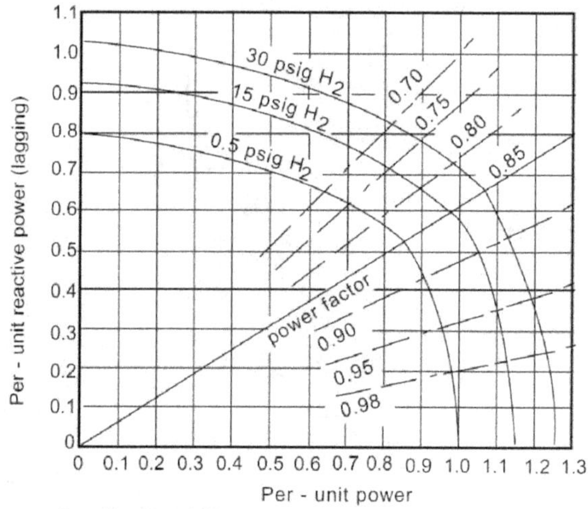


**Figure 3.12** Generator Reactive-Capability Curves.

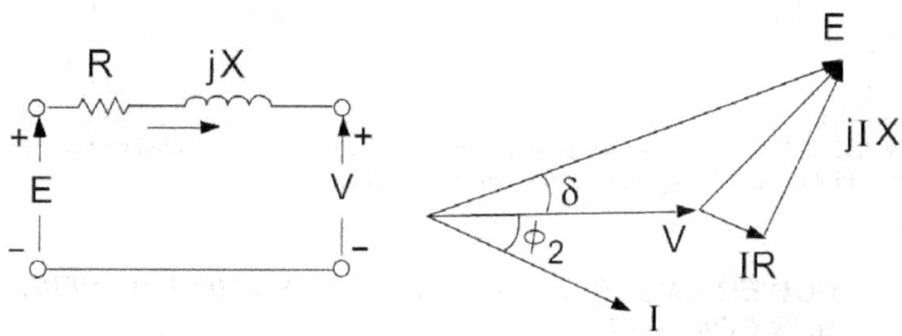


**Figure 3.13** Equivalent Circuit and Phasor Diagram for a Simple Link.

$$S_2^* = V^* I$$

Therefore,

$$S_1^* = \left(\frac{E^2}{Z}\angle -\psi\right) - \left(\frac{EV}{Z}\angle -\psi - \delta\right) \tag{3.12}$$

$$S_2^* = \left(\frac{EV}{Z}\angle \delta - \psi\right) - \left(\frac{V^2}{Z}\angle -\psi\right) \tag{3.13}$$

Recall that

$$S^* = P - jQ$$

When the resistance is negligible; then

$$\psi = 90^\circ$$
$$Z = X$$

and the active and reactive power equations are obtained as:

$$P_1 = P_2 = \frac{EV}{X}\sin\delta \tag{3.14}$$

$$Q_1 = \frac{E^2 - EV\cos\delta}{X} \tag{3.15}$$

$$Q_2 = \frac{EV\cos\delta - V^2}{X} \tag{3.16}$$

In large-scale power systems, a three-phase synchronous machine is represented as a voltage source connected through an equivalent system reactance ($X_e$) to the network which has a high generation capacity relative to any single unit. We often refer to the network or system as an *infinite bus* when a change in input mechanical power or in field excitation to the unit does not cause an appreciable change in system frequency or terminal voltage. Figure 3.14 shows such a situation, where $V$ is the infinite bus voltage.

The previous analysis shows that in the present case we have for power transfer,

$$P = P_{\max}\sin\delta \tag{3.17}$$

with

$$P_{\max} = \frac{EV}{X_t} \tag{3.18}$$

and

$$X_t = X_s + X_e \tag{3.19}$$

If we try to advance $\delta$ further than 90° (corresponding to maximum power transfer) by increasing the mechanical power input, the electrical power output would decrease from the $P_{\max}$ point. Therefore the angle $\delta$ increases further as the machine accelerates. This drives the machine and system apart electrically. The value $P_{\max}$ is called the *steady-state stability limit* or *pull-out power*.

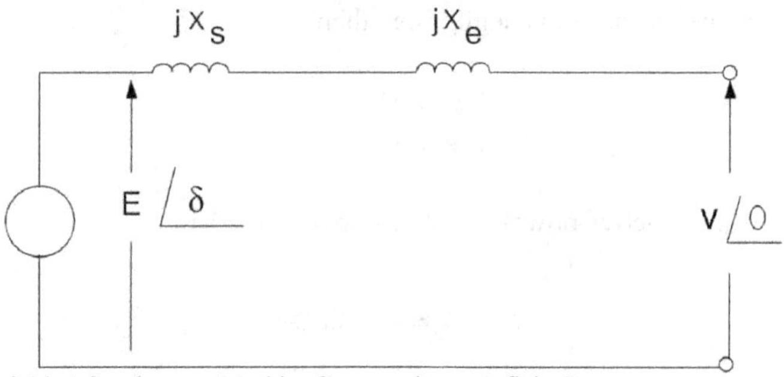


**Figure 3.14** A Synchronous Machine Connected to an Infinite Bus.

## Example 3.3

A synchronous generator with a synchronous reactance of 1.15 p.u. is connected to an infinite bus whose voltage is one p.u. through an equivalent reactance of 0.15 p.u. The maximum permissible active power output is 1.2 p.u.

A. Compute the excitation voltage $E$.
B. The power output is gradually reduced to 0.7 p.u. with fixed field excitation. Find the new current and power angle $\delta$.

*Solution*

A. The total reactance is

$$X_t = 1.15 + 0.15 = 1.3$$

Thus we have,

$$1.2 = \frac{EV}{X_t} = \frac{(E)(1)}{1.3}$$

Therefore,

$$E = 1.56 \text{ p.u.}$$

B. We have for any angle $\delta$,

$$P = P_{\max} \sin\delta$$

Therefore,

$$0.7 = 1.2 \sin\delta$$

This results in

$$\delta = 35.69^\circ$$

The current is

$$I = \frac{E - V}{jX_t}$$

Substituting the given values, we obtain

$$I = \frac{1.56\angle 35.69^\circ - 1.0}{j1.3}$$
$$= 0.7296\angle -16.35^\circ \text{ A}$$

The following is a MATLAB™ script to solve problems of the type presented in Example 3.3.

```
% example 3.3
% enter the data
Xs=1.25; % synchronous reactance
Xe=0.25; % equivalent reactance
Pm=1.2; % max permisible output
V=1; % infinite bus voltage
% to find the total reactance
Xt=Xs+Xe;
% A. To compute the exitation voltage
% from Pm=E*V/Xt
E=Pm*Xt/V
% B. The power output is gradually
reduced to 0.7 p.u.
% with fixed field excitation.
% to find power angle delta
P=0.7;  % power output
% from P=Pm*sin(delta)
delta=asin(P/Pm);
delta_deg=delta*180/pi
E_complex=E*(cos(delta)+i*sin(delta));
% To find the new current
% modulus and argumen
I=(E_complex-V)/Xt*i;
modulus_I=abs(I)
eta=atan(imag(I)/real(I));
argumen_I=eta*180/pi
```

The solution is obtained by running the script as follows

```
EDU»
E = 1.5600
delta_deg = 35.6853
modulus_I = 0.7296
argumen_I = -16.3500°
```

## Reactive Power Generation

Eq. (3.16) suggests that the generator produces reactive power ($Q_2 > 0$) if

$$E\cos\delta > V$$

In this case, the generator appears to the network as if it is a capacitor or a reactive power source. This condition applies for high magnitude *E*, and the machine is said to be overexcited. On the other hand, the machine is called underexcited if it consumes reactive power ($Q_2 < 0$). Here we have

$$E\cos\delta < V$$

In this case the machine absorbs reactive power from the system. Figure 3.15 shows phasor diagrams for both cases. The overexcited synchronous machine is normally employed to provide synchronous condenser action, where usually no real load is carried by the machine ($\delta = 0$). In this case we have

$$Q_2 = \frac{V(E-V)}{X} \tag{3.20}$$

Control of reactive power generation is carried out by simply changing *E*, by varying the dc excitation.

### Example 3.4

Compute the reactive power generated by the machine of Example 3.3 under the conditions in part (b). If the machine is required to generate a reactive power of 0.4 p.u. while supplying the same active power by changing the field excitation, find the new excitation voltage and power angle $\delta$.

*Solution*

The reactive power generated is obtained according to Eq. (3.16) as

$$Q_2 = \frac{1(1.56\cos 35.69 - 1)}{1.3} = 0.205$$

With a new excitation voltage and stated active and reactive powers, we have

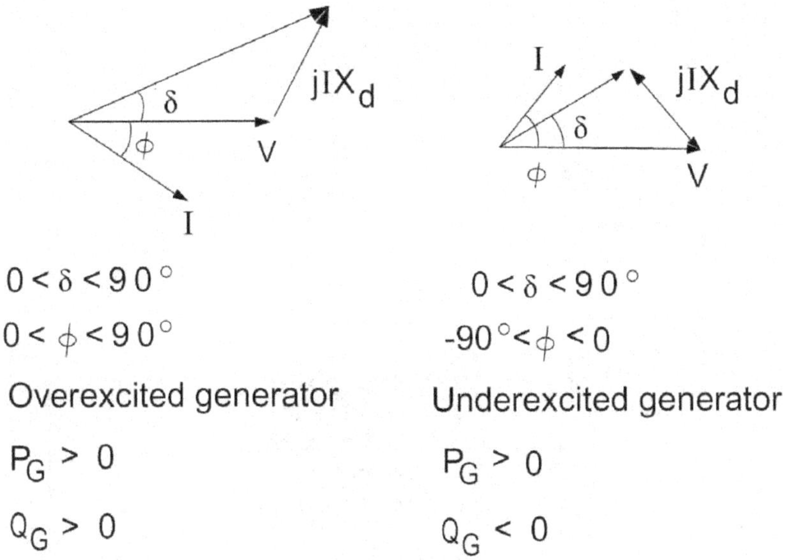


**Figure 3.15** Phasor Diagrams for Overexcited and Underexcited Synchronous Machines.

using Eq. (3.14) and (3.16)

$$0.7 = \frac{(E)(1)}{(1.3)}\sin\delta$$

$$0.4 = \frac{1(E\cos\delta - 1)}{1.3}$$

We thus obtain

$$\tan\delta = \frac{(1.3)(0.7)}{(1.52)}$$

$$\delta = 30.9083^\circ$$

From the above we get

$$E = \frac{(1.3)(0.7)}{\sin(30.9083)} = 1.7716$$

The following script implements the solution of this example in MATLAB™ environment.

```
% example 3.4
% enter the data
Xs=1.15; % synchronous reactance
Xe=0.15; % equivalent reactance
Pm=1.2; % max permisible output
V=1; % infinite bus voltage
%
% to find the total reactance
Xt=Xs+Xe;
% A. To compute the exitation voltage
% from Pm=E*V/Xt
E=Pm*Xt/V;
P=0.7;  % power output
% from P=Pm*sin(delta)
delta=asin(P/Pm);
% to compute reactive power generated
Q2=(E*V*cos(delta)-V^2)/Xt;
% If the machine is required to
generate a reactive power
% of 0.4 p.u. while supplying the same
active power
% to find the new power angle (delta1)
Q2_required=0.4;
% with a new excitation voltage
% and stated active and reactive powers
% using the equation
% P=(E*V/Xt)sin(delta1) and
Q2=(E*V*cos(delta1)-V^2)/Xt
delta1=atan(P/(Q2_required+V^2/Xt));
delta1_deg=delta1*180/pi
% to find the new field exitation
E_new=P*Xt/sin(delta1)
%
```

The solution is obtained as

```
EDU»
delta1_deg = 30.9083

E_new = 1.7716
```

## 3.7 ACCOUNTING FOR SALIENCY

Field poles in a salient-pole machine cause nonuniformity of the

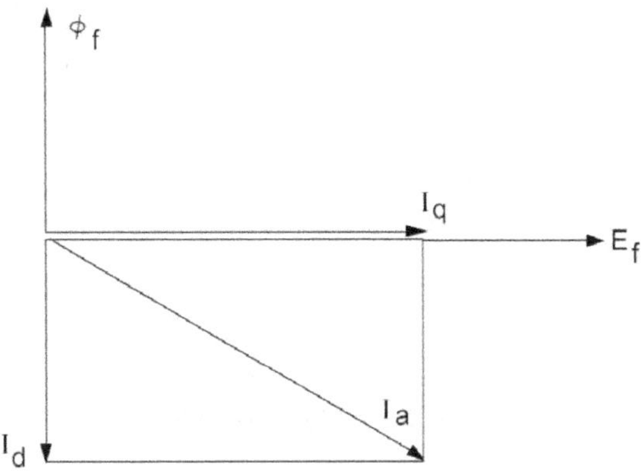


**Figure 3.16** Resolution of Armature Current in Two Components.

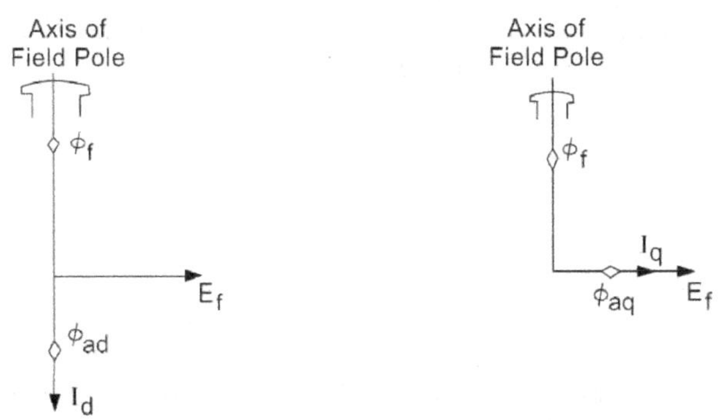


**Figure 3.17** Direct-Axis and Quadrature-Axis Air-Gap Fluxes in a Salient-Pole Synchronous Machine.

magnetic reluctance of the air gap. The reluctance along the polar axis is appreciably less than that along the interpolar axis. We often refer to the polar axis as the *direct axis* and the interpolar as the *quadrature axis*. This effect can be taken into account by resolving the armature current $I_a$ into two components, one in time phase and the other in time quadrature with the excitation voltage as shown in Figure 3.16. The component $I_d$ of the armature current is along the direct axis (the axis of the field poles), and the component $I_q$ is along the quadrature axis.

Let us consider the effect of the direct-axis component alone. With $I_d$ lagging the excitation EMF $E_f$ by 90°, the resulting armature-reaction flux $\phi_{ad}$ is directly opposite the field poles as shown in Figure 3.17. The effect of the quadrature-axis component is to produce an armature-reaction flux $\phi_{aq}$, which is in the quadrature-axis direction as shown in Figure 3.17. The phasor diagram with both components present is shown in Figure 3.18.

In the cylindrical rotor machine, we employed the synchronous reactance $x_s$ to account for the armature-reaction EMF in an equivalent circuit. The same

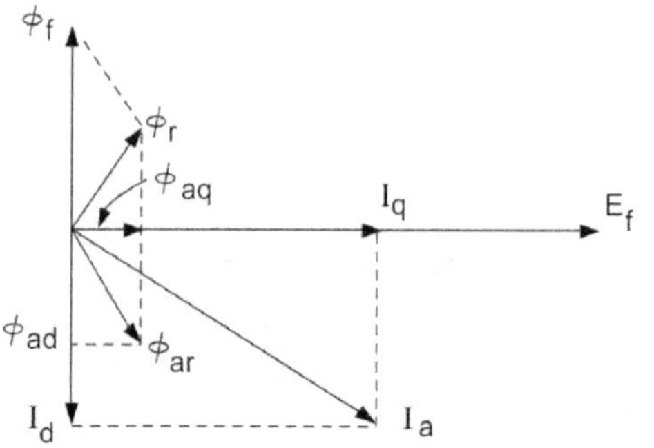

**Figure 3.18** Phasor Diagram for a Salient-Pole Synchronous Machine.

argument can be extended to the salient-pole case. With each of the component currents $I_d$ and $I_q$, we associated component synchronous-reactance voltage drops, $jI_dx_d$ and $jI_qx_q$ respectively. The direct-axis synchronous reactance $x_d$ and the quadrature-axis synchronous reactance $x_q$ are given by

$$x_d = x_l + x_{\phi d}$$

$$x_q = x_l + x_{\phi q}$$

where $x_l$ is the armature leakage reactance and is assumed to be the same for direct-axis and quadrature-axis currents. The direct-axis and quadrature-axis magnetizing reactances $x_{\phi d}$ and $x_{\phi q}$ account for the inductive effects of the respective armature-reaction flux. Figure 3.19 shows a phasor diagram implementing the result.

$$E_f = V_t + I_a r_a + jI_d x_d + jI_q x_d \tag{3.21}$$

In many instances, the power factor angle $\phi$ at the machine terminals is explicitly known rather than the internal power factor angle $(\phi + \delta)$, which is required for the resolution of $I_a$ into its direct-axis and quadrature-axis components. We can avoid this difficulty by recalling that in phasor notation,

$$I_a = I_q + I_d \tag{3.22}$$

Substitution of Eq. (3.22) into Eq. (3.21) for $I_q$ and rearranging, we obtain

$$E_f = V_t + I_a(r_a + jx_q) + jI_d(x_d - x_q) \tag{3.23}$$

Let us define

$$E'_f = V_t + I_a(r_a + jx_q) \tag{3.24}$$

$E'_f$ as defined is in the same direction as $E_f$ since $jI_d$ is also along the same

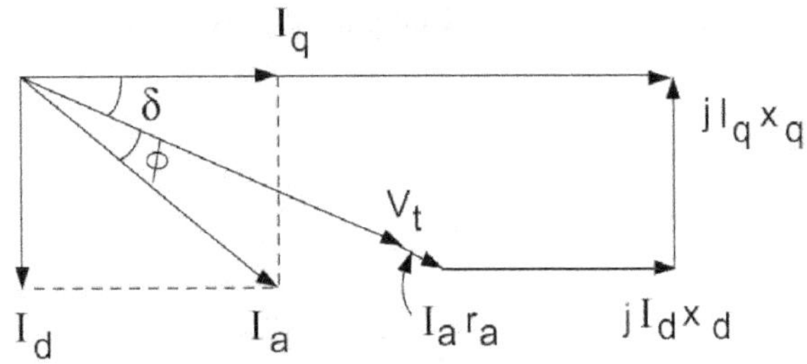


**Figure 3.19** Phasor Diagram for a Synchronous Machine.

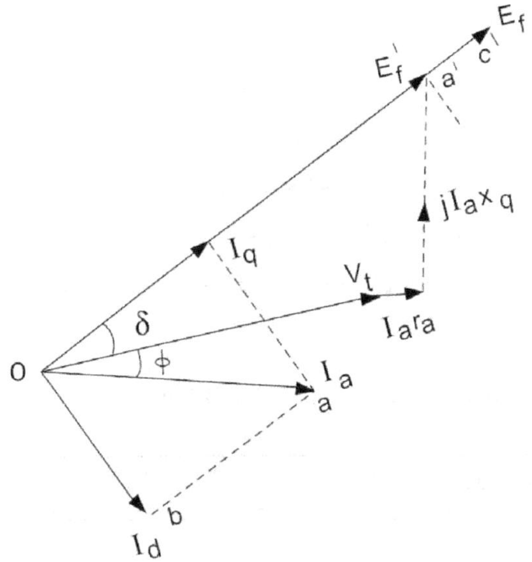


**Figure 3.20** A Modified Phasor Diagram for a Salient-Pole Synchronous Machine.

angle.

direction. Our procedure then is to obtain $E_f'$ as given by Eq. (3.24) and then obtain the component $I_d$ based on the phase angle of $E_f'$. Finally, we find $E_f$ as a result of

$$E_f = E_f' + jI_d(x_d - x_q) \tag{3.25}$$

This is shown in Figure 3.20.

## Example 3.5

A 5-kVA, 220-V, Y-connected, three-phase, salient-pole synchronous generator is used to supply power to a unity PF load. The direct-axis synchronous reactance is 12 ohms and the quadrature-axis synchronous reactance is 7 ohms. Assume that rated current is delivered to the load at rated voltage and that armature resistance is negligible. Compute the excitation voltage and power angle

*Solution*

$$V_t = 127.02 \text{ V}$$

$$I_a = \frac{5\times 10^3}{220\sqrt{3}} = 13.12A$$

We calculate

$$\begin{aligned} E_f' &= V_t + jI_a x_q \\ &= 127.02 + j(13.12)(7) = 156.75\angle 35.87^\circ \end{aligned}$$

Moreover,

$$\begin{aligned} I_d &= I_a \sin 35.87 = 7.69 \text{ A} \\ |E_f| &= |E_f'| + |I_d(x_d - x_q)| \\ &= 156.75 + 7.69(12-7) = 195.20 \text{ V} \\ \delta &= 35.87^\circ \end{aligned}$$

The following script uses MATLAB™ to solve Example 3.5.

```
% Example 3.5
% A 5 kVA, 220 Volts, Y connected, 3 phase,
% salient pole synchronous generator
PF=1;
VL=220;  % Volts
xd=12;
xq=7;
P=5*10^3; % VA
Vt=VL/3^.5;
Ia=P/(VL*3^.5)
% We calculate
Ef_prime=Vt+i*Ia*xq;
abs(Ef_prime)
angle(Ef_prime)*180/pi
Id=Ia*sin(angle(Ef_prime));
Ef=abs(Ef_prime)+abs(Id*(xd-xq))
delta=angle(Ef_prime)*180/pi
```

The solution is

```
EDU»
Ia = 13.1216
ans = 156.7481
ans = 35.8722
Ef = 195.1931
delta =35.8722
```

## 3.8 SALIENT-POLE MACHINE POWER ANGLE CHARACTERISTICS

The power angle characteristics for a salient-pole machine connected to an infinite bus of voltage $V$ through a series reactance of $x_e$ can be arrived at by considering the phasor diagram shown in Figure 3.21. The active power delivered to the bus is

$$P = \left(I_d \sin\delta + I_q \cos\delta\right)V \tag{3.26}$$

Similarly, the delivered reactive power $Q$ is

$$Q = \left(I_d \cos\delta - I_q \sin\delta\right)V \tag{3.27}$$

To eliminate $I_d$ and $I_q$, we need the following identities obtained from inspection of the phasor diagram:

$$I_d = \frac{E_f - V\cos\delta}{X_d} \tag{3.28}$$

$$I_q = \frac{V\sin\delta}{X_q} \tag{3.29}$$

where

$$X_d = x_d + x_e \tag{3.30}$$

$$X_q = x_q + x_e \tag{3.31}$$

Substitution of Eqs. (3.28) and (3.29) into Eqs. (3.26) and (3.27) yields equations that contain six quantities – the two variables $P$ and $\delta$ and the four parameters $E_f$, $V$, $X_d$, and $X_q$ – and can be written in many different ways. The following form illustrates the effect of saliency. Define $P_d$ and $Q_d$ as

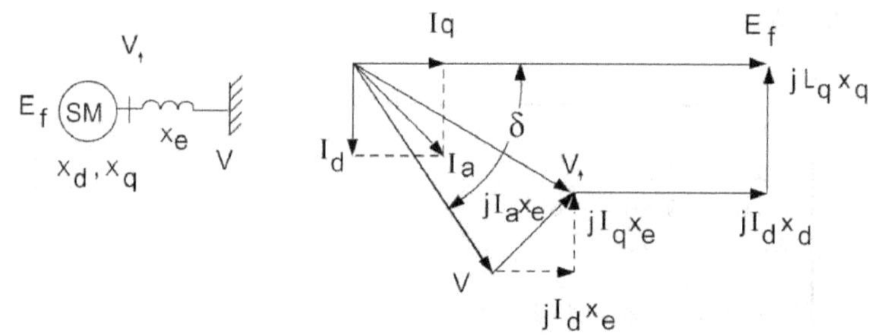


**Figure 3.21** A Salient-Pole Machine Connected to an Infinite Bus through an External Impedance.

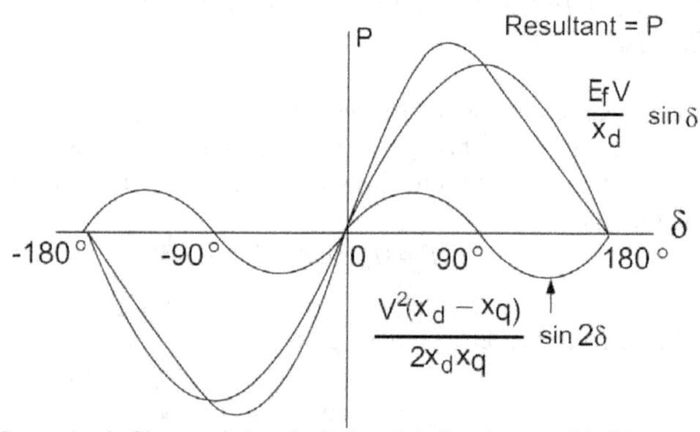


**Figure 3.22** Power Angle Characteristics of a Salient-Pole Synchronous Machine.

$$P_d = \frac{VE_f}{X_d}\sin\delta \tag{3.32}$$

and

$$Q_d = \frac{VE_f}{X_d}\cos\delta - \frac{V^2}{X_d} \tag{3.33}$$

The above equations give the active and reactive power generated by a round rotor machine with synchronous reactance $X_d$. We thus have

$$P = P_d + \frac{V^2}{2}\left(\frac{1}{X_q} - \frac{1}{X_d}\right)\sin 2\delta \tag{3.34}$$

$$Q = Q_d - V^2\left(\frac{1}{X_q} - \frac{1}{X_d}\right)\sin^2\delta \tag{3.35}$$

The second term in the preceding two equations introduces the effect of salient poles, and in the power equation the term corresponds to reluctance torque. Note that if $X_d = X_q$, as in a uniform air-gap machine, the second terms in both equations are zero. Figure 3.22 shows the power angle characteristics of a typical salient-pole machine.

The pull-out power and power angle $\delta$ for the salient-pole machine can be obtained by solving equation (3.36) requiring the partial derivative of $P$ with respect to $\delta$ to be equal to zero.

$$\frac{\partial P}{\partial \delta} = 0 \tag{3.36}$$

The actual value of pull-out power can be shown to be higher than that obtained assuming nonsaliency.

### Example 3.6

A salient-pole synchronous machine is connected to an infinite bus through a link with reactance of 0.2 p.u. The direct-axis and quadrature-axis reactances of the machine are 0.9 and 0.65 p.u., respectively. The excitation voltage is 1.3 p.u., and the voltage of the infinite bus is maintained at 1 p.u. For a power angle of 30°, compute the active and reactive power supplied to the bus.

*Solution*

We calculate $X_d$ and $X_q$ as

$$X_d = x_d + x_e = 0.9 + 0.2 = 1.1$$
$$X_q = x_q + x_e = 0.65 + 0.2 = 0.85$$

Therefore,

$$P = \frac{(1.3)(1)}{1.1}\sin 30^\circ + \frac{1}{2}\left(\frac{1}{0.85} - \frac{1}{1.1}\right)\sin 60^\circ$$
$$= 0.7067 \text{ p.u.}$$

Similarly, the reactive power is obtained using Eq. (3.32) as:

$$Q = \frac{(1.3)(1)}{1.1}\cos 30^\circ - \left(\frac{\cos^2 30^\circ}{1.1} + \frac{\sin^2 30^\circ}{0.85}\right)$$
$$= 0.0475 \text{ p.u.}$$

## *PROBLEMS*

### Problem 3.1

A 5-k VA, 220-V, 60-Hz, six-pole, Y-connected synchronous generator has a leakage reactance per phase of 0.78 ohms and negligible armature resistance. The armature-reaction EMF for this machine is related to the armature current by

$$E_{ar} = -j16.88(I_a)$$

Assume that the generated EMF is related to field current by

$$E_f = 25I_f$$

A. Compute the field current required to establish rated voltage across the terminals of a unity power factor load that draws rated generator armature current.
B. Determine the field current needed to provide rated terminal voltage to a load that draws 125 percent of rated current at 0.8 PF lagging.

**Problem 3.2**

A 9375 kVA, 13,800 kV, 60 Hz, two pole, Y-connected synchronous generator is delivering rated current at rated voltage and unity PF. Find the armature resistance and synchronous reactance given that the field excitation voltage is 11935.44 V and leads the terminal voltage by an angle 47.96°.

**Problem 3.3**

The magnitude of the field excitation voltage for the generator of Problem (3.2) is maintained constant at the value specified above. Find the terminal voltage when the generator is delivering rated current at 0.8 PF lagging.

**Problem 3.4**

A 180 kVA, three-phase, Y-connected, 440 V, 60 Hz synchronous generator has a synchronous reactance of 1.6 ohms and a negligible armature resistance. Find the full load generated voltage per phase at 0.8 PF lagging.

**Problem 3.5**

The synchronous reactance of a cylindrical rotor synchronous generator is 0.9 p.u. If the machine is delivering active power of 1 p.u. to an infinite bus whose voltage is 1 p.u. at unity PF, calculate the excitation voltage and the power angle.

**Problem 3.6**

The synchronous reactance of a cylindrical rotor machine is 1.2 p.u. The machine is connected to an infinite bus whose voltage is 1 p.u. through an equivalent reactance of 0.3 p.u. For a power output of 0.7 p.u., the power angle is found to be 30°.

A. Find the excitation voltage $E_f$ and the pull-out power.
B. For the same power output the power angle is to be reduced to 25°. Find the value of the reduced equivalent reactance connecting the machine to the bus to achieve this. What would be the new pull-

out power?

**Problem 3.7**

Solve Problem 3.5 using MATLAB™.

**Problem 3.8**

A cylindrical rotor machine is delivering active power of 0.8 p.u. and reactive power of 0.6 p.u. at a terminal voltage of 1 p.u. If the power angle is 20°, compute the excitation voltage and the machine's synchronous reactance.

**Problem 3.9**

A cylindrical rotor machine is delivering active power of 0.8 p.u. and reactive power of 0.6 p.u. when the excitation voltage is 1.2 p.u. and the power angle is 25°. Find the terminal voltage and synchronous reactance of the machine.

**Problem 3.10**

A cylindrical rotor machine is supplying a load of 0.8 PF lagging at an infinite bus. The ratio of the excitation voltage to the infinite bus voltage is found to be 1.25. Compute the power angle $\delta$.

**Problem 3.11**

The synchronous reactance of a cylindrical rotor machine is 0.8 p.u. The machine is connected to an infinite bus through two parallel identical transmission links with reactance of 0.4 p.u. each. The excitation voltage is 1.4 p.u. and the machine is supplying a load of 0.8 p.u.

A. Compute the power angle $\delta$ for the outlined conditions.
B. If one link is opened with the excitation voltage maintained at 1.4 p.u. Find the new power angle to supply the same load as in (a).

**Problem 3.12**

The synchronous reactance of a cylindrical rotor generator is 1 p.u. and its terminal voltage is 1 p.u. when connected to an infinite bus through a reactance 0.4 p.u. Find the minimum permissible output vars for zero output active power and unity output active power.

**Problem 3.13**

The apparent power delivered by a cylindrical rotor synchronous machine to an infinite bus is 1.2 p.u. The excitation voltage is 1.3 p.u. and the power angle is 20°. Compute the synchronous reactance of the machine, given that the infinite bus voltage is 1 p.u.

**Problem 3.14**

The synchronous reactance of a cylindrical rotor machine is 0.9 p.u. The machine is connected to an infinite bus through two parallel identical transmission links with reactance of 0.6 p.u. each. The excitation voltage is 1.5 p.u., and the machine is supplying a load of 0.8 p.u.

A. Compute the power angle $\delta$ for the given conditions.
B. If one link is opened with the excitation voltage maintained at 1.5 p.u., find the new power angle to supply the same load as in part (a).

**Problem 3.15**

A round rotor synchronous machine is connected to an infinite bus whose voltage is kept constant at 1.00 p.u. The synchronous reactance of the machine is 0.375. The table given below relates to three operating conditions of the machine. ($Q_2$ is the reactive power at machine terminals) Complete the table neglecting armature reaction.

| | $P$ | $Q_2$ | E | $\delta$ |
|---|---|---|---|---|
| **Condition A** | ? | 0.0 | 1.12 | ? |
| **Condition B** | 2.25 | 0.0 | ? | ? |
| **Condition C** | ? | ? | 1.25 | 37.5° |

**Problem 3.16**

The reactances $x_d$ and $x_q$ of a salient-pole synchronous generator are 0.95 and 0.7 per unit, respectively. The armature resistance is negligible. The generator delivers rated kVA at unity PF and rated terminal voltage. Calculate the excitation voltage.

**Problem 3.17**

The machine of Problem 3.16 is connected to an infinite bus through a link with reactance of 0.2 p.u. The excitation voltage is 1.3 p.u. and the infinite bus voltage is maintained at 1 p.u. For a power angle of 25°, compute the active and reactive power supplied to the bus.

**Problem 3.18**

A salient pole machine supplies a load of 1.2 p.u. at unity power factor to an infinite bus whose voltage is maintained at 1.05 p.u. The machine excitation voltage is computed to be 1.4 p.u. when the power angle is 25°. Evaluate the direct-axis and quadrature-axis synchronous reactances.

**Problem 3.19**

Solve Problem 3.18 using MATLAB™.

**Problem 3.20**

The reactances $x_d$ and $x_q$ of a salient-pole synchronous generator are 1.00 and 0.6 per unit respectively. The excitation voltage is 1.77 p.u. and the infinite bus voltage is maintained at 1 p.u. For a power angle of 19.4°, compute the active and reactive power supplied to the bus.

**Problem 3.21**

For the machine of Problem 3.18, assume that the active power supplied to the bus is 0.8 p.u. compute the power angle and the reactive power supplied to the bus. (Hint: assume cos $\delta \cong 1$ for an approximation).

**Problem 3.22**

A salient-pole synchronous machine is connected to an infinite bus whose voltage is kept constant at 1.05 p.u. The direct axis reactance of the machine is 0.95 p.u. and the quadrature axis reactance is 0.4 p.u. respectively. Neglect armature reaction. The table given relates to two operating conditions of the machine. Complete the table.

| | $P$ | $Q_2$ | $E$ | $\delta$ |
|---|---|---|---|---|
| **Condition A** | ? | ? | 1.3 | 40° |
| **Condition B** | 2.0 | 0 | ? | ? |

# Chapter 4

# THE TRANSFORMER

## 4.1 INTRODUCTION

The transformer is a valuable apparatus in electrical power systems, for it enables us to utilize different voltage levels across the system for the most economical value. Generation of power at the synchronous machine level is normally at a relatively low voltage, typically 20 kV, which is most desirable economically. Stepping up of this generated voltage to high voltage, extra-high voltage, or even to ultra-high voltage is done through power transformers to suit the power transmission requirement to minimize losses and increase the transmission capacity of the lines. This transmission voltage level, typically ranging from 115 to 750 kV, is then stepped down in many stages for distribution and utilization purposes in the range of 115/230 v for home use to 13,200 v in large industrial equipment.

## 4.2 GENERAL THEORY OF TRANSFORMER OPERATION

A transformer contains two or more windings linked by a mutual field. The primary winding is connected to an alternating voltage source, which results in the flow of an alternating flux whose magnitude depends on the voltage and number of turns of the primary winding. The alternating flux links the secondary winding and induces a voltage in it with a value that depends on the number of turns of the secondary winding. The operation of the transformer depends on Faraday's voltage law and Ampere's law. If the primary voltage is $\upsilon_1$, the core flux $\phi$ is established such that the counter EMF $e$ equals the impressed voltage (neglecting winding resistance). Thus,

$$\upsilon_1 = e_1 = N_1\left(\frac{d\phi}{dt}\right) \tag{4.1}$$

Here $N_1$ is the number of turns of the primary winding. The EMF $e_2$ is induced in the secondary by the alternating core flux $\phi$:

$$\upsilon_2 = e_2 = N_2\left(\frac{d\phi}{dt}\right) \tag{4.2}$$

Taking the ratio of Eqs. (4.1) to (4.2), we obtain

$$\frac{\upsilon_1}{\upsilon_2} = \frac{N_1}{N_2} \tag{4.3}$$

Neglecting losses, the instantaneous power is equal on both sides of the transformer, as shown below:

$$v_1 i_i = v_2 i_2 \tag{4.4}$$

Combining Eqs. (4.3) and (4.4), we get

$$\frac{i_1}{i_2} = \frac{N_2}{N_1} \tag{4.5}$$

Thus the current ratio is the inverse of the voltage ratio. We can conclude that almost any desired voltage ratio, or ratio of transformation, can be obtained by adjusting the number of turns of the transformer winding.

Transformer action requires a flux to link the two windings. This will be obtained more effectively if an iron core is used because an iron core confines the flux to a definite path linking both windings. A magnetic material such as iron undergoes a loss of energy due to the application of alternating voltage to its *B-H* loop. The losses are composed of two parts. The first is called the *eddy-current loss*, and the second is the *hysteresis loss*. Eddy-current loss is basically an $I^2R$ loss due to the induced currents in the magnetic material. To reduce these losses, the magnetic circuit is usually made of a stack of thin iron-alloy laminations. Hysteresis loss is caused by the energy used in orienting the magnetic domains of the material along the field. The loss depends on the material used.

Two types of construction are used, as shown in Figure 4.1. The first is called the *core type*, which is a single ring encircled by one or more groups of windings. The mean length of the magnetic circuit for this type is long,

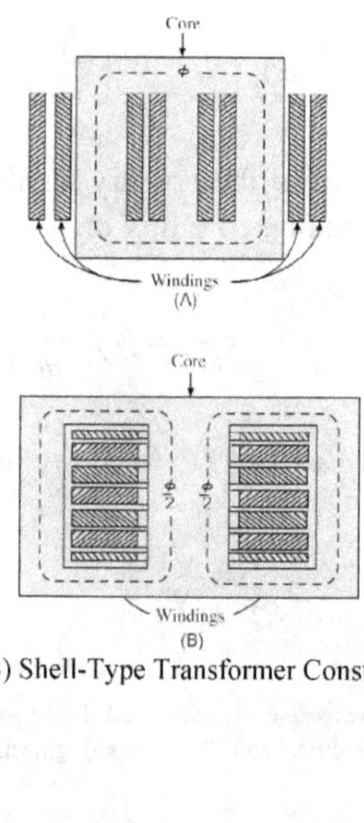


**Figure 4.1** (A) Core-Type and (B) Shell-Type Transformer Construction.

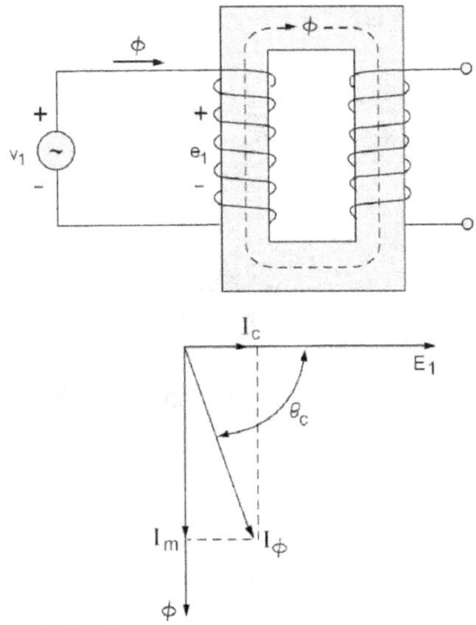


**Figure 4.2** Transformer on No-Load.

whereas the mean length of windings is short. The reverse is true for the *shell type*, where the magnetic circuit encloses the windings.

Due to the nonlinearity of the *B-H* curve of the magnetic material, the primary current on no-load (for illustration purposes) will not be a sinusoid but rather a certain distorted version, which is still periodic. For analysis purposes, a Fourier analysis shows that the fundamental component is out of phase with the applied voltage. This fundamental primary current is basically made of two components. The first is in phase with the voltage and is attributed to the active power taken by eddy-current and hysteresis losses and is called the *core-loss component* $I_c$ of the exciting current $I_\phi$. The current component that lags $e$ by 90° is called the magnetizing current $I_m$. Higher harmonics are neglected. Figure 4.2 shows the no-load phasor diagram for a single-phase transformer.

Consider an ideal transformer (with negligible winding resistances and reactances and no exciting losses) connected to a load as shown in Figure 4.3. Clearly Eqs. (4.1)-(4.5) apply. The dot markings indicate terminals of corresponding polarity in the sense that both windings encircle the core in the same direction if we begin at the dots. Thus comparing the voltage of the two windings shows that the voltages from a dot-marked terminal to an unmarked terminal will be of the same polarity for the primary and secondary windings (i.e., $v_1$ and $v_2$ are in phase). From Eqs. (4.3) and (4.5) we can write for sinusoidal steady state operation

$$\frac{V_1}{I_1} = \left(\frac{N_1}{N_2}\right)^2 \frac{V_2}{I_2}$$

But the load impedance $Z_2$ is

$$\frac{V_2}{I_2} = Z_2$$

Thus,

$$\frac{V_1}{I_1} = \left(\frac{N_1}{N_2}\right)^2 Z_2$$

The result is that $Z_2$ can be replaced by an equivalent impedance $Z_2'$ in the primary circuit. Thus,

$$Z_2' = \left(\frac{N_1}{N_2}\right)^2 Z_2 \tag{4.6}$$

The equivalence is shown in Figure 4.3.

Representations of the transformer that are more realistic must account for winding parameters as well as the exciting current. The equivalent circuit of the transformer can be visualized by following the chain of events as we proceed

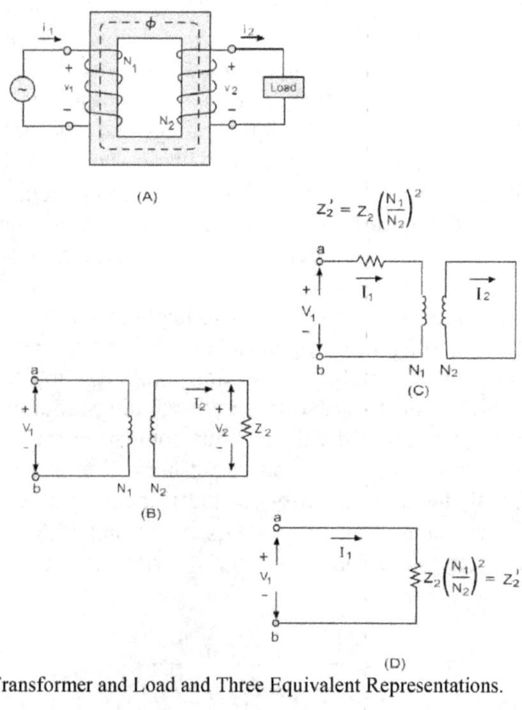


**Figure 4.3** Ideal Transformer and Load and Three Equivalent Representations.

from the primary winding to the secondary winding in Figure 4.4. First the impressed voltage $V_1$ will be reduced by a drop $I_1R_1$ due to the primary winding resistance as well as a drop $jI_1X_1$ due to the primary leakage represented by the inductive reactance $X_1$. The resulting voltage is denoted $E_1$. The current $I_1$ will supply the exciting current $I_\phi$ as well as the current $I_2'$, which will be transformed through to the secondary winding. Thus

$$I_1 = I_\phi + I_2'$$

Since $I_\phi$ has two components ($I_c$ in phase with $E_1$ and $I_m$ lagging $E_1$ by 90°), we can model its effect by the parallel combination of the elements $G_c$ and $B_m$ as shown in the circuit. Next $E_1$ and $I_1$ are transformed by an ideal transformer with turns ratio $N_1/N_2$. As a result, $E_2$ and $I_2$ emerge on the secondary side. $E_2$ undergoes drops $I_2'R_2$ and $jI_2'X_2$ in the secondary winding to result in the terminal voltage $V_2$.

Figure 4.4(B) shows the transformer's equivalent circuit in terms of variables referred to the primary. This circuit is called "circuit referred to the primary side." Note that

$$V_2' = \frac{N_1}{N_2}(V_2) \tag{4.7}$$

$$I_2' = \frac{N_2}{N_1}(I_2) \tag{4.8}$$

$$R_2' = R_2\left(\frac{N_1}{N_2}\right)^2 \tag{4.9}$$

$$X_2' = X_2\left(\frac{N_1}{N_2}\right)^2 \tag{4.10}$$

Although the equivalent circuit illustrated above is simply a *T*-network, it is customary to use approximate circuits such as shown in Figure 4.5. In the first two circuits we move the shunt branch either to the secondary or primary sides to form inverted *L*-circuits. Further simplifications are shown where the shunt branch is neglected in Figure 4.5(C) and finally with the resistances neglected in Figure 4.5(D). These last two circuits are of sufficient accuracy in most power system applications. In Figure 4.5 note that

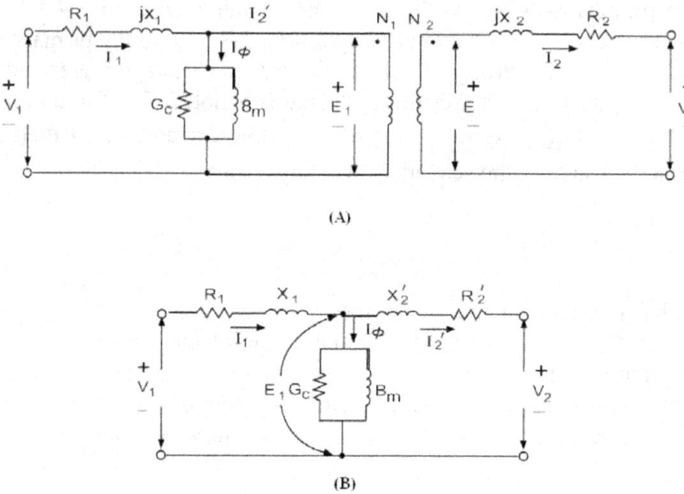


**Figure 4.4** Equivalent Circuits of Transformer.

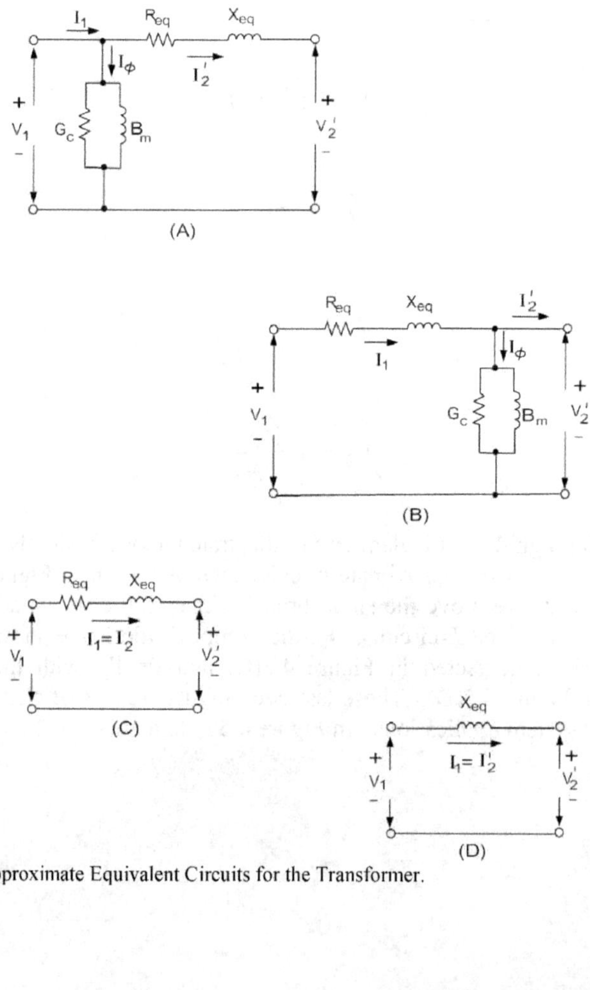


**Figure 4.5** Approximate Equivalent Circuits for the Transformer.

$$R_{\text{eq}} = R_1 + R_2'$$
$$X_{\text{eq}} = X_1 + X_2'$$

**Example 4.1**

A 100-kVA, 400/2000 V, single-phase transformer has the following parameters

$R_1 = 0.01$ $R_2 = 0.25$ ohms
$X_1 = 0.03$ ohms $X_2 = 0.75$ ohms
$G_c = 2.2$ mS $B_m = 6.7$ mS

Note that $G_c$ and $B_m$ are given in terms of primary reference. The transformer supplies a load of 90 kVA at 2000 V and 0.8 PF lagging. Calculate the primary voltage and current using the equivalent circuits shown in Figure 4.5. Verify your solution using MATLAB™.

*Solution*

Let us refer all the data to the primary (400 V) side:

$R_1 = 0.01$ ohm $X_1 = 0.03$ ohms

$$R_2' = 0.25\left(\frac{400}{2000}\right)^2 \qquad X_2' = 0.75\left(\frac{400}{2000}\right)^2$$
$$= 0.01 \text{ ohms} \qquad = 0.03 \text{ ohm}$$

Thus, the equivalent series resistance $R_{\text{eq}}$ and series reactance $X_{\text{eq}}$ are

$$R_{\text{eq}} = R_1 + R_2' \qquad X_{\text{eq}} = X_1 + X_2'$$
$$= 0.02 \text{ ohm} \qquad = 0.06 \text{ ohm}$$

The voltage $V_2 = 2000$ V; thus

$$V_2' = 2000\left(\frac{400}{2000}\right) = 400 \text{ V}$$

The current $I_2'$ is obtained from the VA rating and voltage as:

$$|I_2'| = \frac{90\times10^3}{400} = 225 \text{ A}$$

The power factor of 0.8 lagging implies that

$$I_2' = 225\angle -36.87^\circ \text{ A}$$

For ease of computation, we start with the simplest circuit of Figure 4.5(D). Let us denote the primary voltage calculated through this circuit by $V_{1_d}$. It is clear then that

$$V_{1_d} = V_2' + jI_2'(X_{\text{eq}})$$
$$= 400\angle 0 + j\left(225\angle -36.87^\circ\right)(0.06)$$

Thus,

$$V_{1_d} = 408.243\angle 1.516^\circ \text{ V}$$
$$I_{1_d} = 225\angle -36.87^\circ \text{ A}$$

Comparing circuits (C) and (D) in Figure 4.5, we deduce that

$$V_{1_c} = V_2' + I_2'\left(R_{\text{eq}} + jX_{\text{eq}}\right) = V_{1_d} + I_2'\left(R_{\text{eq}}\right)$$

Thus,

$$V_{1_c} = 408.243\angle 1.516^\circ + (225\angle -36.87)(0.04)$$
$$= 411.78\angle 1.127^\circ \text{ V}$$
$$I_{1_c} = I_2' = 225\angle -36.87^\circ \text{ A}$$

Let us consider circuit (A) in Figure 4.5. We can see that

$$V_{1_a} = V_{1_c} = 411.78\angle 1.127^\circ \text{ V}$$

But

$$I_{1_a} = I_2' + (G_c + jB_m)V_{1_a}$$
$$= 225\angle -36.87^\circ + \left(2.2\times 10^{-3} - j6.7\times 10^{-3}\right)(411.78\angle 1.127)$$
$$= 227.418\angle -37.277^\circ \text{ A}$$

Circuit (B) is a bit different since we start with $V_2'$ impressed on the shunt branch. Thus,

$$I_{1_b} = I_2' + (G_c + jB_m)V_2'$$
$$= 225\angle -36.87^\circ + \left(2.2\times 10^{-3} - j6.7\times 10^{-3}\right)(400\angle 0)$$
$$= 227.37\angle -37.277^\circ$$

Now

$$V_{1_b} = V_2' + I_{1_b}\left(R_{\text{eq}} + jX_{\text{eq}}\right)$$
$$= 411.96\angle 1.1265^{\circ}$$

The following is a MATLAB™ script implementing Example 4.1.

```
% Example 4-1
% To enter the data
R1=0.01;
X1=0.03;
Gc=2.2*10^(-3);
R2=0.25;
X2=0.75;
Bm=-6.7*10^(-3);
V2=2000;
N1=400;
N2=2000;
pf=0.8;
S=90*10^3;
% To refer all the data to the primary
side
R2_prime=R2*(N1/N2)^2;
X2_prime=X2*(N1/N2)^2;
Req=R1+R2_prime;
Xeq=X1+X2_prime;
%The voltage V2=2000 V; thus
V2_prime=V2*(N1/N2);
% To find I2' complex
% Power factor of 0.8 lagging
theta=acos(pf);
theta_deg=theta*180/pi;
I2_prime=abs(S/V2_prime);
I2_primecom=I2_prime*(cos(-
theta)+i*sin(-theta));
I2_primearg=-theta_deg;
% From the figure 5-5(d). The primary
voltage V1d is
V1d=V2_prime+i*I2_primecom*Xeq;
delta=angle(V1d);
delta_deg=delta*180/pi;
I1d=I2_primecom;
% Comparing circuit (c) and (d) in
figure 4.5 we have
V1c_compl=V2_prime+I2_primecom*(Req+i*X
eq);
V1c_mod=abs(V1c_compl);
V1c_arg=angle(V1c_compl);
```

MATLAB™ con't.

```
V1c_argdeg=V1c_arg*180/pi;
I1c=I2_primecom;
% Consider to the circuit (a) in figure
4.5
V1a_compl=V1c_compl;
I1a_compl=I2_primecom+(Gc+i*Bm)*V1a_com
pl;
I1a_mod=abs(I1a_compl);
I1a_arg=angle(I1a_compl);
I1a_argdeg=I1a_arg*180/pi;
% Consider to the circuit (b) in figure
4.5
I1b_compl=I2_primecom+(Gc+i*Bm)*V2_prim
e;
I1b_mod=abs(I1b_compl);
I1b_arg=angle(I1b_compl);
I1b_argdeg=I1b_arg*180/pi;
V1b_compl=V2_prime+I1b_compl*(Req+i*Xeq
);
V1b_mod=abs(V1b_compl);
V1b_arg=angle(V1b_compl);
V1b_argdeg=V1b_arg*180/pi;
% The exact equivalent circuit is now
considered
% as shown in Figure 4.4 (b. First we
calculate E1
E1_compl=V2_prime+I2_primecom*(R2_prime
+i*X2_prime);
E1_mod=abs(E1_compl);
E1_arg=angle(E1_compl);
E1_argdeg=E1_arg*180/pi;
% Now we calculate I1
I1_compl=I2_primecom+E1_compl*(Gc+i*Bm)
I1_mod=abs(I1_compl)

I1_arg=angle(I1_compl);
I1_argdeg=I1_arg*180/pi
% Thus, we have
V1_compl=E1_compl+I1_compl*(R1+i*X1)
V1_mod=abs(V1_compl)
V1_arg=angle(V1_compl);
V1_argdeg=V1_arg*180/pi
```

The solution is

```
EDU»
I1_compl = 1.8092e+002 - 1.3771e+002i

I1_mod = 227.3679

I1_argdeg = -37.2772

V1_compl = 4.1179e+002 + 8.1005e+000i

V1_mod = 411.8702

V1_argdeg = 1.1269
```

## Transformer Performance Measures

Two important performance measures are of interest when selecting transformers. These are the voltage regulation and efficiency of the transformer. The voltage regulation measures the variation in the secondary voltage when the load is varied from zero to rated value at a constant power factor. The percentage voltage regulation (P.V.R) is thus given by

$$\text{P.V.R.} = 100 \frac{\left|V_{2\,(\text{no load})}\right| - \left|V_{2\,\text{rated}}\right|}{\left|V_{2\,\text{rated}}\right|} \tag{4.11}$$

If we neglect the exciting current and refer the equivalent circuit to the secondary side, we have by inspection of Figure 4.6,

$$\text{P.V.R.} = 100 \frac{\left(\frac{V_1}{a}\right) - |V_2|}{|V_2|}$$

where $a$ is the transformer ratio:

$$a = \frac{N_1}{N_2}$$

From the phasor diagram we have approximately in terms of transformer constants:

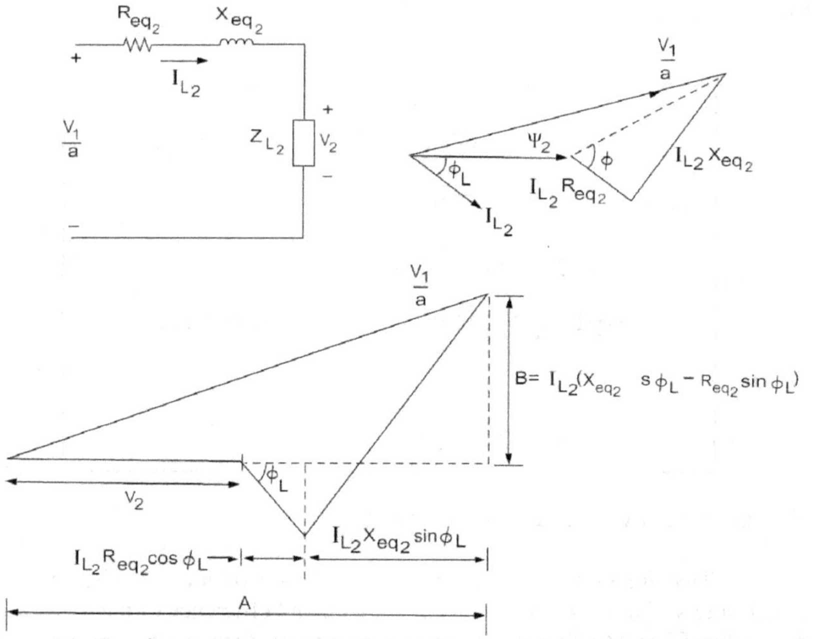


**Figure 4.6** Transformer Approximate Equivalent Circuit and Associated Phasor Diagrams for Voltage Regulation Derivation.

$$\text{P.V.R.} \cong 100\left[\frac{I_{L_2}\left(R_{\text{eq}_2}\cos\phi_L + X_{\text{eq}_2}\sin\phi_L\right)}{V_2} + \frac{1}{2}\left\{\frac{I_{L_2}\left(X_{eq_2}\cos\phi_L - R_{eq_2}\sin\phi_L\right)}{V_2}\right\}^2\right] \quad \textbf{(4.12)}$$

The efficiency of the transformer is the ratio of output (secondary) power to the input (primary) power. Formally the efficiency is $\eta$:

$$\eta = \frac{P_2}{P_1} \quad \textbf{(4.13)}$$

Let $I_L$ be the load current. The input power is the sum of output power $P_2$ and losses $P_l$.

$$P_1 = P_2 + P_l$$

The power loss in the transformer is made of two parts: the $I^2R$ loss and the core loss $P_c$.

As a result, the efficiency is obtained as:

$$\eta = \frac{|V_2||I_L|\cos\phi_L}{|V_2||I_L|\cos\phi_L + P_c + |I_L|^2 (R_{eq})} \tag{4.14}$$

The following example utilizes results of Example 4.1 to illustrate the computations involved in evaluating the efficiency.

**Example 4.2**

Find the P.V.R. and efficiency for the transformer of Example 4.1.

*Solution*

Let us apply the basic formula of Eq. (4.12). We have from Example 4.1:

$$\begin{aligned} V_2 &= 2000 \text{ V} \\ I_{L_2} &= 45 \text{ A} \\ R_{eq_2} &= 0.02\left(\frac{2000}{400}\right)^2 = 0.5 \text{ ohm} \\ X_{eq_2} &= 0.06\left(\frac{2000}{400}\right)^2 = 1.5 \text{ ohm} \end{aligned}$$

Thus substituting in Eq. (4.12), we get

$$\begin{aligned} \text{P.V.R.} &= 100\left\{\frac{45[0.5(0.8)+1.5(0.6)]}{2000} + \frac{1}{2}\left[\frac{45[1.5(0.8)-0.5(0.6)]}{2000}\right]^2\right\} \\ &= 2.9455 \text{ percent} \end{aligned}$$

To calculate the efficiency of the transformer, we need only to apply the basic definition. Take the results of the exact circuit. The input power is

$$\begin{aligned} P_1 &= V_1 I_1 \cos\phi_1 \\ &= (411.77)(227.418)(\cos 38.404) \\ &= 73{,}385.66 \text{ W} \\ P_2 &= V_2 I_2 \cos\phi_2 \\ &= 90 \times 10^3 \times 0.8 \\ &= 72{,}000 \text{ W} \end{aligned}$$

Thus,

$$\eta = \frac{72{,}000}{73{,}385.66} = 0.98112$$

The efficiency of a transformer varies with the load current $I_L$. It attains a maximum when

$$\frac{\partial \eta}{\partial I_L} = 0$$

The maximum efficiency can be shown to occur for

$$P_c = |I_L|^2 (R_{eq}) \tag{4.15}$$

That is, when the $I^2R$ losses equal the core losses, maximum efficiency is attained.

**Example 4.3**

Find the maximum efficiency of the transformer of Example 4.1 under the same power factor and voltage conditions.

*Solution*

We need first the core losses. These are obtained from the exact equivalent circuit of Figure 4.4

$$\begin{aligned} E_1 &= V_2' + I_2'[R_2' + jX_2'] \\ &= 400 + 225\angle -36.87(0.01 + j0.03) \\ &= 405.87\angle 0.57174^\circ \end{aligned}$$

$$\begin{aligned} P_c &= |E_1|^2 (G_c) \\ &= (405.87)^2 (2.2\times 10^{-3}) \\ &= 362.407 \text{ W} \end{aligned}$$

For maximum efficiency,

$$P_c = I_L^2 (R_{eq})$$

Referred to the primary, we thus have

$$362.407 = I_L^2 (0.02)$$

Thus for maximum efficiency,

$$I_L = 134.612 \text{ A}$$

$$\eta_{max} = \frac{V_2'|I_L|\cos\phi_L}{V_2'|I_L|\cos\phi_L + 2P_c}$$

$$= \frac{(400)(134.612)(0.8)}{(400)(134.612)(0.8) + 2(362.407)}$$

$$= 0.98345$$

## 4.3 TRANSFORMER CONNECTIONS

Single-phase transformers can be connected many ways. To start with, consider two single-phase transformers *A* and *B*. They can be connected in four different combinations provided that the polarities are observed. Figure 4.7 illustrates a series-series connection where the primaries of the two transformers are connected in series whereas the secondaries are connected in series. Figure 4.8 illustrates the series-parallel connection and the parallel-series connection. Note that when windings are connected in parallel, those having the same voltage and polarity are paralleled. When connected in series, windings of opposite polarity are joined in one junction. Coils of unequal voltage ratings may be series-connected with polarities either aiding or opposing.

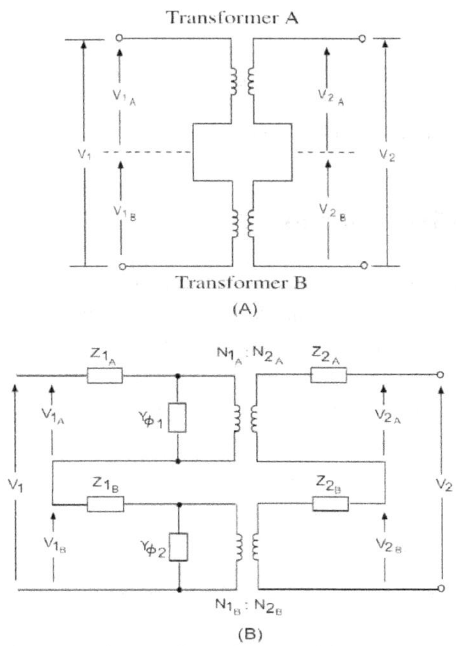


**Figure 4.7** Two Transformers with Primaries in Series and Secondaries in Series. (A) Connection Diagram, and (B) Exact Equivalent Circuit.

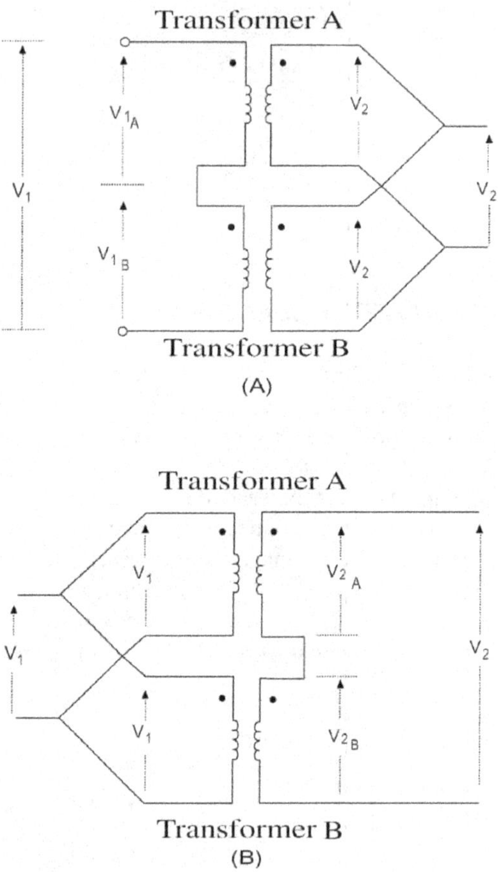


**Figure 4.8** Series-Parallel and Parallel-Series Connections for Single-Phase Transformers.

## Three-Winding Transformers

In many parts of the power system, the three winding transformer is used. The three windings are housed on a common core to achieve economic savings Figure 4.9 shows a three-winding transformer with a practical equivalent circuit. The impedances $Z_1$, $Z_2$, and $Z_3$ are calculated from the three impedances obtained by considering each pair of windings separately with

$$Z_1 = \frac{Z_{12} + Z_{13} - Z_{23}}{2} \quad \textbf{(4.16)}$$

$$Z_2 = \frac{Z_{12} + Z_{23} - Z_{13}}{2} \quad \textbf{(4.17)}$$

$$Z_3 = \frac{Z_{13} + Z_{23} - Z_{12}}{2} \quad \textbf{(4.18)}$$

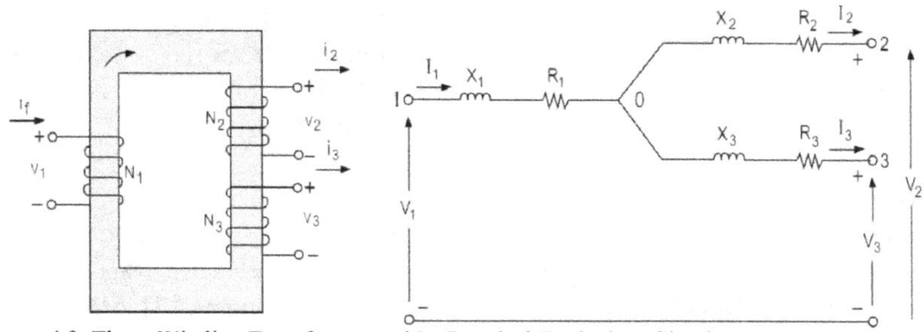


**Figure 4.9** Three-Winding Transformer and Its Practical Equivalent Circuit.

The $I^2R$ or load loss for a three-winding transformer can be obtained from analysis of the equivalent circuit shown.

## The Autotransformer

The basic idea of the autotransformer is permitting the interconnection of the windings electrically. Figure 4.10 shows a two-winding transformer connected in an autotransformer step-up configuration. We will assume that each winding will have the same voltage per turn, i.e.,

$$\frac{V_1}{N_1} = \frac{V_2}{N_2}$$

The rating of the transformer when connected in a two-winding configuration is

$$S_{\text{rated}} = V_1 I_1 = V_2 I_2 \tag{4.19}$$

In the configuration chosen, the apparent power into the load is given by

$$\begin{aligned} S_0 &= (V_1 + V_2)I_2 \\ &= V_2 I_2 \left(1 + \frac{N_1}{N_2}\right) \end{aligned} \tag{4.20}$$

The input apparent power is

$$\begin{aligned} S_i &= V_1(I_1 + I_2) \\ &= V_1 I_1 \left(1 + \frac{N_1}{N_2}\right) \end{aligned}$$

Thus the rating of the autotransformer is higher than the original rating of the two-winding configuration. Note that in both configurations, the same current passes in each winding, and as a result the losses remain the same. Due to the increased power rating, the efficiency is thus improved.

Autotransformers are generally used when the transformation ratio is 3:1 or less. Two disadvantages of the autotransformer are the lack of electric isolation between primary and secondary and the increased short-circuit current over that of the corresponding two-winding configuration.

## Example 4.4

A 50-kVA, 2.4/0.6-kV transformer is connected as a step-up autotransformer from a 2.4-kV supply. Calculate the currents in each part of the transformer and the load rating. Neglect losses. Verify your solution using MATLAB™.

*Solution*

With reference to Figure 4.10, the primary winding rated current is

$$I_1 = \frac{50}{2.4} = 20.83\text{ A}$$

The secondary rated current is

$$I_2 = \frac{50}{0.6} = 83.33\text{ A}$$

Thus the load current is

$$I_L = 83.33\text{ A}$$

The load voltage is

$$V_L = V_1 + V_2 = 3\text{ kV}$$

As a result, the load rating is

$$S_L = V_L I_L = 250\text{ kVA}$$

Note that

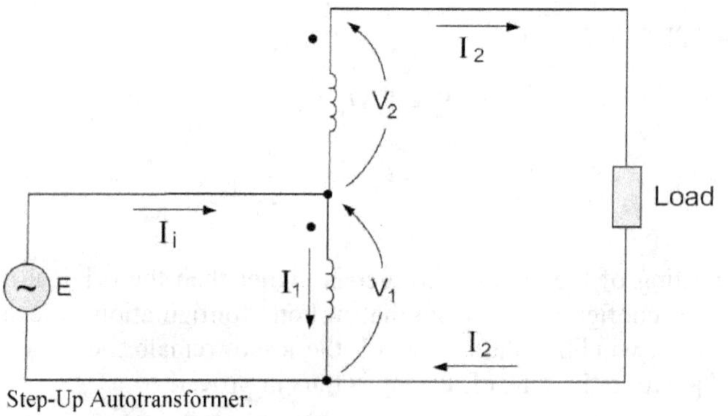


**Figure 4.10** Step-Up Autotransformer.

$$I_i = I_1 + I_2$$
$$= 104.16 \text{ A}$$
$$V_i = V_1 = 2.4 \text{ kV}$$

Thus,

$$S_i = (2.4)(104.16) = 150 \text{ kVA}$$

A MATLAB™ script implementing Example 4.4 is shown here

```
% Example 4-4
% Autotransformer
KVA=30;
KVp=2.4;
KVs=0.6;
% The primary winding rated current is
I1=KVA/KVp
% The secondary rated current is
I2=KVA/KVs
% The load current is
IL=I2
% the load voltage is
VL = KVp+KVs
% The load rating is
SL=VL*IL
```

The solution is obtained as

```
EDU»
I1 = 12.5000

I2 = 50

IL = 50

VL = 3

SL = 150
```

## Three-Phase Transformer Connections

For three-phase system applications it is possible to install three-phase transformer units or banks made of three single-phase transformers connected in the desired three-phase configurations. The latter arrangement is advantageous from a reliability standpoint since it allows installing a single standby single-phase transformer instead of a three-phase unit. This provides a considerable cost saving. We have seen that there are two possible three-phase connections; the Y-connection and the Δ-connection. We thus see that three-phase transformers can be connected in four different ways. In the Y/Y connection, both primary and secondary windings are connected in Y. In addition, we have Δ/Δ, Y/Δ, or Δ/Y connections. The Y-connected windings may or may not be grounded.

The Y/Δ configuration is used for stepping down from a high voltage to a medium or low voltage. This provides a grounding neutral on the high-voltage side. Conversely, the Δ/Y configuration is used in stepping up to a high voltage. The Δ/Δ connection enables one to remove one transformer for maintenance while the other two continue to function as a three-phase bank (with reduced rating) in an open-delta or V-connection. The difficulties arising from the harmonic contents of the exciting current associated with the Y/Y connection make it seldom used.

In Figure 4.11, the four common three-phase transformer connections are shown along with the voltage and current relations associated with the transformation. It is important to realize that the line-to-ground voltages on the Δ side lead the corresponding Y-side values by 30° and that the line currents on the Δ side also lead the currents on the Y side by 30°.

Consider the Y/Δ three-phase transformer shown in Figure 4.12. We can show that

$$V_{An} = \frac{N_1}{N_2}\sqrt{3}V_{an}\angle -30^\circ$$

That is the Δ-side line-to-ground secondary voltage $V_{an}$ leads the Y-side line-to-ground primary voltage $V_{An}$ by 30°.

Turning our attention now to the current relations, we have

$$I_a = \frac{N_1}{N_2}\sqrt{3}I_A\angle 30^\circ$$

Thus the secondary line current leads the primary current by 30°.

Three-phase autotransformers are usually Y-Y connected with the neutral grounded. A third (tertiary) Δ-connected set of windings is included to carry the third harmonic component of the exciting current. A schematic diagram of a three-phase autotransformer with a Δ-tertiary is shown in Figure 4.13.

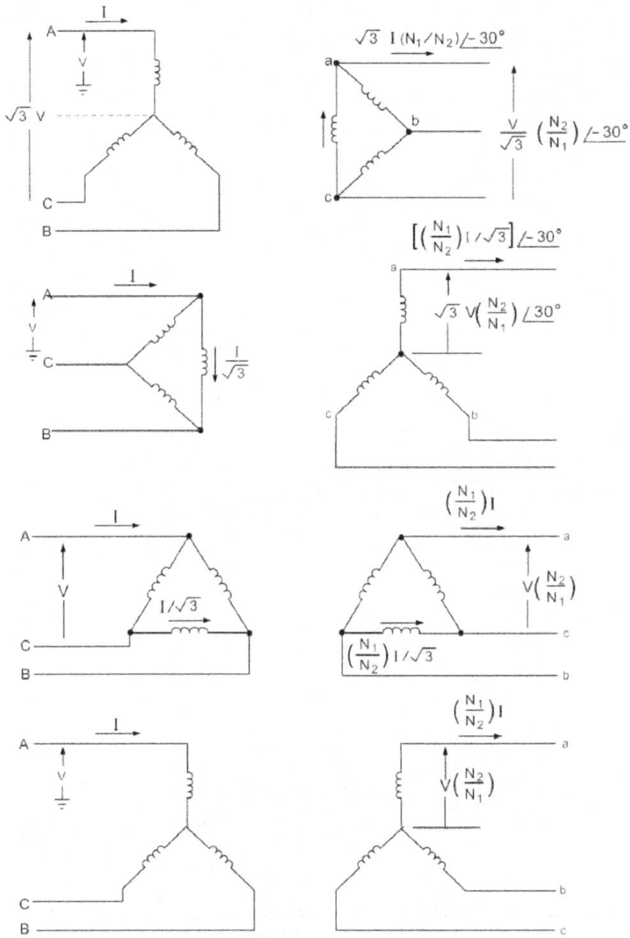


**Figure 4.11** Three-Phase Transformer Connections.

## Control Transformers

Transformers are used not only to step up or step down bulk power voltages but also to control the operations of the power system in terms of voltage levels and hence power flow. Two examples of control transformer applications involve (1) tap changing under load (TCUL) transformers, and (2) the regulating transformer.

### *Tap Changing under Load*

The TCUL transformer maintains a prescribed voltage at a point in the system by changing the transformation ratio by increasing or decreasing the number of active turns in one winding with respect to another winding. This is performed while not interfering with the load. In practice, a voltage measuring device actuates the motor that drives the tap changer. If the actual voltage is higher than a desired upper limit, the motor will change to the next lower tap

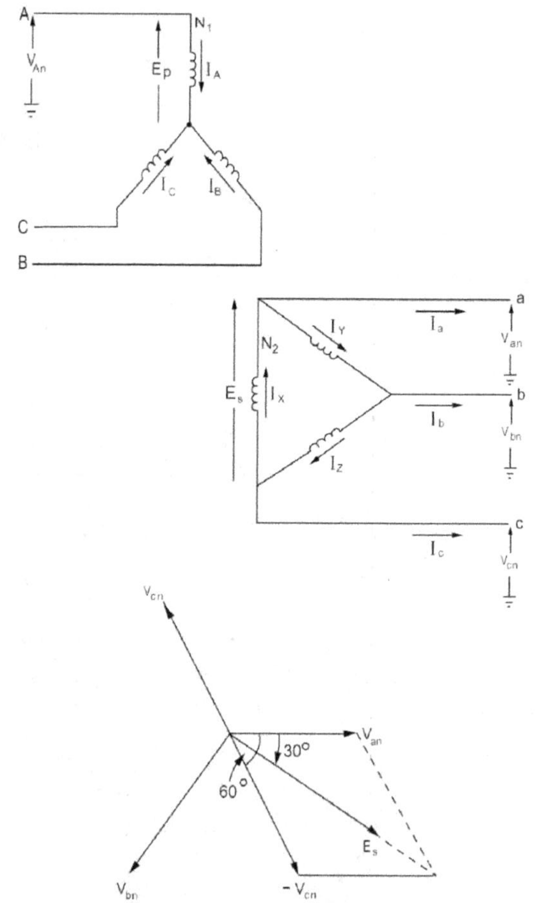

**Figure 4.12** A Y-Δ Transformer and a Phasor Diagram.

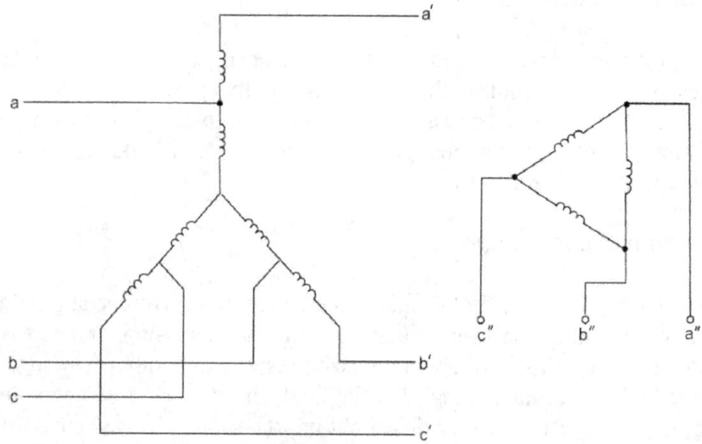

**Figure 4.13** Schematic Diagram of a Three-Winding Autotransformer.

voltage; similarly, a voltage lower than the desired will cause a change to the next higher up.

### *The Regulating Transformer*

The regulating transformer changes (by a small amount) the voltage magnitude and phase angle at a certain point in the system. Figure 4.14 shows the arrangement of a regulating transformer. Assume that:

$$V_{an} = V\angle 0$$

$$V_{bn} = V\angle -120^\circ$$

$$V_{cn} = V\angle +120^\circ$$

The primary windings of the transformers *A*, *B*, and *C* are connected in Δ. The secondary windings 1, 3, and 5 are connected in Y with their voltages adjustable. From the phase-shift property in Δ-Y transformers, we have

$$V_{ko} = \frac{V_m}{\sqrt{3}}\angle 30^\circ$$

$$V_{lo} = \frac{V_m}{\sqrt{3}}\angle -90^\circ$$

$$V_{mo} = \frac{V_m}{\sqrt{3}}\angle 150^\circ$$

The magnitude of $V_m$ can be controlled in a small range and is utilized for adjusting the magnitude of the three-phase voltage $V_{a'}$, $V_{b''}$, and $V_{c'}$. The tertiary windings 2, 4, and 6 have voltages

$$V_{rl} = V_\phi\angle 30^\circ \tag{4.21}$$

$$V_{sm} = V_\phi\angle -90^\circ \tag{4.22}$$

$$V_{tk} = V_\phi\angle 150^\circ \tag{4.23}$$

The magnitude $V_\phi$ is adjustable and is used for control of the phase angle of the voltages $V_{a'}$, $V_{b''}$, and $V_{c'}$.

We can derive the voltages $V_{km}$, $V_{lk}$, $V_{ml}$ from $V_{ko}$, $V_{lo}$, $V_{mo}$ as

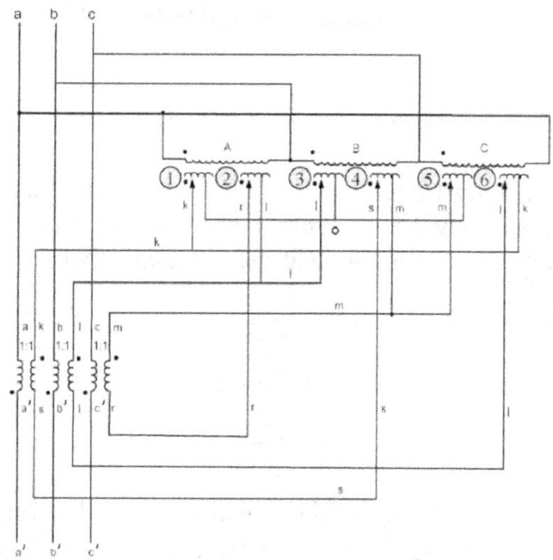


**Figure 4.14** Schematic of Regulating Transformer.

$$V_{km} = V_m \angle 0$$
$$V_{lk} = V_m \angle -120^\circ$$
$$V_{ml} = V_m \angle +120^\circ$$

Note that $V_{km}$, $V_{lk}$, and $V_{ml}$ are in phase with the system voltages $V_{an}$, $V_{bn}$, and $V_{cn}$. The voltages $V_{rl}$, $V_{sm}$, and $V_{tk}$ are 90° out of phase with the same voltages. The incremental voltages $\Delta V_a$, $\Delta V_b$, and $\Delta V_c$ are given by

$$\Delta V_a = V_{ks}$$
$$\Delta V_b = V_{lt}$$
$$\Delta V_c = V_{mr}$$

or

$$\Delta V_a = V_{km} - V_{sm} = V_m \angle 0 - V_\phi \angle -90^\circ \tag{4.24}$$

$$\Delta V_b = V_{lk} - V_{tk} = V_m \angle -120^\circ - V_\phi \angle 150^\circ \tag{4.25}$$

$$\Delta V_c = V_{ml} - V_{rl} = V_m \angle +120^\circ - V_\phi \angle 30^\circ \tag{4.26}$$

The $\Delta V$ values are added in series in each phase to give

$$V_{a'n} = V_{an} + \Delta V_a \tag{4.27}$$

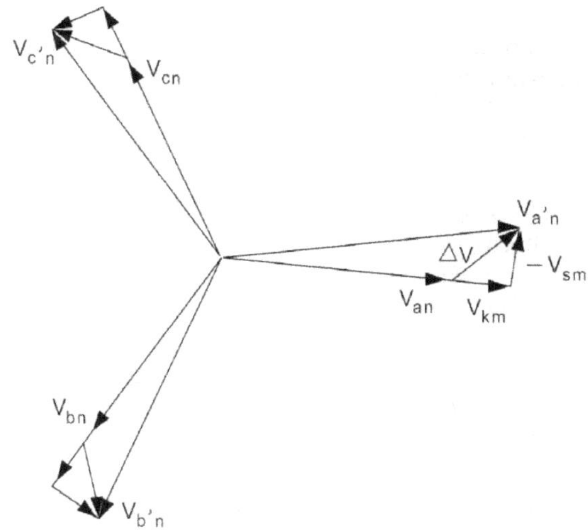


**Figure 4.15** Output Voltages of Regulating Transformer.

$$V_{b'n} = V_{bn} + \Delta V_b \tag{4.28}$$

$$V_{c'n} = V_{cn} + \Delta V_c \tag{4.29}$$

A phasor diagram of the voltages in the system is shown in Figure 4.15.

## PROBLEMS

### Problem 4.1

A 50-kVA, 400/2000 V, single-phase transformer has the following parameters:

| | |
|---|---|
| $R_1 = 0.02$ ohm | $R_2 = 0.5$ ohm |
| $X_1 = 0.06$ ohm | $X_2 = 1.5$ ohm |
| $G_c = 2$ mS | $Bm = -6$ mS |

Note that $G_c$ and $B_m$ are given in terms of primary reference. The transformer supplies a load of 40 kVA at 2000 V and 0.8 PF lagging. Calculate the primary voltage and current using the equivalent circuits shown in Figure 4.5 and that of Figure 4.4. Verify your solution using MATLAB™.

### Problem 4.2

Find the P.V.R. and efficiency for the transformer of Problem 4.1.

### Problem 4.3

Find the maximum efficiency of the transformer of Problem 4.1, under the same conditions. Verify your solution using MATLAB™.

**Problem 4.4**

The equivalent impedance referred to the primary of a 2300/230-V, 500-kVA, single-phase transformer is

$$Z = 0.2 + j0.6 \text{ ohm}$$

Calculate the percentage voltage regulation (P.V.R.) when the transformer delivers rated capacity at 0.8 power factor lagging at rated secondary voltage. Find the efficiency of the transformer at this condition given that core losses at rated voltage are 2 kW.

**Problem 4.5**

A 500/100 V, two-winding transformer is rated at 5 kVA. The following information is available:

A. The maximum efficiency of the transformer occurs when the output of the transformer is 3 kVA.
B. The transformer draws a current of 3 A, and the power is 100 W when a 100-V supply is impressed on the low-voltage winding with the high-voltage winding open-circuit.

Find the rated efficiency of the transformer at 0.8 PF lagging. Verify your solution using MATLAB™.

**Problem 4.6**

The no-load input power to a 50-kVA, 2300/230-V, single-phase transformer is 200 VA at 0.15 PF at rated voltage. The voltage drops due to resistance and leakage reactance are 0.012 and 0.018 times rated voltage when the transformer is operated at rated load. Calculate the input power and power factor when the load is 30 kW at 0.8 PF lagging at rated voltage. Verify your solution using MATLAB™.

**Problem 4.7**

A 500 KVA, 2300/230 V single phase transformer delivers full rated KVA at 0.8 p.f. lagging to a load at rated secondary voltage. The primary voltage magnitude is 2400 V under these conditions and the efficiency is 0.97. Find the equivalent circuit parameters of this transformer neglecting the no load circuit.

**Problem 4.8**

Solve Problem 4.7 using MATLAB™ for a power factor of 0.7.

**Problem 4.9**

A single phase transformer has a turns ratio of 2:1, and an equivalent reactance $X_{eq} = 4$ ohms. The primary voltage is 2020 V at 0.75 p.f. lagging. The voltage regulation for this power factor is found to be 0.09, and the efficiency is 95% under these conditions. Neglect the no load circuit.

A. Find the transformer's equivalent circuit resistance $R_{eq}$ referred to the primary side.

B. Find the current drawn by the transformer referred to the primary side.

C. If the load power factor is changed to 0.9 lagging with the load's active power and voltage magnitude unchanged, find the required primary voltage.

**Problem 4.10**

Two 2400/600 V single phase transformers are rated at 300 and 200 KVA respectively. Find the rating of the transformers' combination if one uses the following connections:

A. Series-series
B. Parallel-series
C. Parallel-parallel

**Problem 4.11**

A 30-kVA, 2.4/0.6-kV transformer is connected as a step-up autotransformer from a 2.4-kV supply. Calculate the currents in each part of the transformer and the load rating. Neglect losses.

**Problem 4.12**

A three-phase bank of three single-phase transformers steps up the three-phase generator voltage of 13.8 kV (line-to-line) to a transmission voltage of 138 kV (line-to-line). The generator rating is 83 MVA. Specify the voltage, current and kVA ratings of each transformer for the following connections:

A. Low-voltage windings Δ, high-voltage windings Y
B. Low-voltage windings Y, high-voltage windings Δ
C. Low-voltage windings Y, high-voltage windings Y
D. Low-voltage windings Δ, high-voltage windings Δ

**Problem 4.13**

The load at the secondary end of a transformer consists of two parallel branches:

*Load 1:* an impedance $Z$ given by

$$Z = 0.75\angle 45^\circ$$

*Load 2:* inductive load with $P = 1.0$ p.u., and $S = 1.5$ p.u.

The load voltage magnitude is an unknown. The transformer is fed by a feeder, whose sending end voltage is kept at 1 p.u. Assume that the load voltage is the reference. The combined impedance of transformer and feeder is given by:

$$Z = 0.02 + j0.08 \text{ p.u.}$$

A. Find the value of the load voltage.

B. If the load contains induction motors requiring at least 0.85 p.u. voltage to start, will it be possible to start the motors? If not, suggest a solution.

**Problem 4.14**

A three phase transformer delivers a load of 66 MW at 0.8 p.f. lagging and 138 KV (line-to-line). The primary voltage under these conditions is 14.34 KV (line-to-line), the apparent power is 86 MVA and the power factor is 0.78 lagging.

A. Find the transformer ratio.
B. Find the series impedance representation of the transformer.
C. Find the primary voltage when the load is 75 MVA at 0.7 p.f. lagging at a voltage of 138 KV.

**Problem 4.15**

A three phase transformer delivers a load of 83 MVA at 0.8 p.f. lagging and 138 KV (line-to-line). The primary voltage under these conditions is 14.34 KV (line-to-line), the apparent power is 86 MVA and the power factor is 0.78 lagging.

A. Find the transformer ratio.
B. Find the series impedance representation of the transformer.
C. Find the primary apparent power and power factor as well as the voltage when the load is 75 MVA at 0.7 p.f. lagging at a voltage of 138 KV.

**Problem 4.16**

A two winding transformer is rated at 50 kVA. The maximum efficiency of the transformer occurs when the output of the transformer is 35 kVA. Find the rated efficiency of the transformer at 0.8 PF lagging given that the no load losses are 200 W.

**Problem 4.17**

The no-load input to a 5 kVA, 500/100-V, single-phase transformer is 100 W at 0.15 PF at rated voltage. The voltage drops due to resistance and leakage reactance are 0.01 and 0.02 times the rated voltage when the transformer operates at rated load. Calculate the input power and power factor when the load is 3 kW at 0.8 PF lagging at rated voltage.

**Problem 4.18**

A 25 kVA, 2200/220 V, 60 Hz, single-phase transformer has the following equivalent-circuit parameters referred to the high-voltage side.

$$R_1 = 3.0\ \Omega \qquad R_2' = 3.0\ \Omega$$

$$X_{l1} = 12\ \Omega \qquad X_{l2}' = 12\ \Omega$$

$X_m = 20{,}000\ \Omega$ $R_c = 50{,}000\ \Omega$

The transformer is supplying 17.5 kVA at 220 V and a lagging power factor of 0.8. Use the equivalent Cantilever model circuit of the transformer shown in Figure 4.16. Determine:

A. The magnitude of the required primary voltage.
B. The power factor of the primary side, the input kVA, and the efficiency of the transformer.

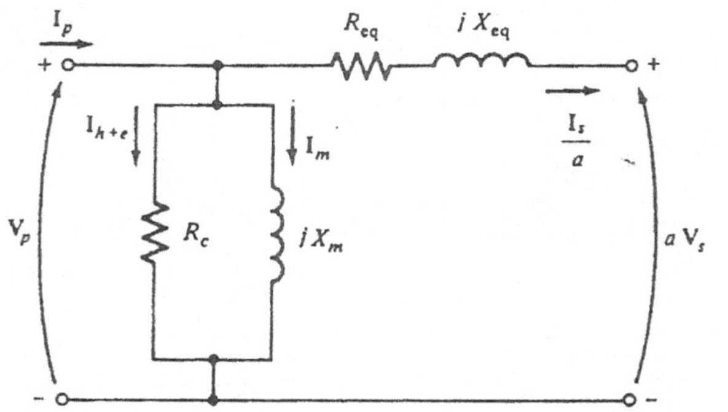


**Figure 4.16** Equivalent Circuit of Transformer for Problem 4.18.

## Problem 4.19

Consider the circuit of Problem 4.18.

A. A short circuit test is conducted on the transformer with 22 volts applied to the secondary side with the primary short circuited. Determine the readings of the ammeter and wattmeter connected to the secondary side for this test.
B. An open circuit test is conducted on the transformer with 2200 volts applied to the primary side with the secondary side left open. Determine the readings of the ammeter and wattmeter connected to the primary for this open circuit test.
C. The transformer is supplying 15 kVA at 220-V and a lagging power factor of 0.85. Determine the primary voltage.

## Problem 4.20

Consider a three-winding transformer, as shown in Figure 4.9, with the following particulars:

$Z_1 = 0.02 + j0.08$ $Z_2 = 0.02 + j0.08$ $Z_3 = 0.02 + j0.08$

$V_2 = 400$ $I_2 = 60\angle -30^\circ$ $I_3 = 50\angle -40^\circ$

Assume that $V_2$ is the reference phasor, calculate:

A. The intermediate voltage $V_0$

B. The primary current $I_1$ and the primary voltage $V_1$
C. The tertiary voltage $V_3$ referred to the primary side
D. The apparent powers and power factors at the primary, secondary and tertiary terminals.
E. The transformer efficiency.

# Chapter 5

# ELECTRIC POWER TRANSMISSION

## 5.1 *INTRODUCTION*

Electric energy is produced at electric power generating stations and transported over high-voltage transmission lines to utilization points. The trend toward using higher voltages is motivated by the increased line capacity while reducing line losses per unit of power transmitted. The reduction in losses is significant and is an important aspect of energy conservation. Better use of land is a benefit of the larger transmission capacity of the lines.

This chapter develops a fundamental understanding of modeling the performance of electric power transmission systems.

## 5.2 *ELECTRIC TRANSMISSION LINE PARAMETERS*

An electric transmission line is modeled using series resistance, series inductance, shunt (parallel to ground) capacitance, and shunt conductance. The line resistance and inductive reactance are important. For some studies it is possible to omit the shunt capacitance and conductance and thus simplify the equivalent circuit considerably.

We deal here with aspects of determining these parameters on the basis of line length, type of conductor used, and the spacing of the conductors as they are mounted on the supporting structure of the transmission towers.

A wire or combination of wires not insulated from one another is called a *conductor*. A stranded conductor consists of a group of wires, usually twisted or braided together. In a concentrically stranded conductor, each successive layer contains six more wires than the preceding one. There are two basic constructions: the one-wire core and the three-wire core.

### Types of Conductors and Conductor Materials

Phase conductors in extra high voltage – ultra high voltage EHV-UHV transmission systems employ aluminum conductors and aluminum or steel conductors for overhead ground wires. Many types of conductors are available. These include:

A. Aluminum Conductors
   There are five designs:
   1. Homogeneous designs: These are denoted as All-Aluminum-Conductors (AAC) or All-Aluminum-Alloy Conductors (AAAC).

2. Composite designs: These are essentially aluminum-conductor-steel-reinforced conductors (ACSR) with steel core material.
3. Expanded ASCR: These use solid aluminum strands with a steel core. Expansion is by open helices of aluminum wire, flexible concentric tubes, or combinations of aluminum wires and fibrous ropes.
4. Aluminum-clad conductor (Alumoweld).
5. Aluminum-coated conductors.

B. Steel Conductors
Galvanized steel conductors with various thicknesses of zinc coatings are used.

## Line Resistance

The resistance of the conductor is the most important cause of power loss in a power line. Direct-current resistance is given by the familiar formula:

$$R_{\text{dc}} = \frac{\rho l}{A} \text{ ohms}$$

where

$\rho$ = resistivity of conductor
$l$ = length
$A$ = cross-sectional area

Any consistent set of units may be used in the calculation of resistance. In the SI system of units, $\rho$ is expressed in ohm-meters, length in meters, and area in square meters. A system commonly used by power systems engineers expresses resistivity in ohms circular mils per foot, length in feet, and area in circular mils.

There are certain limitations in the use of this equation for calculating the resistance of transmission line conductors. The following factors need to be considered:

1. Effect of conductor stranding.
2. When ac flows in a conductor, the current is not distributed uniformly over the conductor cross-sectional area. This is called *skin effect* and is a result of the nonuniform flux distribution in the conductor. This increases the resistance of the conductor.
3. The resistance of magnetic conductors varies with current magnitude.
4. In a transmission line there is a nonuniformity of current distribution caused by a higher current density in the elements of adjacent conductors nearest each other than in the elements farther

apart. The phenomenon is known as *proximity effect*. It is present for three-phase as well as single-phase circuits. For the usual spacing of overhead lines at 60 Hz, the proximity effect is neglected.

## 5.3 LINE INDUCTANCE

In transmission lines the inductive reactance is by far the most dominating impedance element.

### Inductance of a Single-Phase Two-Wire Line

The inductance of a simple two-wire line consisting of two solid cylindrical conductors of radii $r_1$ and $r_2$ shown in Figure 5.1 is considered first.

The total inductance of the circuit due to the current in conductor 1 only is given by:

$$L_1 = (2\times 10^{-7}) ln\left(\frac{D}{r_1'}\right) \tag{5.1}$$

Similarly, the inductance due to the current in conductor 2 is

$$L_2 = (2\times 10^{-7}) ln\left(\frac{D}{r_2'}\right) \tag{5.2}$$

Thus $L_1$ and $L_2$ are the phase inductances. For the complete circuit we have

$$L_t = L_1 + L_2 \tag{5.3}$$

$$L_t = (4\times 10^{-7}) ln\left(\frac{D}{\sqrt{r_1' r_2'}}\right) \tag{5.4}$$

where

$$\begin{aligned} r_i' &= r_i e^{-1/4} \\ &= 0.7788 r_i \end{aligned} \tag{5.5}$$

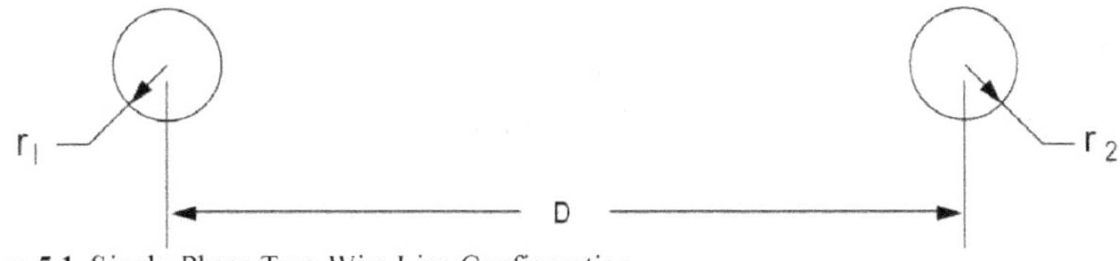


**Figure 5.1** Single-Phase Two-Wire Line Configuration.

We compensate for the internal flux by using an adjusted value for the radius of the conductor. The quantity $r'$ is commonly referred to as the solid conductor's *geometric mean radius* (GMR).

An inductive voltage drop approach can be used to get the same results

$$V_1 = j\omega(L_{11}I_1 + L_{12}I_2) \tag{5.6}$$

$$V_2 = j\omega(L_{12}I_1 + L_{22}I_2) \tag{5.7}$$

where $V_1$ and $V_2$ are the voltage drops per unit length along conductors 1 and 2 respectively. The self-inductances $L_{11}$ and $L_{22}$ correspond to conductor geometries:

$$L_{11} = (2\times10^{-7})ln\left(\frac{1}{r_1'}\right) \tag{5.8}$$

$$L_{22} = (2\times10^{-7})ln\left(\frac{1}{r_2'}\right) \tag{5.9}$$

The mutual inductance $L_{12}$ corresponds to the conductor separation $D$. Thus

$$L_{12} = (2\times10^{-7})ln\left(\frac{1}{D}\right) \tag{5.10}$$

Now we have

$$I_2 = -I_1$$

The complete circuit's voltage drop is

$$V_1 - V_2 = j\omega(L_{11} + L_{22} - 2L_{12})I_1 \tag{5.11}$$

In terms of the geometric configuration, we have

$$\Delta V = V_1 - V_2$$

$$\Delta V = j\omega(2\times10^{-7})\left[ln\left(\frac{1}{r_1'}\right) + ln\left(\frac{1}{r_2'}\right) - 2ln\left(\frac{1}{D}\right)\right]I_1$$

$$= j\omega(4\times10^{-7})ln\left(\frac{D}{\sqrt{r_1'r_2'}}\right)I_1$$

Thus

$$L_t = (4\times10^{-7})ln\left(\frac{D}{\sqrt{r_1' r_2'}}\right) \tag{5.12}$$

where

$$L_t = L_{11} + L_{22} - 2L_{12}$$

We recognize this as the inductance of two series-connected magnetically coupled coils, each with self-inductance $L_{11}$ and $L_{22}$, respectively, and having a mutual inductance $L_{12}$.

The phase inductance expressions given in Eqs. (5.1) and (5.2) can be obtained from the voltage drop equations as follows:

$$V_1 = j\omega(2\times10^{-7})\left[I_1 ln\left(\frac{1}{r_1'}\right) + I_2 ln\left(\frac{1}{D}\right)\right]$$

However,

$$I_2 = -I_1$$

Thus,

$$V_1 = j\omega(2\times10^{-7})\left[I_1 ln\left(\frac{D}{r_1'}\right)\right]$$

In terms of phase inductance we have

$$V_1 = j\omega L_1 I_1$$

Thus for phase one,

$$L_1 = (2\times10^{-7})ln\left(\frac{D}{r_1'}\right) \text{henries/meter} \tag{5.13}$$

Similarly, for phase two,

$$L_2 = (2\times10^{-7})ln\left(\frac{D}{r_2'}\right) \text{henries/meter} \tag{5.14}$$

Normally, we have identical line conductors.

In North American practice, we deal with the inductive reactance of the line per phase per mile and use the logarithm to the base 10. Performing this conversion, we obtain

$$X = k \log \frac{D}{r'} \text{ ohms per conductor per mile} \tag{5.15}$$

where

$$\begin{aligned} k &= 4.657 \times 10^{-3} f \\ &= 0.2794 \text{ at } 60 \text{ Hz} \end{aligned} \tag{5.16}$$

assuming identical line conductors.

Expanding the logarithm in the expression of Eq. (5.15), we get

$$X = k \log D + k \log \frac{1}{r'} \tag{5.17}$$

The first term is called $X_d$ and the second is $X_a$. Thus

$$X_d = k \log D \text{ inductive reactance spacing factor in ohms per mile} \tag{5.18}$$

$$X_a = k \log \frac{1}{r'} \text{ inductive reactance at 1-ft spacing in ohms per mile} \tag{5.19}$$

Factors $X_a$ and $X_d$ may be obtained from tables available in many handbooks.

### Example 5.1

Find the inductive reactance per mile per phase for a single-phase line with phase separation of 25 ft and conductor radius of 0.08 ft.

*Solution*

We first find $r'$, as follows:

$$\begin{aligned} r' &= re^{-1/4} \\ &= (0.08)(0.7788) \\ &= 0.0623 \text{ ft} \end{aligned}$$

We therefore calculate

$$X_a = 0.2794 \log \frac{1}{0.0623}$$
$$= 0.3368$$
$$X_d = 0.2794 \log 25$$
$$= 0.3906$$
$$X = X_a + X_d = 0.7274 \text{ ohms per mile}$$

The following MATLAB™ script implements Example 5.1 based on Eqs. (5.17) to (5.19)

```
% Example 5-1
r=0.08
D=25
r_prime=0.7788*r
Xa=0.2794*(log10(1/(r_prime)))
Xb=0.2794*(log10 (D))
X=Xa+Xb
```

The answers obtained from MATLAB™ are as follows:

```
EDU»
r = 0.0800
D = 25
r_prime = 0.0623
Xa = 0.3368
Xb = 0.3906
X = 0.7274
```

## Bundle Conductors

At voltages above 230 kV (extra high voltage) and with circuits with only one conductor per phase, the corona effect becomes more excessive. Associated with this phenomenon is a power loss as well as interference with communication links. Corona is the direct result of high-voltage gradient at the conductor surface. The gradient can be reduced considerably by using more than one conductor per phase. The conductors are in close proximity compared with the spacing between phases. A line such as this is called a *bundle-conductor line*. The bundle consists of two or more conductors (subconductors) arranged on the perimeter of a circle called the *bundle circle* as shown in Figure 5.2. Another important advantage of bundling is the attendant reduction in line reactances, both series and shunt. The analysis of bundle-conductor lines is a specific case of the general multiconductor configuration problem.

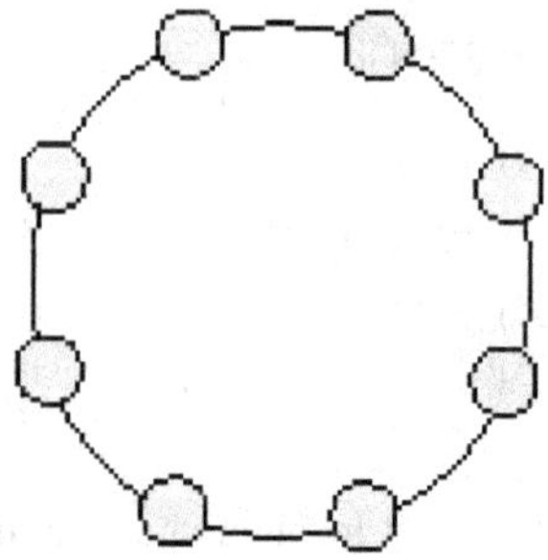

**Figure 5.2** Bundle Conductor.

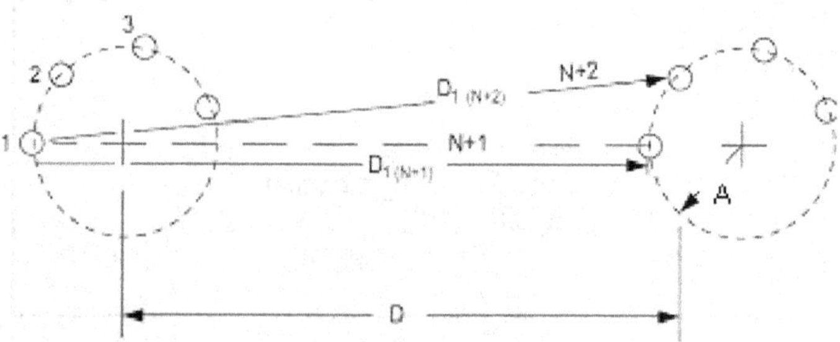


**Figure 5.3** Single-Phase Symmetrical Bundle-Conductor Circuit.

## Inductance of a Single-Phase Symmetrical Bundle-Conductor Line

Consider a symmetrical bundle with $N$ subconductors arranged in a circle of radius $A$. The angle between two subconductors is $2\pi/N$. The arrangement is shown in Figure 5.3.

We define the geometric mean distance (GMD) by

$$\text{GMD} = \left\{\left[D_{1(N+1)}\right]\left[D_{1(N+2)}\right]\cdots\left[D_{1(2N)}\right]\right\}^{1/N} \tag{5.20}$$

Let us observe that practically the distances $D_{1(N+1)}$, $D_{1(N+2)}$, . . . , are all almost equal in value to the distance $D$ between the bundle centers. As a result,

$$\text{GMD} \cong D \tag{5.21}$$

Also, define the geometric mean radius as

$$\text{GMR} = \left[Nr'(A)^{N-1}\right]^{1/N} \tag{5.22}$$

The inductance is then obtained as

$$L = (2\times 10^{-7}) ln\left(\frac{\text{GMD}}{\text{GMR}}\right) \tag{5.23}$$

In many instances, the subconductor spacing $S$ in the bundle circle is given. It is easy to find the radius $A$ using the formula

$$S = 2A\sin\left(\frac{\pi}{N}\right) \tag{5.24}$$

which is a consequence of the geometry of the bundle as shown in Figure 5.4.

### Example 5.2

Figure 5.5 shows a 1000-kv, single-phase, bundle-conductor line with eight subconductors per phase. The phase spacing is $D_1 = 18$ m, and the subconductor spacing is $S = 50$ cm. Each subconductor has a diameter of 5 cm. Calculate the line inductance.

*Solution*

We first evaluate the bundle radius $A$. Thus,

$$0.5 = 2A\sin\left(\frac{\pi}{8}\right)$$

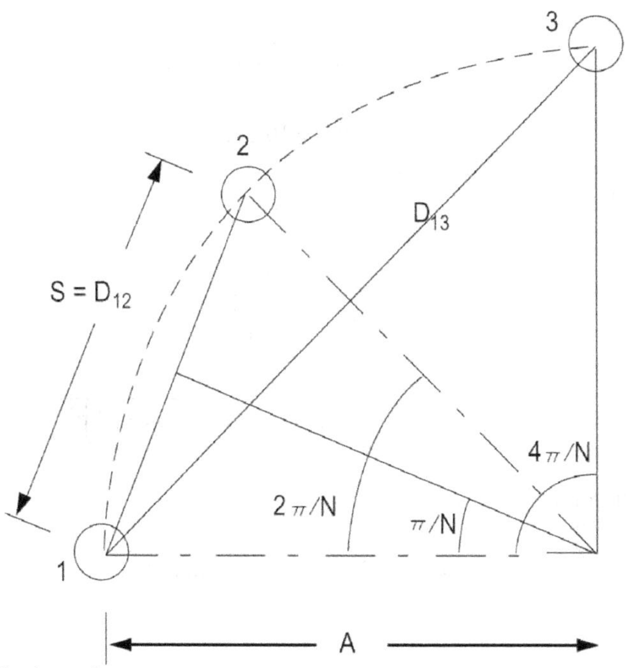


**Figure 5.4** Conductor Geometry.

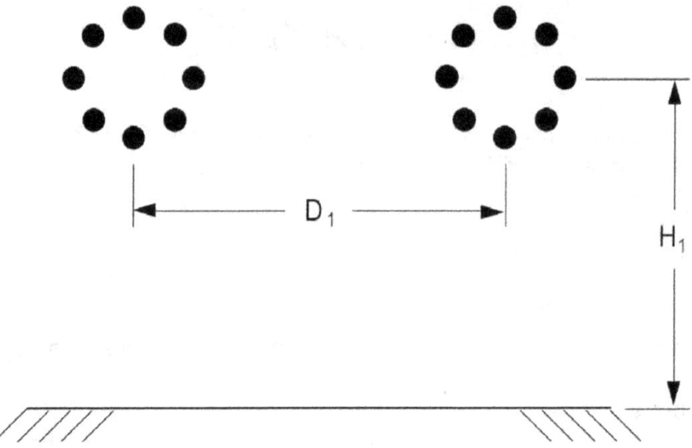


**Figure 5.5** 1000-kV Single-Phase Bundle-Conductor Line.

Therefore,

$$A = 0.6533 \text{ m}$$

Assume that the following practical approximation holds:

$$\text{GMD} = D_1 = 18 \text{ m}$$

The subconductor's geometric mean radius is

$$r_1' = 0.7788\left(\frac{5}{2}\times 10^{-2}\right)$$
$$= 1.947\times 10^{-2} \text{ m}$$

Thus we have

$$L = (2\times 10^{-7})ln\left\{\frac{\text{GMD}}{\left[Nr_1'(A)^{N-1}\right]^{1/N}}\right\}$$
$$= (2\times 10^{-7})ln\left\{\frac{18}{\left[(8)(1.947\times 10^{-2})(0.6533)^7\right]^{1/8}}\right\}$$

The result of the above calculation is

$$L = 6.99\times 10^{-7} \text{ henries/meter}$$

The following MATLAB™ script implements Example 5.2 based on Eqs. (5.21) to (5.24)

```
% Example 5-2
%
N=8
S=0.5
d=0.05
r=d/2
r_prime=0.7788*r
GMD=18
A=(S/2)/sin(pi/N)
GMR=(N * r_prime *(A)^(N-1))^(1/N)
L=2*1e-7*log(GMD/GMR)
```

The answers obtained from MATLAB™ are as follows:

```
EDU»
N = 8
S = 0.5000
d = 0.0500
r = 0.0250
r_prime = 0.0195
GMD = 18
A = 0.6533
GMR = 0.5461
L = 6.9907e-007
```

## Inductance of a Balanced Three-Phase Single-Circuit Line

We consider a three-phase line whose phase conductors have the general arrangement shown in Figure 5.6. We use the voltage drop per unit length concept. This is a consequence of Faraday's law. In engineering practice we have a preference for this method. In our three-phase system, we can write

$$V_1 = j\omega(L_{11}I_1 + L_{12}I_2 + L_{13}I_3)$$
$$V_2 = j\omega(L_{12}I_1 + L_{22}I_2 + L_{23}I_3)$$
$$V_3 = j\omega(L_{13}I_1 + L_{23}I_2 + L_{33}I_3)$$

Here we generalize the expressions of Eqs. (5.8) and (5.10) to give

$$L_{ii} = (2\times10^{-7})ln\left(\frac{1}{r_i'}\right) \tag{5.25}$$

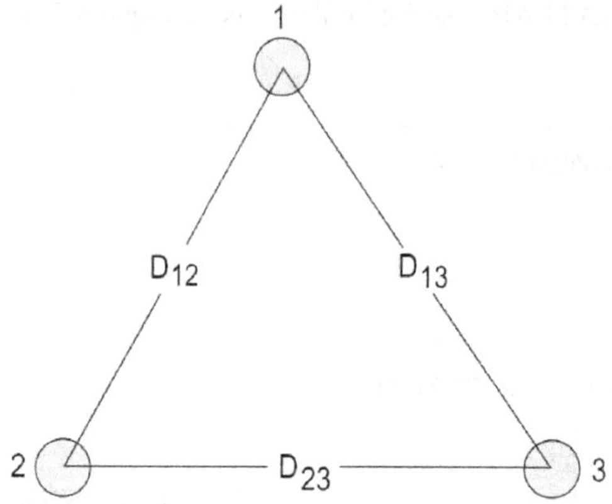


**Figure 5.6** A Balanced Three-Phase Line.

$$L_{kj} = (2\times10^{-7}) ln\left(\frac{1}{D_{kj}}\right) \tag{5.26}$$

We now substitute for the inductances in the voltage drops equations and use the condition of balanced operation to eliminate one current from each equation. The result is

$$\begin{aligned} V_1' &= I_1 ln\left(\frac{D_{13}}{r_1'}\right) + I_2 ln\left(\frac{D_{13}}{D_{12}}\right) \\ V_2' &= I_1 ln\left(\frac{D_{23}}{D_{12}}\right) + I_2 ln\left(\frac{D_{23}}{r_2'}\right) \\ V_3' &= I_2 ln\left(\frac{D_{13}}{D_{23}}\right) + I_3 ln\left(\frac{D_{13}}{r_3'}\right) \end{aligned} \tag{5.27}$$

Here,

$$V_i' = \frac{V_i}{j\omega(2\times10^{-7})}$$

We note that for this general case, the voltage drop in phase one, for example, depends on the current in phase two in addition to its dependence on $I_1$. Thus the voltage drops will not be part of a balanced three phase system. This situation is undesirable.

Consider the case of equilaterally spaced conductors generally referred to as the *delta* configuration; that is

$$D_{12} = D_{13} = D_{23} = D$$
$$r_1' = r_2' = r_3' = r'$$

The voltage drops will thus be given by

$$V_1' = I_1 ln\left(\frac{D}{r'}\right)$$
$$V_2' = I_2 ln\left(\frac{D}{r'}\right)$$
$$V_3' = I_3 ln\left(\frac{D}{r'}\right) \quad \textbf{(5.28)}$$

And in this case the voltage drops will form a balanced three-phase system.

Consider the so often called H-type configuration. The conductors are placed in one horizontal plane as shown in Figure 5.7. The distances between conductors are thus

$$D_{12} = D_{23} = D$$
$$D_{13} = 2D$$

and the voltage drops are given by

$$V_1' = I_1 ln\left(\frac{2D}{r'}\right) + I_2 ln2$$
$$V_2' = I_2 ln\left(\frac{D}{r'}\right)$$
$$V_3' = I_2 ln2 + I_3 ln\left(\frac{2D}{r'}\right) \quad \textbf{(5.29)}$$

We note that only conductor number two has a voltage drop proportional to its current.

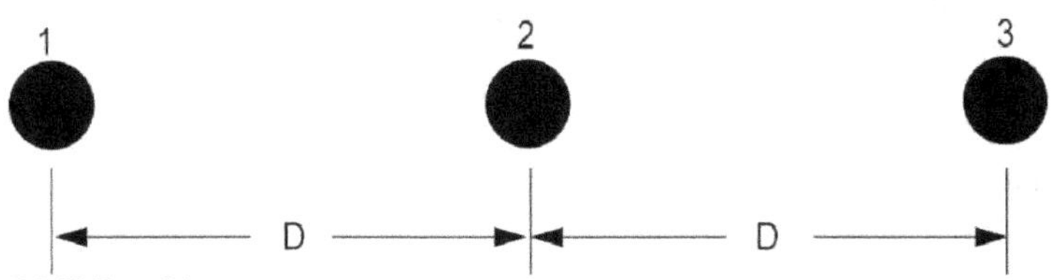


**Figure** 5.7 H-Type Line.

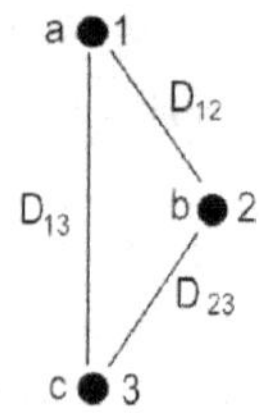


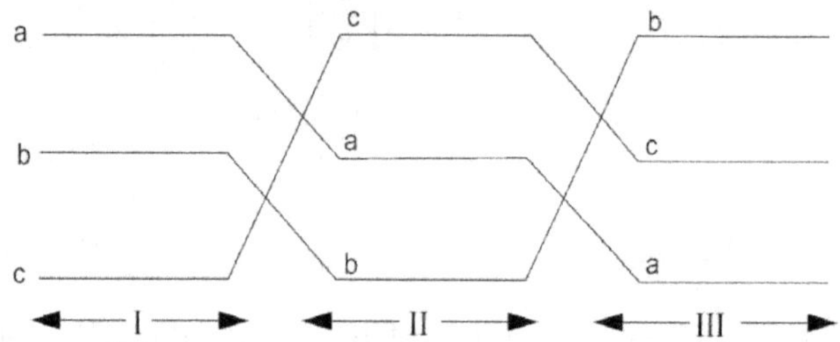


**Figure 5.8** Transposed Line.

## Transposition of Line Conductors

The equilateral triangular spacing configuration is not the only configuration commonly used in practice. Thus the need exists for equalizing the mutual inductances. One means for doing this is to construct transpositions or rotations of overhead line wires. A transposition is a physical rotation of the conductors, arranged so that each conductor is moved to occupy the next physical position in a regular sequence such as *a-b-c*, *b-c-a*, *c-a-b*, etc. Such a transposition arrangement is shown in Figure 5.8. If a section of line is divided into three segments of equal length separated by rotations, we say that the line is "completely transposed."

Consider a completely transposed three-phase line. We can demonstrate that by completely transposing a line, the mutual inductance terms disappear, and the voltage drops are proportional to the current in each phase.

Define the geometric mean distance GMD as

$$\text{GMD} = (D_{12}D_{13}D_{23})^{1/3} \tag{5.30}$$

and the geometric mean radius GMR as

$$\text{GMR} = r' \tag{5.31}$$

we attain

$$L = (2\times10^{-7})ln\left(\frac{\text{GMD}}{\text{GMR}}\right)\text{henries/meter} \tag{5.32}$$

**Example 5.3**

Calculate the inductance per phase of the three-phase solid conductor line shown in Figure 5.9. Assume that the conductor diameter is 5 cm and the phase separation $D_1$ is 8 m. Assume that the line is transposed.

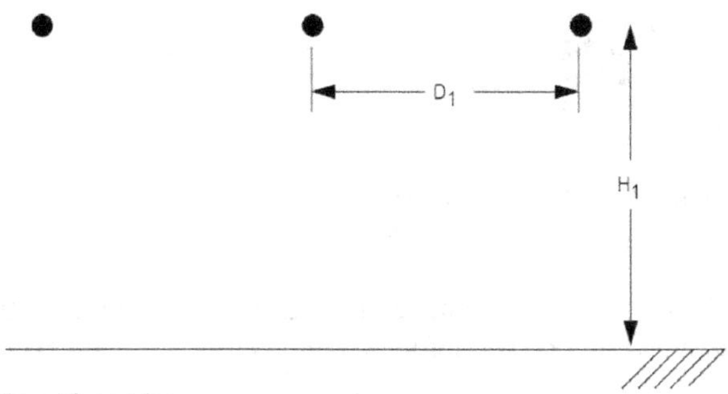


**Figure 5.9** A Three-Phase Line.

*Solution*

The geometric mean distance is given by

$$\begin{aligned}\text{GMD} &= [D_1 D_1 (2D_1)]^{1/3}\\ &= 1.2599 D_1\\ &= 10.08\text{ m}\end{aligned}$$

The geometric mean radius is

$$\begin{aligned}r' &= (e^{-1/4})\frac{5\times10^{-2}}{2}\\ &= 0.0195\text{ m}\end{aligned}$$

Therefore,

$$\begin{aligned}L &= (2\times10^{-7})ln\left(\frac{10.08}{0.0195}\right)\\ &= 1.25\times10^{-6}\text{ henries/meter}\end{aligned}$$

**Inductance of Multiconductor Three-Phase Systems**

Consider a single-circuit, three-phase system with multiconductor-configured phase conductors as shown in Figure 5.10. Assume equal current

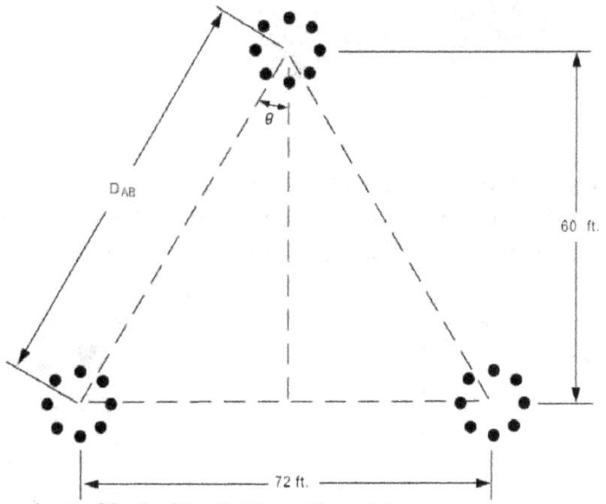


**Figure 5.10** Multiconductor Single-Circuit Three-Phase Line.

distribution in the phase subconductors and complete transposition. We can show that the phase inductance for the system is the following expression:

$$L = (2\times10^{-7})ln\left(\frac{\text{GMD}}{\text{GMR}}\right) \tag{5.33}$$

In this case, the geometric mean distance is given by

$$\text{GMD} = (D_{AB}D_{BC}D_{CA})^{1/3} \tag{5.34}$$

where $D_{AB}$, $D_{BC}$, and $D_{CA}$ are the distances between phase centers. The geometric mean radius (GMR) is obtained using the same expression as that for the single-phase system. Thus,

$$\text{GMR} = \left[\prod_{i=1}^{N}(D_{si})\right]^{1/N} \tag{5.35}$$

For the case of symmetrical bundle conductors, we have

$$\text{GMR} = \left[Nr'(A)^{N-1}\right]^{1/N} \tag{5.36}$$

The inductive reactance per mile per phase $X_L$ in the case of a three-phase, bundle-conductor line can be obtained using

$$X_L = X_a + X_d \tag{5.37}$$

where as before for 60 Hz operation,

$$X_a = 0.2794 \log \frac{1}{\text{GMR}} \tag{5.38}$$

$$X_d = 0.2794 \log \text{GMD} \tag{5.39}$$

The GMD and GMR are defined by Eqs. (5.34) and (5.36).

**Example 5.4**

Consider a three-phase line with an eight subconductor-bundle delta arrangement with a 42 in. diameter. The subconductors are ACSR 84/19 (Chukar) with $r' = 0.0534$ ft. The horizontal phase separation is 75 ft, and the vertical separation is 60 ft. Calculate the inductive reactance of the line in ohms per mile per phase.

*Solution*

From the geometry of the phase arrangements, we have

$$\tan\theta = \frac{36}{60}$$

$$\theta = 30.96^\circ$$

$$D_{AB} = \frac{60}{\cos 30.96^\circ}$$

$$= 69.97 \text{ ft}$$

Thus,

$$\text{GMD} = [(69.97)(69.97)(75)]^{1/3} = 71.577 \text{ ft}$$

For the subconductors we have $r' = 0.0534$ ft. The bundle particulars are $N = 8$ and $A = (42/2)$ in. Therefore,

$$\text{GMR} = \left[8(0.0534)\left(\frac{21}{12}\right)^7\right]^{1/8}$$

$$= 1.4672 \text{ ft}$$

Thus,

$$X_a = 0.2794 \log \frac{1}{1.4672}$$

$$= -0.0465$$

$$X_d = 0.2794 \log 71.577$$

$$= 0.518$$

As a result,

$$X_L = X_a + X_d = 0.4715 \text{ ohms per mile}$$

## Inductance of Three-Phase, Double-Circuit Lines

A three-phase, double-circuit line is essentially two three-phase circuits connected in parallel. Normal practice calls for identical construction for the two circuits. If the two circuits are widely separated, then we can obtain the line reactance as simply half that of one single-circuit line. For the situation where the two circuits are on the same tower, the above approach may not produce results of sufficient accuracy. The error introduced is mainly due to neglecting the effect of mutual inductance between the two circuits. Here we give a simple but more accurate expression for calculating the reactance of double-circuit lines.

We consider a three-phase, double-circuit line with full line transposition such that in segment I, the relative phase positions are as shown in Figure 5.11.

The inductance per phase per unit length is given by

$$L = \left(2\times10^{-7}\right) ln\left(\frac{\text{GMD}}{\text{GMR}}\right) \tag{5.40}$$

where the double-circuit geometric mean distance is given by

$$\text{GMD} = \left(D_{AB_{eq}} D_{BC_{eq}} D_{AC_{eq}}\right)^{1/3} \tag{5.41}$$

with mean distances defined by

$$\begin{aligned} D_{AB_{eq}} &= \left(D_{12} D_{1'2'} D_{12'} D_{1'2}\right)^{1/4} \\ D_{BC_{eq}} &= \left(D_{23} D_{2'3'} D_{23'} D_{2'3}\right)^{1/4} \\ D_{AC_{eq}} &= \left(D_{13} D_{1'3'} D_{13'} D_{1'3}\right)^{1/4} \end{aligned} \tag{5.42}$$

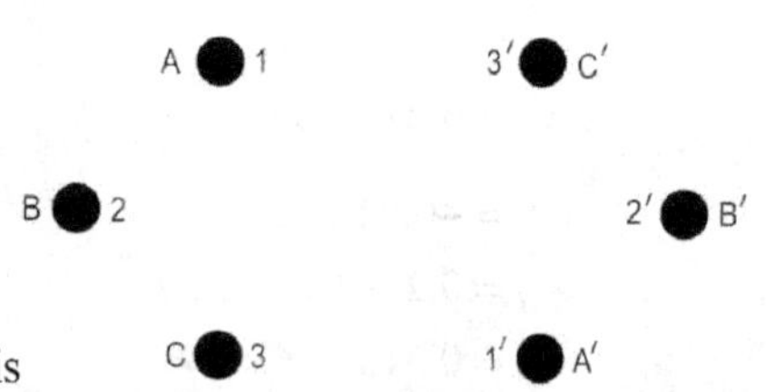


is

**Figure 5.11** Double-Circuit Conductors' Relative Positions in Segment *I* of Transposition.

where subscript eq. refers to equivalent spacing. The GMR

$$\mathrm{GMR} = [(\mathrm{GMR}_A)(\mathrm{GMR}_B)(\mathrm{GMR}_C)]^{1/3} \tag{5.43}$$

with phase GMR's defined by

$$\begin{aligned} \mathrm{GMR}_A &= [r'(D_{11'})]^{1/2} \\ \mathrm{GMR}_B &= [r'(D_{22'})]^{1/2} \\ \mathrm{GMR}_C &= [r'(D_{33'})]^{1/2} \end{aligned} \tag{5.44}$$

We see from the above result that the same methodology adopted for the single-circuit case can be utilized for the double-circuit case.

**Example 5.5**

Calculate the inductance per phase for the three-phase, double-circuit line whose phase conductors have a GMR of 0.06 ft, with the horizontal conductor configuration as shown in Figure 5.12.

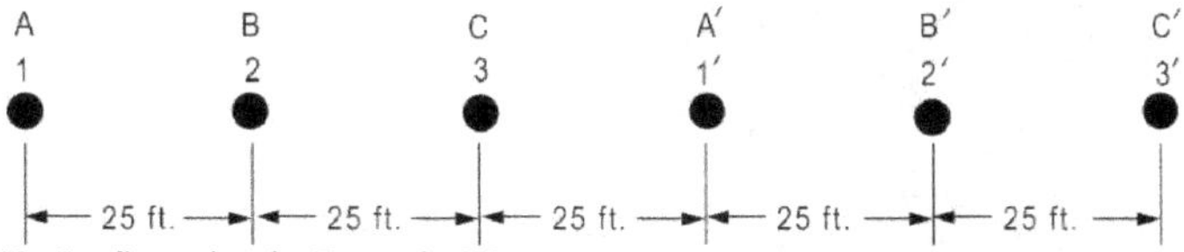


**Figure 5.12** Configuration for Example 5.5.

*Solution*

We use Eq. (5.42):

$$\begin{aligned} D_{AB_{\mathrm{eq}}} &= [(25)(25)(50)(100)]^{1/4} \\ &= 42.04\ \mathrm{ft} \\ D_{BC_{\mathrm{eq}}} &= [(25)(25)(50)(100)]^{1/4} \\ &= 42.04\ \mathrm{ft} \\ D_{AC_{\mathrm{eq}}} &= [(50)(50)(125)(25)]^{1/4} \\ &= 52.87\ \mathrm{ft} \end{aligned}$$

As a result,

$$\begin{aligned} \mathrm{GMD} &= [(42.04)(42.04)(52.87)]^{1/3} \\ &= 45.381\ \mathrm{ft} \end{aligned}$$

The equivalent GMR is obtained using Eq. (5.44) as

$$r_{eq} = \left[(0.06)^3(75)^3\right]^{1/6}$$
$$= 2.121\,\text{ft}$$

As a result,

$$L = (2\times 10^{-7})ln\left(\frac{45.381}{2.121}\right)$$
$$= 0.6126\times 10^{6} \text{ henries/meter}$$

The following MATLAB™ script implements Example 5.5 based on Eqs. (5.40) to (5.44)

```
% Example 5-5
%
r_prime=0.06;
D_AAprime=75;
D_BBprime=75;
D_CCprime=75;
D_AB=25;
D_BC=D_AB;
D_CAprime=D_AB;
D_AprimeBprime=D_AB;
D_BprimeCprime=D_AB;
D_BCprime=D_BC+D_CAprime+D_AprimeBprime
+D_BprimeCprime;
D_CBprime=D_CAprime+D_AprimeBprime;
D_ABprime=D_AB+D_BC+D_CAprime+D_AprimeB
prime;
D_BAprime=D_BC+D_CAprime;
D_CA=D_AB+D_BC;
D_CprimeAprime=D_AprimeBprime+D_BprimeC
prime;
D_ACprime=D_ABprime+D_BprimeCprime;
D_ABeq=(D_BC*D_BCprime*D_BprimeCprime*D
_CBprime)^(1/4)
D_BCeq=(D_AprimeBprime*D_ABprime*D_AB*D
_BAprime)^(1/4)
D_ACeq=(D_CA*D_CprimeAprime*D_CAprime*D
_ACprime)^(1/4)
GMD=(D_ABeq*D_BCeq*D_ACeq)^(1/3)
% The equivalent GMR
r_eq=(r_prime^3*D_AAprime^3)^(1/6)
L=(2*10^-7)*log(GMD/r_eq)
```

The results of running the script are shown below:

```
EDU»
D_ABeq =      42.0448
D_BCeq =      42.0448
D_ACeq =    52.8686
GMD =    45.3810
r_eq =       2.1213
L =   6.1261e-007
```

## 5.4 LINE CAPACITANCE

The previous sections treated two line parameters that constitute the series impedance of the transmission line. The line inductance normally dominates the series resistance and determines the power transmission capacity of the line. There are two other line-parameters whose effects can be appreciable for high transmission voltages and line length. The line's shunt admittance consists of the conductance ($g$) and the capacitive susceptance ($b$). The conductance of a line is usually not a major factor since it is dominated by the capacitive susceptance $b = \omega C$. The line capacitance is a leakage (or charging) path for the ac line currents.

The capacitance of a transmission line is the result of the potential differences between the conductors themselves as well as potential differences between the conductors and ground. Charges on conductors arise, and the capacitance is the charge per unit potential difference. Because we are dealing with alternating voltages, we would expect that the charges on the conductors are also alternating (i.e., time varying). The time variation of the charges results in what is called line-charging currents. In this section we treat line capacitance for a number of conductor configurations.

### Capacitance of Single-Phase Line

Consider a single-phase, two-wire line of infinite length with conductor radii of $r_1$ and $r_2$ and separation $D$ as shown in Figure 5.13. The potential at an arbitrary point $P$ at distances $r_a$ and $r_b$ from $A$ and $B$, respectively, is given by

$$V_p = \frac{q}{2\pi\varepsilon_0} ln\left(\frac{r_b}{r_a}\right) \tag{5.45}$$

where $q$ is the charge density in coulombs per unit length.

The potential $V_A$ on the conductor $A$ of radius $r_1$ is therefore obtained by setting $r_a = r_1$ and $r_b = D$ to yield

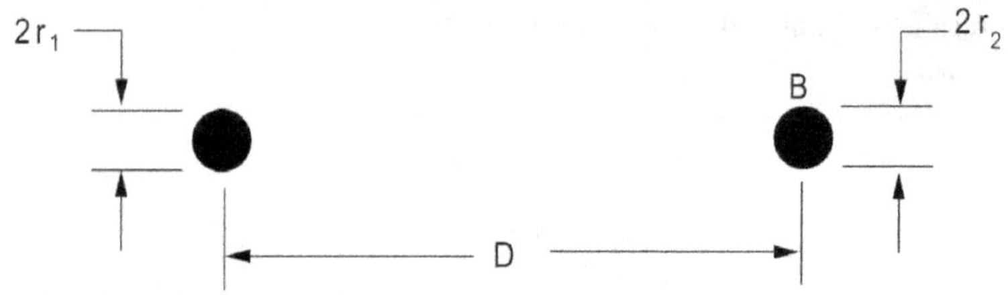


**Figure 5.13** Single-Phase, Two-Wire Line.

$$V_A = \frac{q}{2\pi\varepsilon_0} ln\left(\frac{D}{r_1}\right) \tag{5.46}$$

Likewise for conductor $B$ of radius $r_2$, we have

$$V_B = \frac{q}{2\pi\varepsilon_0} ln\left(\frac{r_2}{D}\right) \tag{5.47}$$

The potential difference between the two conductors is therefore

$$V_{AB} = V_A - V_B = \frac{q}{\pi\varepsilon_0} ln\left(\frac{D}{\sqrt{r_1 r_2}}\right) \tag{5.48}$$

The capacitance between the two conductors is defined as the charge on one conductor per unit of potential difference between the two conductors. As a result,

$$C_{AB} = \frac{q}{V_{AB}} = \frac{\pi\varepsilon_0}{ln\left(\frac{D}{\sqrt{r_1 r_2}}\right)} \text{ farads per meter} \tag{5.49}$$

If $r_1 = r_2 = r$, we have

$$C_{AB} = \frac{\pi\varepsilon_0}{ln\left(\frac{D}{r}\right)} \tag{5.50}$$

Converting to microfarads ($\mu$F) per mile and changing the base of the logarithmic term, we have

$$C_{AB} = \frac{0.0388}{2\log\left(\frac{D}{r}\right)} \ \mu\text{F per mile} \tag{5.51}$$

Equation (5.51) gives the line-to-line capacitance between the conductors. The capacitance to neutral for conductor $A$ is defined as

$$C_{AN} = \frac{q}{V_A} = \frac{2\pi\varepsilon_0}{ln\left(\frac{D}{r_1}\right)} \tag{5.52}$$

Likewise, observing that the charge on conductor $B$ is $-q$, we have

$$C_{BN} = \frac{-q}{V_B} = \frac{2\pi\varepsilon_0}{ln\left(\frac{D}{r_2}\right)} \tag{5.53}$$

For $r_1 = r_2$, we have

$$C_{AN} = C_{BN} = \frac{2\pi\varepsilon_0}{ln\left(\frac{D}{r}\right)} \tag{5.54}$$

Observe that

$$C_{AN} = C_{BN} = 2C_{AB} \tag{5.55}$$

Converting to $\mu$F per mile, we obtain

$$C_{AN} = \frac{0.0388}{\log\frac{D}{r}} \ \mu\text{F per mile to neutral} \tag{5.56}$$

The capacitive reactance $X_c$ is given by

$$X_c = \frac{1}{2\pi fC} = k'\log\frac{D}{r} \ \text{ohms}\cdot\text{mile to neutral} \tag{5.57}$$

where the factor $k'$ is given by

$$k' = \frac{4.1\times10^6}{f} \tag{5.58}$$

Expanding the logarithm, we have

$$X_c = k' \log D + k' \log \frac{1}{r} \tag{5.59}$$

The first term is called $X_{d'}$, the capacitive reactance spacing factor, and the second is called $X_{a'}$, the capacitive reactance at 1-ft spacing.

$$X_{d'} = k' \log D \tag{5.60}$$

$$X_{a'} = k' \log \frac{1}{r} \tag{5.61}$$

$$X_c = X_{d'} + X_{a'} \tag{5.62}$$

The last relationships are very similar in form to those given for the inductance case. One difference that should be noted is that the conductor radius for the capacitance formula is the actual outside radius of the conductor and not the modified value $r'$.

**Example 5.6**

Find the capacitive reactance in ohms · mile per phase for a single-phase line with phase separation of 25 ft and conductor radius of 0.08 ft for 60-Hz operation.

*Solution*

Note that this is the same line of Example 5.1. We have for $f = 60$ Hz:

$$k' = \frac{4.1\times10^6}{f} = 0.06833\times10^6$$

We calculate

$$\begin{aligned} X_{d'} &= k' \log 25 \\ &= 95.52\times10^3 \\ X_{a'} &= k' \log \frac{1}{0.08} \\ &= 74.95\times10^3 \end{aligned}$$

As a result,

$$X_c = X_{d'} + X_{a'}$$
$$X_c = 170.47\times10^3 \text{ ohms} \cdot \text{mile to neutral}$$

The following MATLAB™ script implements Example 5.6 based on equations (5.60) to (5.62)

```
% Example 5-6
% Data
f=60; % Hz
D=25; % phase separation (ft)
r=0.08;% conductor radius (ft)
% To calculate the capacitive reactance
% in ohms.mile per phase
kp=4.1*10^6/f;
Xdp=kp*log10(D)
Xap=kp*log10(1/r)
Xc=Xdp+Xap
```

The results of running the script are as shown below:

```
EDU»
Xdp =    9.5526e+004
Xap =    7.4956e+004
Xc =   1.7048e+005
```

### Including the Effect of Earth

The effect of ground needs to be considered if the conductors are not high enough above ground. This can be done using the theory of image charges. These are imaginary charges of the same magnitude as the physical charges but of opposite sign and are situated below the ground at a distance equal to that between the physical charge and ground. The potential at ground due to the charge and its image is zero, which is consistent with the usual assumption that ground is a plane of zero potential.

### General Multiconductor Configurations

Considering a system of $n$ parallel and very long conductors with charges $q_1, q_2, \ldots, q_n$, respectively, we can state that the potential at point $P$ having distances $r_1, r_2, \ldots, r_n$ to the conductor as shown in Figure 5.14 is given by

$$V_P = \frac{q_1}{2\pi\varepsilon_0} ln\left(\frac{1}{r_1}\right) + \frac{q_2}{2\pi\varepsilon_0} ln\left(\frac{1}{r_2}\right) + \cdots + \frac{q_n}{2\pi\varepsilon_0} ln\left(\frac{1}{r_n}\right) \tag{5.63}$$

This is a simple extension of the two-conductor case.

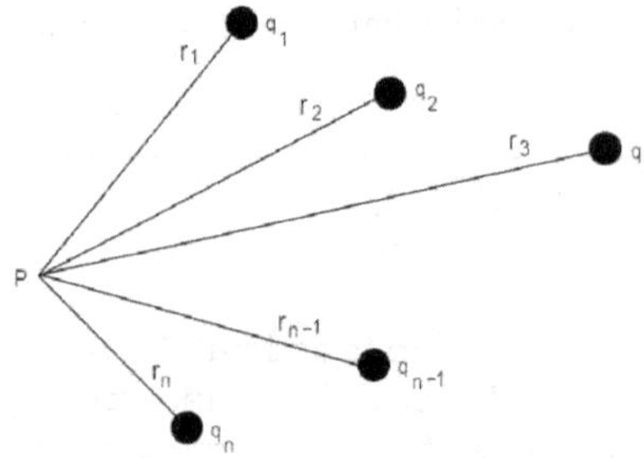


**Figure 5.14** A Multiconductor Configuration.

If we consider the same $n$ parallel long conductors and wish to account for the presence of ground, we use the theory of images. As a result, we will have $n$ images charges $-q_1, -q_2, \ldots, -q_n$ situated below the ground at distance $r_1^-, r_2^-, \ldots, r_n^-$ from $P$. This is shown in Figure 5.15. The potential at $P$ is therefore

$$V_P = \frac{q_1}{2\pi\varepsilon_0} ln\left(\frac{r_1^-}{r_1}\right) + \frac{q_2}{2\pi\varepsilon_0} ln\left(\frac{r_2^-}{r_2}\right) + \cdots + \frac{q_n}{2\pi\varepsilon_0} ln\left(\frac{r_n^-}{r_1}\right) \quad \textbf{(5.64)}$$

The use of this relationship in finding the capacitance for many systems will be treated next.

## Capacitance of a Single-Phase Line Considering the Effect of Ground

Consider a single-phase line with conductors $A$ and $B$ as before. To account for ground effects, we introduce the image conductors $A'$ and $B'$. The situation is shown in Figure 5.16.

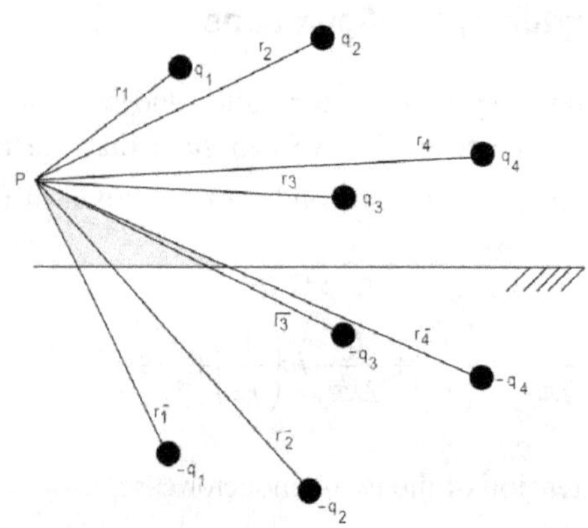


**Figure 5.15** A Multiconductor Configuration Accounting for Ground Effect.

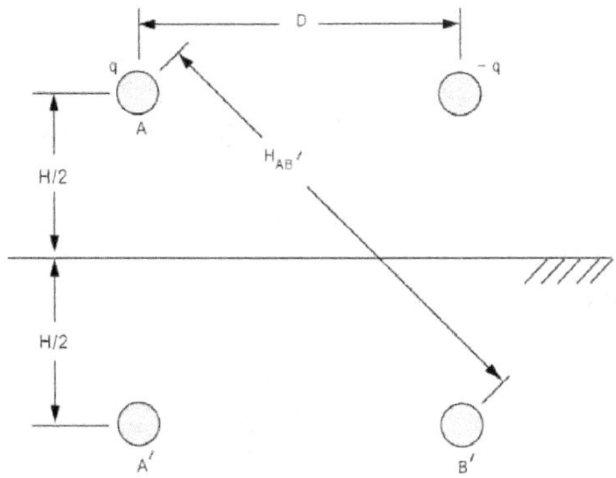


**Figure 5.16** Single-Phase Line and Its Image.

The voltage of phase $A$ is given according to Eq. (5.64) by

$$V_A = \frac{q}{2\pi\varepsilon_0} ln\left(\frac{H}{r} \cdot \frac{D}{H_{AB'}}\right) \tag{5.65}$$

The voltage of phase $B$ is

$$V_B = \frac{q}{2\pi\varepsilon_0} ln\left(\frac{H_{AB'}}{D} \cdot \frac{r}{H}\right) \tag{5.66}$$

The voltage difference between $A$ and $B$ is thus

$$\begin{aligned} V_{AB} &= V_A - V_B \\ &= \frac{q}{\pi\varepsilon_0} ln\left(\frac{H}{r} \cdot \frac{D}{H_{AB'}}\right) \end{aligned} \tag{5.67}$$

The capacitance between the two conductors is thus

$$C_{AB} = \frac{\pi\varepsilon_0}{ln\left(\frac{D}{r} \cdot \frac{H}{H_{AB'}}\right)} \tag{5.68}$$

The capacitance to neutral is obtained using

$$C_{AN} = \frac{q}{V_A}$$

$$= \frac{2\pi\varepsilon_0}{ln\left(\frac{D}{r}\cdot\frac{H}{H_{AB'}}\right)} \text{ farads per meter} \quad \textbf{(5.69)}$$

Observe that again

$$C_{AB} = \frac{C_{AN}}{2}$$

Let us examine the effect of including ground on the capacitance for a single-phase line in the following example.

### Example 5.7

Find the capacitance to neutral for a single-phase line with phase separation of 20 ft and conductor radius of 0.075 ft. Assume that the height of the conductor above ground is 80 ft.

*Solution*

We have

$$D = 20 \text{ ft}$$
$$r = 0.075 \text{ ft}$$
$$H = 160 \text{ ft}$$

As a result,

$$H_{AB'} = \sqrt{(160)^2 + (20)^2} = 161.2452 \text{ ft}$$

Therefore we have

$$C_{AN_1} = \frac{2\pi\varepsilon_0}{ln\left(\frac{20}{0.075}\cdot\frac{160}{161.2452}\right)}$$

$$= \frac{2\pi\varepsilon_0}{5.578} \text{ farads per meter}$$

If we neglect earth effect, we have

$$C_{AN_2} = \frac{2\pi\varepsilon_0}{ln\left(\frac{20}{0.075}\right)}$$

$$= \frac{2\pi\varepsilon_0}{5.586} \text{ farads per meter}$$

The relative error involved if we neglect earth effect is:

$$\frac{C_{AN_1} - C_{AN_2}}{C_{AN_1}} = 0.0014$$

which is clearly less than 1 %.

## Capacitance of a Single-Circuit, Three-Phase Line

We consider the case of a three-phase line with conductors not equilaterally spaced. We assume that the line is transposed and as a result can assume that the capacitance to neutral in each phase is equal to the average value. This approach provides us with results of sufficient accuracy for our purposes. This configuration is shown in Figure 5.17.

We use the three-phase balanced condition that the sinusoidally varying charges satisfy

$$q_a + q_b + q_c = 0$$

The average potential on phase $A$ is given by

$$V_A = \frac{q_a}{2\pi\varepsilon_0} ln\left[\frac{(D_{12}D_{23}D_{13})^{1/3}}{r}\right] \tag{5.70}$$

The capacitance to neutral is therefore given by

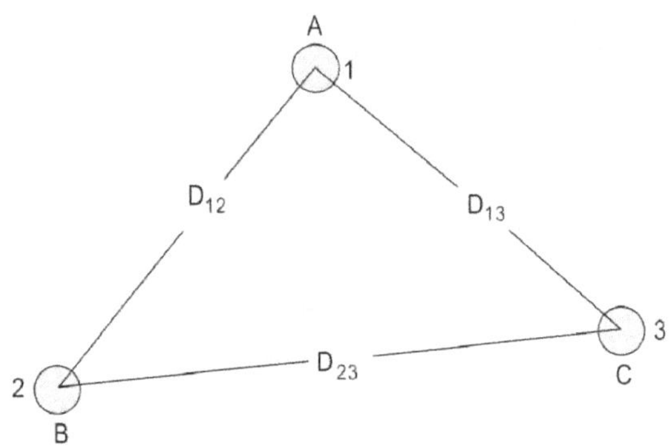


**Figure 5.17** Three-Phase Line with General Spacing.

$$C_{AN} = \frac{q_a}{V_A} = \frac{2\pi\varepsilon_0}{ln\left(\frac{D_{eq}}{r}\right)} \tag{5.71}$$

where

$$D_{eq} = \sqrt[3]{D_{12}D_{23}D_{13}} \tag{5.72}$$

Observe that $D_{eq}$ is the same as the geometric mean distance obtained in the case of inductance. Moreover, we have the same expression for the capacitance as that for a single-phase line. Thus,

$$C_{AN} = \frac{2\pi\varepsilon_0}{ln\left(\frac{\text{GMD}}{r}\right)} \text{ farad per meter} \tag{5.73}$$

If we account for ground effect (the influence of earth), we come up with a slightly modified expression for the capacitance. Consider the same three-phase line with the attendant image line shown in Figure 5.18. The line is assumed to be transposed. As a result, the average phase $A$ voltage will be given by

$$V_A = \frac{q_a}{3(2\pi\varepsilon_0)} ln\left[\frac{(D_{12}D_{23}D_{13})(H_1H_2H_3)}{r^3(H_{12}H_{13}H_{23})}\right] \tag{5.74}$$

From the above,

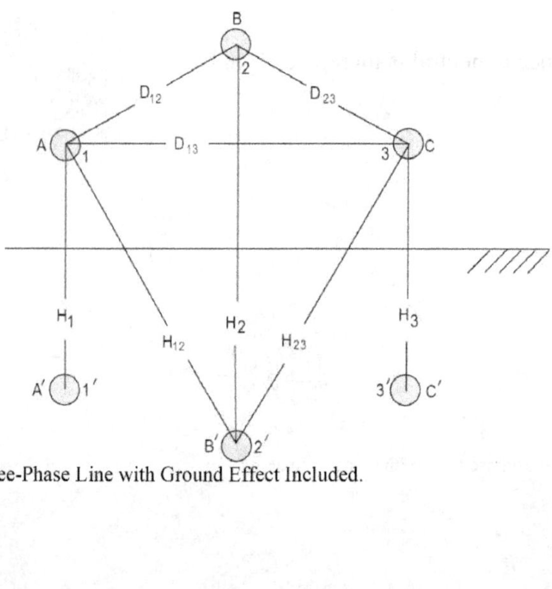


**Figure 5.18** Three-Phase Line with Ground Effect Included.

$$C_{AN} = \frac{2\pi\varepsilon_0}{ln\left[\frac{D_{eq}}{r}\left(\frac{H_1H_2H_3}{H_{12}H_{13}H_{23}}\right)^{1/3}\right]}$$

or

$$C_{AN} = \frac{2\pi\varepsilon_0}{ln\frac{D_{eq}}{r} + ln\left(\frac{H_1H_2H_3}{H_{12}H_{13}H_{23}}\right)^{1/3}} \tag{5.75}$$

We define the mean distances

$$H_s = (H_1H_2H_3)^{1/3} \tag{5.76}$$

$$H_m = (H_{12}H_{23}H_{13})^{1/3} \tag{5.77}$$

Then the capacitance expression reduces to

$$C_{AN} = \frac{2\pi\varepsilon_0}{ln\left(\frac{D_{eq}}{r}\right) - ln\left(\frac{H_m}{H_s}\right)} \tag{5.78}$$

We can thus conclude that including the effect of ground will give a higher value for the capacitance than that obtained by neglecting the ground effect.

### Example 5.8

Find the capacitance to neutral for the single-circuit, three-phase, 345-kV line with conductors having an outside diameter of 1.063 in. with phase configuration as shown in Figure 5.19. Repeat including the effect of earth, assuming the height of the conductors is 50 ft.

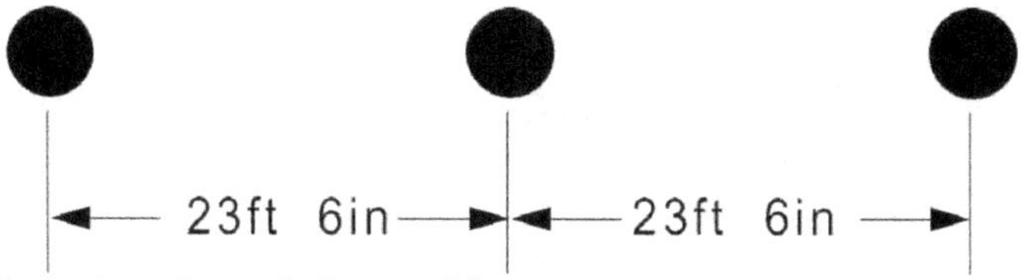


**Figure 5.19** Conductor Layout for Example 5.8.

*Solution*

$$\begin{aligned} \text{GMD} &= [(23.5)(23.5)(47)]^{1/3} \\ &= 29.61\ \text{ft} \end{aligned}$$

$$r = \frac{1.063}{(2)(12)} = 0.0443\ \text{ft}$$

$$\begin{aligned} C_{AN} &= \frac{2\pi\varepsilon_0}{ln\left(\dfrac{\text{GMD}}{r}\right)} \\ &= 8.5404\times10^{-12}\ \text{farads per meter} \end{aligned}$$

$$H_1 = H_2 = H_3 = 2\times50 = 100\ \text{ft}$$

$$H_{12} = H_{23} = \sqrt{(23.5)^2 + (100)^2} = 102.72$$

$$H_{13} = \sqrt{(47)^2 + (100)^2} = 110.49$$

$$\begin{aligned} ln\left(\frac{H_s}{H_m}\right) &= ln\left[\frac{(100)(100)(100)}{(102.72)(102.72)(110.49)}\right]^{1/3} \\ &= -0.0512 \end{aligned}$$

Thus,

$$\begin{aligned} C_{AN} &= \frac{1}{(18\times10^9)(6.505 - 0.0512)} \\ &= 8.6082\times10^{-12}\ \text{farads per meter} \end{aligned}$$

The following MATLAB™ script implements Example 5.8.

```
% Example 5-8
% data
D12=23.5; % ft
D23=23.5; % ft
D13=47; % ft
r=0.0443; % ft
eo=(1/(36*pi))*10^-9;
% To find the capacitance to neutral in
farads/m
GMD=(D12*D23*D13)^(1/3)
CAN=(2*pi*eo)/(log(GMD/r))
% To calculate the capacitance to
neutral,
% including the effect of earth
H1=2*50; % ft
H2=H1;
```

MATLAB™ con't.

```
H3=H1;
H12=(D12^2+H1^2)^.5;
H23=H12;
H13=(D13^2+H3^2)^.5;
Hs=(H1*H2*H3)^(1/3);
Hm=(H12*H23*H13)^(1/3);
CAN=(2*pi*eo)/(log(GMD/r)-log(Hm/Hs))
```

The results of running the script are as shown below:

```
EDU»
GMD =     29.6081

CAN =    8.5407e-012

CAN =    8.6084e-012
```

## Capacitance of Double-Circuit Lines

The calculation of capacitance of a double-circuit line can be quite involved if rigorous analysis is followed. In practice, however, sufficient accuracy is obtained if we assume that the charges are uniformly distributed and that the charge $q_a$ is divided equally between the two phase $A$ conductors. We further assume that the line is transposed. As a result, we obtain capacitance formulae similar in nature to those for the single-circuit line.

Consider a double-circuit line with phases, $A$, $B$, $C$, $A'$, $B'$, and $C'$ placed in positions 1, 2, 3, 1′, 2′, and 3′, respectively, in segment $I$ of the transposition cycle. The situation is shown in Figure 5.20.

The average voltage of phase $A$ can be shown to be given by

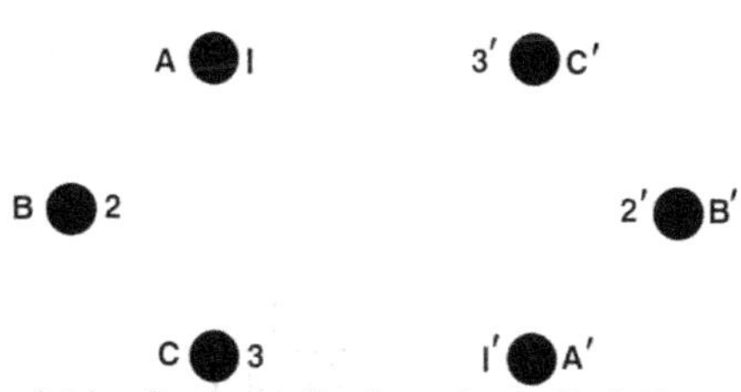


**Figure 5.20** Double-Circuit Line Conductor Configuration in Cycle Segment *I* of Transposition.

$$V_A = \frac{q_a}{12(2\pi\varepsilon_0)} ln\left[\frac{(D_{12}D_{1'2'}D_{12'}D_{1'2})(D_{23}D_{2'3'}D_{2'3}D_{23'})(D_{13}D_{1'3'}D_{1'3}D_{13'})}{(r^6)(D_{11'}^2 D_{22'}^2 D_{33'}^2)}\right] \tag{5.79}$$

As a result,

$$C_{AN} = \frac{2\pi\varepsilon_0}{ln\left(\frac{\text{GMD}}{\text{GMR}}\right)} \tag{5.80}$$

As we have done for the inductance case, we define

$$\text{GMD} = \left(D_{AB_{eq}} D_{BC_{eq}} D_{AC_{eq}}\right)^{1/3} \tag{5.81}$$

$$D_{AB_{eq}} = (D_{12}D_{1'2'}D_{12'}D_{1'2})^{1/4} \tag{5.82}$$

$$D_{BC_{eq}} = (D_{23}D_{2'3'}D_{2'3}D_{23'})^{1/4} \tag{5.83}$$

$$D_{AC_{eq}} = (D_{13}D_{1'3'}D_{13'}D_{1'3})^{1/4} \tag{5.84}$$

The GMR is given by

$$GMR = (r_A r_B r_C)^{1/3} \tag{5.85}$$

with

$$r_A = (rD_{11'})^{1/2} \tag{5.86}$$

$$r_B = (rD_{22'})^{1/2} \tag{5.87}$$

$$r_C = (rD_{33'})^{1/2} \tag{5.88}$$

If we wish to include the effect of the earth in the calculation, a simple extension will do the job.

As a result, we have

$$C_{AN} = \frac{2\pi\varepsilon_0}{ln\left(\frac{\text{GMD}}{\text{GMR}}\right) + \alpha} \tag{5.89}$$

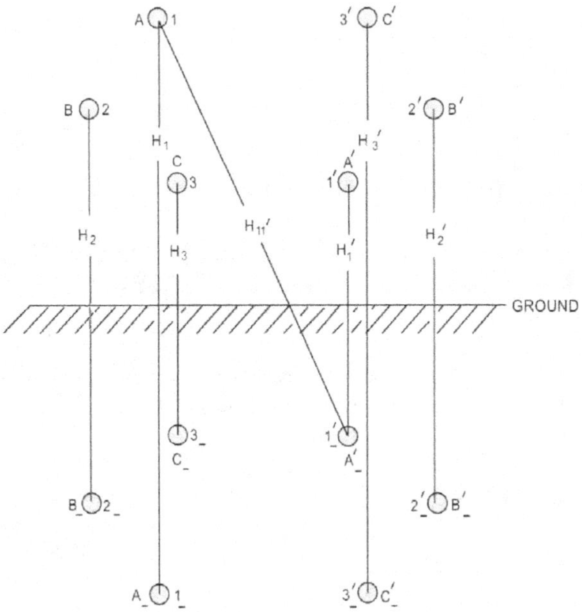


**Figure 5.21** Double-Circuit Line with Ground Effect.

where GMD and GMR are as given by Eqs. (5.81) and (5.85). Also, we defined

$$\alpha = ln\left(\frac{H_s}{H_m}\right) \tag{5.90}$$

$$H_s = \left(H_{s_1} H_{s_2} H_{s_3}\right)^{1/3} \tag{5.91}$$

with

$$H_{s_1} = \left(H_1 H_{1'} H_{11'}^2\right)^{1/4} \tag{5.92}$$

$$H_{s_2} = \left(H_2 H_{2'} H_{22'}^2\right)^{1/4} \tag{5.93}$$

$$H_{s_3} = \left(H_3 H_{3'} H_{33'}^2\right)^{1/4} \tag{5.94}$$

and

$$H_m = \left(H_{m_{12}} H_{m_{32}} H_{m_{23}}\right)^{1/3} \tag{5.95}$$

$$H_{m_{12}} = \left(H_{12} H_{1'2'} H_{12'} H_{1'2}\right)^{1/4} \tag{5.96}$$

$$H_{m_{13}} = \left(H_{13}H_{1'3'}H_{13'}H_{1'3}\right)^{1/4} \tag{5.97}$$

$$H_{m_{23}} = \left(H_{23}H_{2'3'}H_{23'}H_{2'3}\right)^{1/4} \tag{5.98}$$

## Capacitance of Bundle-Conductor Lines

It is sufficient to consider a single-phase line to reach conclusions that can be readily extended to the three-phase case. We use this in the present discussion pertaining to bundle-conductor lines.

Consider a single-phase line with bundle conductor having $N$ subconductors on a circle of radius $A$. Each subconductor has a radius of $r$.

We have

$$C_{AN} = \frac{2\pi\varepsilon_0}{ln\left\{\frac{D}{\left[rN(A)^{N-1}\right]^{1/N}}\right\}} \text{ farads per meter} \tag{5.99}$$

The extension of the above result to the three-phase case is clearly obtained by replacing $D$ by the GMD. Thus

$$C_{AN} = \frac{2\pi\varepsilon_0}{ln\left\{\frac{\text{GMD}}{\left[rN(A)^{N-1}\right]^{1/N}}\right\}} \tag{5.100}$$

with

$$\text{GMD} = \left(D_{AB}D_{BC}D_{AC}\right)^{1/3} \tag{5.101}$$

The capacitive reactance in megaohms calculated for 60 Hz and 1 mile of line using the base 10 logarithm would be as follows:

$$X_c = 0.0683 \log\left\{\frac{\text{GMD}}{\left[rN(A)^{N-1}\right]^{1/N}}\right\} \tag{5.102}$$

$$X_c = X_a' + X_d' \tag{5.103}$$

This capacitive reactive reactance can be divided into two parts

$$X'_a = 0.0683 \log\left\{\frac{1}{\left[rN(A)^{N-1}\right]^{1/N}}\right\} \quad \textbf{(5.104)}$$

and

$$X'_d = 0.0683 \log(\text{GMD}) \quad \textbf{(5.105)}$$

If the bundle spacing $S$ is specified rather than the radius $A$ of the circle on which the conductors lie, then as before,

$$A = \frac{S}{2\sin\left(\frac{\pi}{N}\right)} \text{ for} N > 1 \quad \textbf{(5.106)}$$

## 5.5 TWO-PORT NETWORKS

A network can have two terminals or more, but many important networks in electric energy systems are those with four terminals arranged in two pairs. A two-terminal pair network might contain a transmission line model or a transformer model, to name a few in our power system applications. The box is sometimes called a *coupling network*, or *four-pole*, or a *two-terminal pair*. The term *two-port network* is in common use. It is a common mistake to call it a four-terminal network. In fact, the two-port network is a *restricted* four-terminal network since we require that the current at one terminal of a pair must be equal and opposite to the current at the other terminal of the pair.

An important problem arises in the application of two-port network theory to electric energy systems, which is called the *transmission problem*. It is required to find voltage and current at one pair of terminals in terms of quantities at the other pair.

*The transmission problem* is handled by assuming a pair of equations of the form

$$V_s = AV_r + BI_r \quad \textbf{(5.107)}$$

$$I_s = CV_r + DI_r \quad \textbf{(5.108)}$$

to represent the two-port network. In matrix form, we therefore have

$$\begin{bmatrix} V_s \\ I_s \end{bmatrix} = \begin{bmatrix} A & B \\ C & D \end{bmatrix} \begin{bmatrix} V_r \\ I_r \end{bmatrix} \quad \textbf{(5.109)}$$

For bilateral networks we have

$$AD - BC = 1 \tag{5.110}$$

Thus, there are but three independent parameters in the *ABCD* set as well.

Symmetry of a two-port network reduces the number of independent parameters to two. The network is *symmetrical* if it can be turned end for end in a system without altering the behavior of the rest of the system. An example is the transmission line, as will be seen later on. To satisfy this definition, a symmetrical network must have

$$A = D \tag{5.111}$$

We consider an important two-port network that plays a fundamental role in power system analysis – this is the symmetrical $\pi$-network. Figure 5.22 shows a symmetrical $\pi$-network. We can show that

$$A = \left(1 + \frac{ZY}{2}\right) \tag{5.112}$$

$$B = Z \tag{5.113}$$

$$C = Y\left(1 + \frac{ZY}{4}\right) \tag{5.114}$$

$$D = A \tag{5.115}$$

One of the most valued aspects of the *ABCD* parameters is that they are readily combined to find overall parameters when networks are connected in cascade. Figure 5.23 shows two cascaded two-part networks. We can write

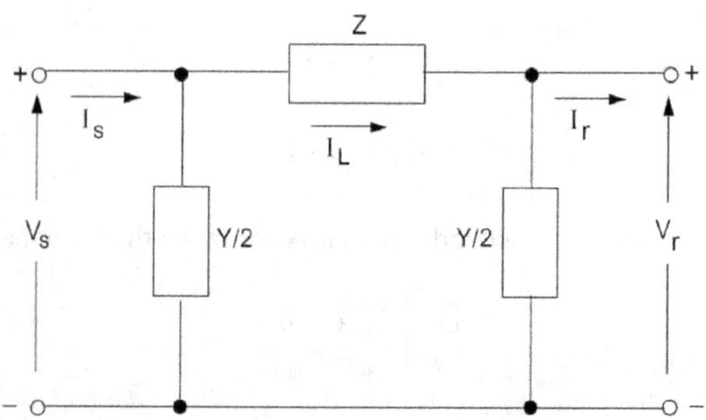


**Figure 5.22** A $\pi$-Network.

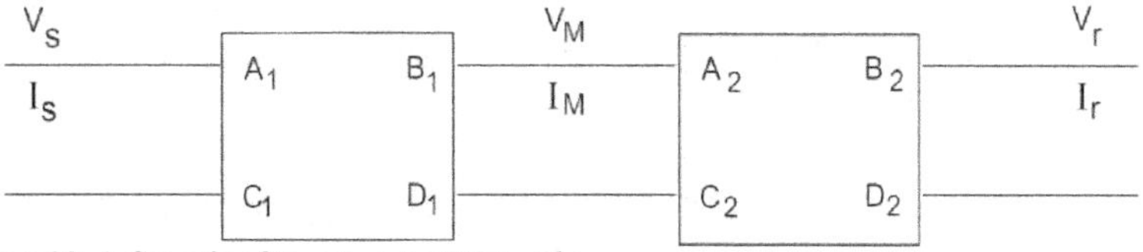


**Figure 5.23** A Cascade of Two two-Port Networks.

$$\begin{bmatrix} V_s \\ I_s \end{bmatrix} = \begin{bmatrix} A_1 & B_1 \\ C_1 & D_1 \end{bmatrix} \begin{bmatrix} V_M \\ I_M \end{bmatrix},$$

$$\begin{bmatrix} V_M \\ I_M \end{bmatrix} = \begin{bmatrix} A_2 & B_2 \\ C_2 & D_2 \end{bmatrix} \begin{bmatrix} V_r \\ I_r \end{bmatrix}$$

From which, eliminating ($V_M$, $I_M$), we obtain

$$\begin{bmatrix} V_s \\ I_s \end{bmatrix} = \begin{bmatrix} A_1 & B_1 \\ C_1 & D_1 \end{bmatrix} \begin{bmatrix} A_2 & B_2 \\ C_2 & D_2 \end{bmatrix} \begin{bmatrix} V_r \\ I_r \end{bmatrix}$$

Thus the equivalent *ABCD* parameters of the cascade are

$$A = A_1 A_2 + B_1 C_2 \tag{5.116}$$

$$B = A_1 B_2 + B_1 D_2 \tag{5.117}$$

$$C = C_1 A_2 + D_1 C_2 \tag{5.118}$$

$$D = C_1 B_2 + D_1 D_2 \tag{5.119}$$

If three networks or more are cascaded, the equivalent *ABCD* parameters can be obtained most easily by matrix multiplications as was done above.

## 5.6 TRANSMISSION LINE MODELS

The line parameters discussed in the preceding sections were obtained on a per-phase, per unit length basis. We are interested in the performance of lines with arbitrary length, say $l$. To be exact, one must take an infinite number of incremental lines, each with a differential length. Figure 5.24 shows the line with details of one incremental portion ($dx$) at a distance ($x$) from the receiving end.

The assumptions used in subsequent analyses are:

1. The line is operating under sinusoidal, balanced, steady-state

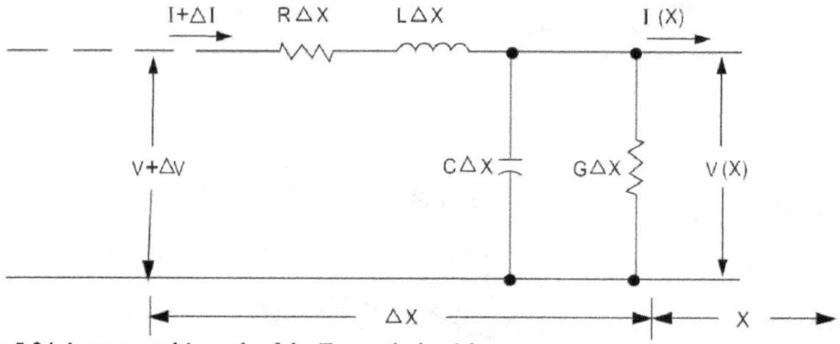


**Figure 5.24** Incremental Length of the Transmission Line.

conditions.

2. The line is transposed.

With these assumptions, we analyze the line on a per phase basis. Application of Kirchhoff's voltage and current relations yields

$$\Delta V = I(x) z \Delta x$$
$$\Delta I = V(x) y \Delta x$$

Let us introduce the propagation constant $\upsilon$ defined as

$$\upsilon = \sqrt{zy} \tag{5.120}$$

The series impedance per-unit length is defined by

$$z = R + j\omega L \tag{5.121}$$

The shunt admittance per-unit length is defined by

$$y = G + j\omega C \tag{5.122}$$

$R$ and $L$ are series resistance and inductance per unit length, and $G$ and $C$ are shunt conductance and capacitance to neutral per unit length.

In the limit, as $\Delta x \to 0$, we can show that

$$\frac{d^2 V}{dx^2} = \upsilon^2 V \tag{5.123}$$

$$\frac{d^2 I}{dx^2} = \upsilon^2 I \tag{5.124}$$

Equation (5.123) can be solved as an ordinary differential equation in

$V$. The solution turns out to be

$$V(x) = A_1 \exp(\upsilon x) + A_2 \exp(-\upsilon x) \tag{5.125}$$

Now taking the derivative of $V$ with respect to $x$ to obtain $I(x)$ as

$$I(x) = \frac{A_1 \exp(\upsilon x) - A_2 \exp(-\upsilon x)}{Z_c} \tag{5.126}$$

Here we introduce

$$Z_c = \sqrt{\frac{z}{y}} \tag{5.127}$$

$Z_c$ is the characteristic (wave) impedance of the line.

The constants $A_1$ and $A_2$ may be evaluated in terms of the initial conditions at $x = 0$ (the receiving end). Thus we have

$$V(0) = A_1 + A_2$$
$$Z_c I(0) = A_1 - A_2$$

from which we can write

$$V(x) = \frac{1}{2}\{[V(0) + Z_c I(0)]\exp(\upsilon x) + [V(0) - Z_c I(0)]\exp(-\upsilon x)\} \tag{5.128}$$

$$I(x) = \frac{1}{2}\left\{\left[I(0) + \frac{V(0)}{Z_c}\right]\exp(\upsilon x) + \left[I(0) - \frac{V(0)}{Z_c}\right]\exp(-\upsilon x)\right\} \tag{5.129}$$

Equations (5.128) and (5.129) can be used for calculating the voltage and current at any distance $x$ from the receiving end along the line. A more convenient form of these equations is found by using hyperbolic functions.

We recall that

$$\sinh\theta = \frac{\exp(\theta) - \exp(-\theta)}{2}$$
$$\cosh\theta = \frac{\exp(\theta) + \exp(-\theta)}{2}$$

By rearranging Eqs. (5.128) and (5.129) and substituting the hyperbolic function for the exponential terms, a new set of equations is found. These are

$$V(x) = V(0)\cosh \upsilon x + Z_c I(0)\sinh \upsilon x \tag{5.130}$$

and

$$I(x) = I(0)\cosh \upsilon x + \frac{V(0)}{Z_c}\sinh \upsilon x \tag{5.131}$$

We define the following *ABCD* parameters:

$$A(x) = \cosh \upsilon x \tag{5.132}$$

$$B(x) = Z_c \sinh \upsilon x \tag{5.133}$$

$$C(x) = \frac{1}{Z_c}\sinh \upsilon x \tag{5.134}$$

$$D(x) = \cosh \upsilon x \tag{5.135}$$

As a result, we have

$$V(x) = A(x)V(0) + B(x)I(0)$$
$$I(x) = C(x)V(0) + D(x)I(0)$$

For evaluation of the voltage and current at the sending end $x = l$, it is common to write

$$V_s = V(l)$$
$$I_s = I(l)$$
$$V_r = V(0)$$
$$I_r = I(0)$$

Thus we have

$$V_s = AV_r + BI_r \tag{5.136}$$

$$I_s = CV_r + DI_r \tag{5.137}$$

The subscripts $s$ and $r$ stand for sending and receiving values, respectively. We have from above:

$$A = A(l) = \cosh \upsilon l \tag{5.138}$$

$$B = B(l) = Z_c \sinh \upsilon l \tag{5.139}$$

$$C = C(l) = \frac{1}{Z_c} \sinh \upsilon l \quad (5.140)$$

$$D = D(l) = \cosh \upsilon l \quad (5.141)$$

It is practical to introduce the complex variable $\theta$ in the definition of the *ABCD* parameters. We define

$$\theta = \upsilon l = \sqrt{ZY} \quad (5.142)$$

As a result,

$$A = \cosh \theta \quad (5.143)$$

$$B = Z_c \sinh \theta \quad (5.144)$$

$$C = \frac{1}{Z_c} \sinh \theta \quad (5.145)$$

$$D = A \quad (5.146)$$

Observe that the total line series impedance and shunt admittance are given by

$$Z = zl \quad (5.147)$$

$$Y = yl \quad (5.148)$$

## Evaluating *ABCD* Parameters

If your calculator can not perform hyperbolic function evaluation with complex arguments, then you may want to use the process discussed here. Two methods can be employed to calculate the *ABCD* parameters of a transmission line exactly. Both assume that $\theta$ is calculated in the rectangular form

$$\theta = \theta_1 + j\theta_2$$

The first method proceeds by expanding the hyperbolic functions as follows:

$$\begin{aligned} A &= \frac{e^{\theta}+e^{-\theta}}{2} \\ &= \frac{1}{2}\left(e^{\theta_1}\angle\theta_2 + e^{-\theta_1}\angle -\theta_2\right) \\ \sinh\theta &= \frac{e^{\theta}-e^{-\theta}}{2} \\ &= \frac{1}{2}\left(e^{\theta_1}\angle\theta_2 - e^{-\theta_1}\angle -\theta_2\right) \end{aligned} \tag{5.149}$$

$$B = \sqrt{\frac{Z}{Y}}\sinh\theta \tag{5.150}$$

$$C = \sqrt{\frac{Y}{Z}}\sinh\theta \tag{5.151}$$

Note that $\theta_2$ is in radians to start with in the decomposition of $\theta$.

The second method uses two well-known identities to arrive at the parameter of interest.

$$\begin{aligned} A &= \cosh(\theta_1 + j\theta_2) \\ \cosh\theta &= \cosh\theta_1\cos\theta_2 + j\sinh\theta_1\sin\theta_2 \end{aligned} \tag{5.152}$$

We also have

$$\sinh\theta = \sinh\theta_1\cos\theta_2 + j\cosh\theta_1\sin\theta_2 \tag{5.153}$$

**Example 5.9**

Find the exact *ABCD* parameters for a 235.92-mile long, 735-kV, bundle-conductor line with four subconductors per phase with subconductor resistance of 0.1004 ohms per mile. Assume that the series inductive reactance per phase is 0.5541 ohms per mile and shunt capacitive susceptance of $7.4722 \times 10^{-6}$ siemens per mile to neutral. Neglect shunt conductance.

*Solution*

The resistance per phase is

$$r = \frac{0.1004}{4} = 0.0251 \text{ ohms/mile}$$

Thus the series impedance in ohms per mile is

$$z = 0.0251 + j0.5541 \text{ ohms/mile}$$

The shunt admittance is

$$y = j7.4722\times10^{-6} \text{ siemens/mile}$$

For the line length,

$$Z = zl = (0.0251 + j0.5541)(235.92) = 130.86\angle 87.41^\circ$$
$$Y = yl = j(7.4722\times10^{-6})(235.92) = 1.7628\times10^{-3}\angle 90^\circ$$

We calculate $\theta$ as

$$\begin{aligned}\theta &= \sqrt{ZY} \\ &= \left[(130.86\angle 87.41^\circ)(1.7628\times10^{-3}\angle 90^\circ)\right]^{1/2} \\ &= 0.0109 + j0.4802\end{aligned}$$

Thus,

$$\theta_1 = 0.0109$$
$$\theta_2 = 0.4802$$

We change $\theta_2$ to degrees. Therefore,

$$\theta_2 = (0.4802)\left(\frac{180}{\pi}\right) = 27.5117^\circ$$

Using Eq. (5.149), we then get

$$\begin{aligned}\cosh\theta &= \frac{1}{2}\left(e^{0.0109}\angle 27.5117^\circ + e^{-0.0109}\angle -27.5117^\circ\right) \\ &= 0.8870\angle 0.3242^\circ\end{aligned}$$

From the above,

$$D = A = 0.8870\angle 0.3242^\circ$$

We now calculate sinh $\theta$ as

$$\begin{aligned}\sinh\theta &= \frac{1}{2}\left(e^{0.0109}\angle 27.5117^\circ - e^{-0.0109}\angle -27.5117^\circ\right) \\ &= 0.4621\angle 88.8033^\circ\end{aligned}$$

We have

$$Z_c = \sqrt{\frac{Z}{Y}} = \left( \frac{130.86\angle 87.41^\circ}{1.7628 \times 10^{-3} \angle 90^\circ} \right)^{1/2}$$
$$= (74234.17\angle -2.59)^{1/2}$$
$$= 272.46\angle -1.295^\circ$$

As a result,

$$B = Z_c \sinh\theta$$
$$= 125.904\angle 87.508^\circ$$

Also,

$$C = \frac{1}{Z_c}\sinh\theta$$
$$= 1.696 \times 10^{-3} \angle 90.098^\circ$$

Let us employ the second method to evaluate the parameters. We find the hyperbolic functions:

$$\cosh\theta_1 = \cosh(0.0109)$$
$$= \frac{e^{0.0109} + e^{-0.0109}}{2}$$
$$= 1.000059$$
$$\sinh\theta_1 = \frac{e^{0.0109} - e^{-0.0109}}{2}$$
$$= 1.09002 \times 10^{-2}$$

(Most calculators have built-in hyperbolic functions, so you can skip the intermediate steps). We also have

$$\cos\theta_2 = \cos\left[ (0.4802)\left( \frac{180}{\pi} \right) \right]$$
$$= 0.8869$$
$$\sin\theta_2 = \sin\left[ (0.4802)\left( \frac{180}{\pi} \right) \right]$$
$$= 0.4619566$$

Therefore, we have

$$\begin{aligned}\cosh\theta &= \cosh\theta_1\cos\theta_2 + j\sinh\theta_1\sin\theta_2\\ &= (1.000059)(0.8869) + j(1.09002\times10^{-2})(0.4619566)\\ &= 0.8869695\angle 0.32527^\circ\\ \sinh\theta &= \sinh\theta_1\cos\theta_2 + j\cosh\theta_1\sin\theta_2\\ &= (1.09002\times10^{-2})(0.8869) + j(1.000059)(0.4619566)\\ &= 0.4620851\angle 88.801^\circ\end{aligned}$$

These results agree with the ones obtained using the first method.

**Example 5.10**

Find the voltage, current, and power at the sending end of the line of Example 5.9 and the transmission efficiency given that the receiving-end load is 1500 MVA at 700 kV with 0.95 PF lagging.

*Solution*

We have the apparent power given by

$$S_r = 1500\times10^6 \text{ VA}$$

The voltage to neutral is

$$V_r = \frac{700\times10^3}{\sqrt{3}} \text{ V}$$

Therefore,

$$\begin{aligned}I_r &= \frac{1500\times10^6}{3\left(\frac{700\times10^3}{\sqrt{3}}\right)}\angle -\cos^{-1}0.95\\ &= 1237.18\angle -18.19^\circ \text{ A}\end{aligned}$$

From Example 5.9 we have the values of the $A$, $B$, and $C$ parameters. Thus the sending-end voltage (to neutral) is obtained as

$$\begin{aligned}V_s &= AV_r + BI_r\\ &= \left(0.8870\angle 0.3253^\circ\right)\left(\frac{700\times10^3}{\sqrt{3}}\right)\\ &\quad + \left(125.904\angle 87.508^\circ\right)\left(1237.18\angle -18.19^\circ\right)\\ &= 439.0938\angle 19.66^\circ \text{ kV}\end{aligned}$$

The line-to-line value is obtained by multiplying the above value by

$\sqrt{3}$ , giving

$$V_{S_L} = 760.533\,\text{kV}$$

The sending-end current is obtained as

$$\begin{aligned} I_s &= CV_r + DI_r \\ &= \left(1.696\times10^{-3}\angle 90.098^\circ\right)\left(\frac{700\times10^3}{\sqrt{3}}\right) \\ &\quad + \left(0.887\angle 0.3253^\circ\right)\left(1237.18\angle -18.19^\circ\right) \\ &= 1100.05\angle 18.49^\circ \end{aligned}$$

The sending-end power factor is

$$\begin{aligned} \cos\phi_s &= \cos(19.66 - 18.49) \\ &= \cos(1.17) = 0.99979 \end{aligned}$$

As a result, the sending-end power is

$$\begin{aligned} P_s &= 3(439.0938\times10^3)(1100.05)(0.99979) \\ &= 1448.77\times10^6\ \text{MW} \end{aligned}$$

The efficiency is

$$\begin{aligned} \eta &= \frac{P_r}{P_s} \\ &= \frac{1500\times10^6\times0.95}{1448.77\times10^6} \\ &= 0.9836 \end{aligned}$$

## Lumped Parameter Transmission Line Models

Lumped parameter representations of transmission lines are needed for further analysis of interconnected electric power systems. Their use enables the development of simpler algorithms for the solution of complex networks that involve transmission lines.

Here we are interested in obtaining values of the circuit elements of a $\pi$ circuit, to represent accurately the terminal characteristics of the line given by

$$V_s = AV_r + BI_r$$
$$I_s = CV_r + DI_r$$

It is easy to verify that the elements of the equivalent circuit are given in terms of the *ABCD* parameters of the line by

$$Z_\pi = B \tag{5.154}$$

and

$$Y_\pi = \frac{A-1}{B} \tag{5.155}$$

The circuit is shown in Figure 5.25.

**Example 5.11**

Find the equivalent $\pi$-circuit elements for the line of Example 5.9.

*Solution*

From Example 5.9, we have

$$A = 0.8870\angle 0.3242^\circ$$
$$B = 125.904\angle 87.508^\circ$$

As a result, we have

$$Z_\pi = 125.904\angle 87.508^\circ \text{ ohms}$$
$$Y_\pi = \frac{0.8870\angle 0.3242^\circ - 1}{125.904\angle 87.508^\circ}$$
$$= 8.9851\times 10^{-4}\angle 89.941^\circ \text{ siemens}$$

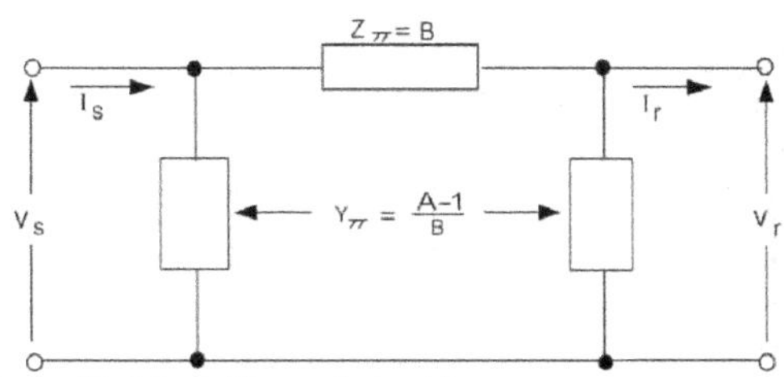


**Figure 5.25** Equivalent $\pi$ Model of a Transmission Line.

The following MATLAB™ Script implements Examples 5.9, 5.10, and 5.11

```
% Example 5-9
% To enter the data
r=0.0251;
x=0.5541;
l=235.92;
y=i*7.4722*10^-6;
Sr=1500*10^6;
Vr=(700*10^3)/3^.5;
% for the line length
z=r+i*x;
Z=z*l;
Y=y*l;
% To calculate theta
theta=(Z*Y)^.5;
theta2=imag(theta)*180/pi;
% To calculate A and D
D=cosh(theta)
A=D
A_mod=abs(A)
delta=angle(A)*180/pi
% To calculate B and C
Zc=(Z/Y)^.5;
B=Zc*sinh(theta)
B_mod=abs(B) delta1=angle(B)*180/pi
C=1/Zc*sinh(theta)
C_mod=abs(C)
delta2=angle(C)*180/pi
% To evaluate the parameters.
% We find the hyperbolic functions
cosh(real(theta));
sinh(real(theta));
%
%Example 5-10
%
Ir=Sr/(3*Vr);
% power factor 0.95 lagging
alpha=acos(0.95);
alpha_deg=alpha*180/pi;
Pr=Sr*cos(alpha);
Ir_compl=Ir*(cos(-alpha)+i*sin(-
alpha));
% To calculate sending end voltage (to
neutral)
Vs=A*Vr+B*Ir_compl
Vs_mod=abs(Vs)
```

MATLAB™ con't.

```
Vs_arg=angle(Vs)*180/pi
% line to line voltage
Vsl=Vs*3^.5
% To calculate sending end current
Is=C*Vr+D*Ir_compl
Is_mod=abs(Is)
Is_arg=angle(Is)*180/pi
% To calculate sending end power factor
pf_sending=cos(angle(Vs)-angle(Is))
% To calculate sending end power
Ps=3*abs(Vs)*abs(Is)*pf_sending
%  To calculate the efficiency
eff=Pr/Ps
%
%example 5-11
%
% To find the equivalent pi-circuit
elements
Zpi=B
Zpi_mod=abs(Zpi)
Zpi_arg=angle(Zpi)*180/pi
Ypi=(A-1)/B
Ypi_mod=abs(Ypi)
Ypi_arg=angle(Ypi)*180/pi
```

The results of running the script are as shown below:

```
EDU»
D =    0.8870+ 0.0050i
A =  0.8870+ 0.0050i
A_mod =      0.8870
delta =     0.3244
B =  5.4745e+000 + 1.2577e+002i
B_mod =    125.8891
delta1 =     87.5076
C =    0.0000+ 0.0017i
C_mod =      0.0017
delta2 =     90.1012
Vs =  4.1348e+005 + 1.4773e+005i
```

## Approximations to the *ABCD* Parameters of Transmission Lines

Consider the series expansion of the hyperbolic functions defining the

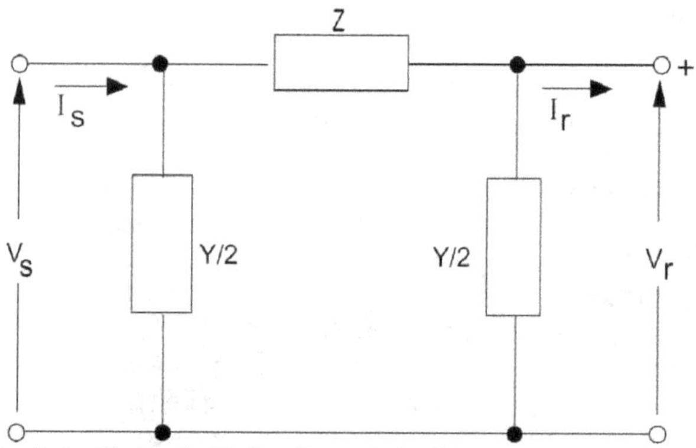


**Figure 5.26** Nominal $\pi$ Model of a Medium Transmission Line.

$A$, $B$, $C$, and $D$ parameters using $\theta = \sqrt{ZY}$.

Usually no more than three terms are required. For overhead lines less than 500 km in length, the following approximate expressions are satisfactory:

$$A = D = 1 + \frac{ZY}{2} \tag{5.156}$$

$$B = Z\left(1 + \frac{ZY}{6}\right) \tag{5.157}$$

$$C = Y\left(1 + \frac{ZY}{6}\right) \tag{5.158}$$

If only the first term of the expansions is used, then

$$B = Z \tag{5.159}$$

$$\frac{A-1}{B} = \frac{Y}{2} \tag{5.160}$$

In this case, the equivalent $\pi$ circuit reduces to the nominal $\pi$, which is used generally for lines classified as medium lines (up to 250 km). Figure 5.26 shows the nominal $\pi$ model of a medium transmission line. The result we obtained analytically could have been obtained easily by the intuitive assumption that the line's series impedance is lumped together and the shunt admittance $Y$ is divided equally with each half placed at each end of the line.

A final model is the short-line (up to 80 km) model, and in this case the shunt admittance is neglected altogether. The line is thus represented only by its series impedance.

### Example 5.12

Find the nominal $\pi$ and short-line representations for the line of Example 5.9. Calculate the sending-end voltage and current of the transmission line using the two representations under the conditions of Example 5.10.

*Solution*

For this line we have

$$Z = 130.86\angle 87.41^\circ$$
$$Y = 1.7628\times 10^{-3}\angle 90^\circ$$

As a result, we have the representations shown in Figure 5.26.

From Example 5.10, we have

$$V_r = \frac{700\times 10^3}{\sqrt{3}}\ \text{V}$$
$$I_r = 1237.18\angle -18.19^\circ\ \text{A}$$

For the short-line representation we have

$$\begin{aligned} V_s &= V_r + I_r Z \\ &= \frac{700\times 10^3}{\sqrt{3}} + (1237.18\angle -18.19^\circ)(130.86\angle 87.41^\circ) \\ &= 485.7682\times 10^3\angle 18.16^\circ\ \text{V} \end{aligned}$$

For the nominal $\pi$ we have

$$\begin{aligned} I_L &= I_r + V_r\left(\frac{Y}{2}\right) \\ &= (1237.18\angle -18.19^\circ) + \frac{700\times 10^3}{\sqrt{3}}(0.8814\times 10^{-3}\angle 90^\circ) \\ &= 1175.74\angle -1.4619^\circ\ \text{A} \end{aligned}$$

Thus,

$$\begin{aligned} V_s &= V_r + I_L Z \\ &= \frac{700\times10^3}{\sqrt{3}} + (1175.74\angle -1.4619^\circ)(130.86\angle 87.41^\circ) \\ &= 442.484\times10^3 \angle 20.2943^\circ \text{ V} \end{aligned}$$

Referring back to the exact values calculated in Example 5.10, we find that the short-line approximation results in an error in the voltage magnitude of

$$\begin{aligned} \Delta V &= \frac{439.0938 - 485.7682}{439.0938} \\ &= -0.11 \end{aligned}$$

For the nominal $\pi$ we have the error of

$$\begin{aligned} \Delta V &= \frac{439.0938 - 442.484}{439.0938} \\ &= -0.00772 \end{aligned}$$

which is less than 1 percent.

The sending-end current with the nominal $\pi$ model is

$$\begin{aligned} I_s &= I_L + V_s\left(\frac{Y}{2}\right) \\ &= 1175.74\angle -1.4619^\circ \\ &\quad + (442.484\angle 20.2943^\circ)(0.8814\times10^3\angle 90^\circ) \\ &= 1092.95\angle 17.89^\circ \text{ A} \end{aligned}$$

The following MATLAB™ Script implements Example 5.12

```
% Example 5-12
% From example 5-9, we have
Z=130.86*(cos(87.41*pi/180)+i*sin(87.41
*pi/180));
Y=i*1.7628*10^-3;
% From example 5-10, we have
Vr=700*10^3/(3^.5);
Ir=1237.18*(cos(-18.19*pi/180)+i*sin(-
18.19*pi/180));
% For the short-line representation we
have
Vs=Vr+Ir*Z;
```

MATLAB™ con't.

```
Vs_mod=abs(Vs)
Vs_arg=angle(Vs)*180/pi
% for the nominal pi, we have
IL=Ir+Vr*(Y/2);
IL_mod=abs(IL)
IL_arg=angle(IL)*180/pi
% Thus
Vs=Vr+IL*Z;
Vs_mod=abs(Vs)
Vs_arg=angle(Vs)*180/pi
% The sending-end current with the
nominal pi model is
Is=IL+Vs*(Y/2)
Is_mod=abs(Is)
Is_arg=angle(Is)*180/pi
```

```
EDU»
Vs_mod =  4.8577e+005
Vs_arg =   18.1557
IL_mod =  1.1757e+003
IL_arg =   -1.4619
Vs_mod =  4.4248e+005
Vs_arg =   20.2943
Is = 1.0401e+003 + 3.3580e+002i
Is_mod =  1.0929e+003
Is_arg =   17.8931
```

## *PROBLEMS*

### Problem 5.1

Determine the inductive reactance in ohms/mile/phase for a 345-kV, single-circuit line with ACSR 84/19 conductor for which the geometric mean radius is 0.0588 ft. Assume a horizontal phase configuration with 26-ft phase separation.

### Problem 5.2

Calculate the inductive reactance in ohms/mile/phase for a 500-kV, single-circuit, two-subconductor bundle line with ACSR 84/19 subconductor for which the GMR is 0.0534 ft. Assume horizontal phase configuration with 33.5-ft phase separation. Assume bundle separation is 18 in.

**Problem 5.3**

Repeat Problem 5.2 for a phase separation of 35 ft.

**Problem 5.4**

Repeat Problem 5.3 with an ACSR 76/19 subconductor for which the GMR is 0.0595 ft.

**Problem 5.5**

Find the inductive reactance in ohms/mile/phase for a 500-kV, single-circuit, two-subconductor bundle line with ACSR 84/19 conductor for which the GMR is 0.0588 ft. Assume horizontal phase configuration with separation of 32 ft. Bundle spacing is 18 in.

**Problem 5.6**

Find the inductive reactance in ohms/mile/phase for the 765-kV, single-circuit, bundle-conductor line with four subconductors per bundle at a spacing of 18 in., given that the subconductor GMR is 0.0385 ft. Assume horizontal phase configuration with 44.5-ft phase separation.

**Problem 5.7**

Repeat Problem 5.6 for bundle spacing of 24 in. and subconductor GMR of 0.0515 ft. Assume phase separation is 45 ft.

**Problem 5.8**

Calculate the inductance in henries per meter per phase for the 1100-kV, bundle-conductor line shown in Figure 5.27. Assume phase spacing $D_1$ = 15.24 m, bundle separation $S$ = 45.72 cm, and conductor diameter is 3.556 cm.

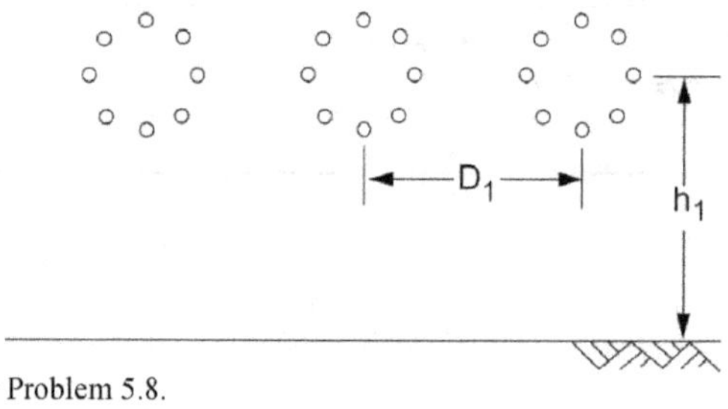


**Figure 5.27** Line for Problem 5.8.

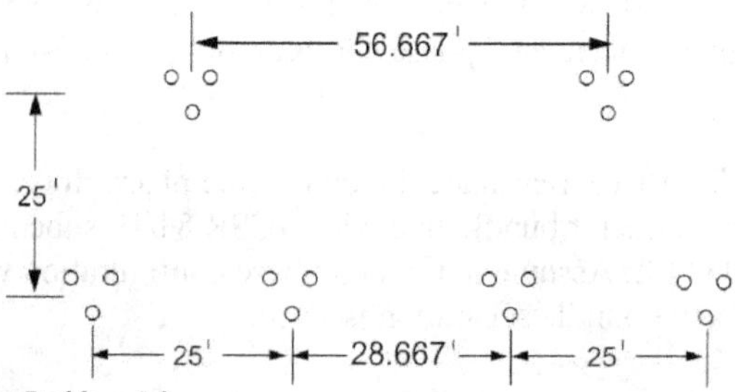


**Figure 5.28** Line for Problem 5.9.

## Problem 5.9

Calculate the inductive reactance in ohms per mile for the 500-kV, double-circuit, bundle-conductor line with three subconductors of 0.0431-ft GMR and with 18-in. bundle separation. Assume conductor configurations as shown in Figure 5.28.

## Problem 5.10

Calculate the inductive reactance in ohms per mile for 345-kV, double-circuit, bundle-conductor line with two subconductors per bundle at 18-in. bundle spacing. Assume subconductor's GMR is 0.0373 ft, and conductor configuration is as shown in Figure 5.29.

## Problem 5.11

Calculate the inductive reactance in ohms per mile for the 345-kV double-circuit, bundle-conductor line with two subconductors per bundle at 18-in. bundle spacing. Assume subconductor's GMR is 0.0497 ft, and conductor configuration is as shown in Figure 5.30.

## Problem 5.12

Determine the capacitive reactance in ohm miles for the line of Problem 5.1. Assume the conductor's outside diameter is 1.76 in. Repeat by including earth effects given that the ground clearance is 45 ft.

## Problem 5.13

Determine the capacitive reactance in ohm miles for the line of Problem 5.2. Assume the conductor's outside diameter is 1.602 in. Repeat by including earth effects given that the ground clearance is 82 ft.

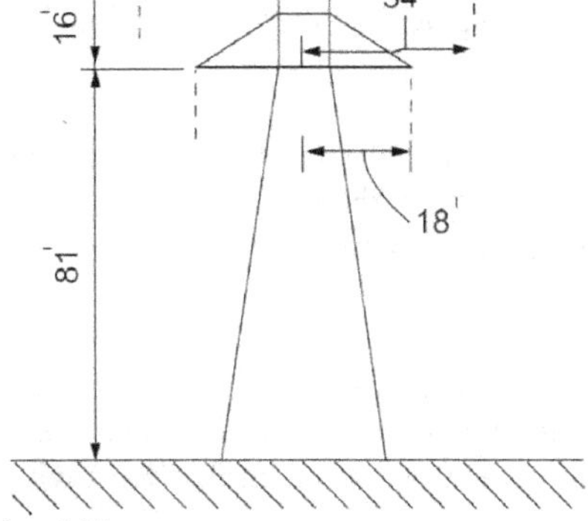


**Figure 5.29** Line for Problem 5.10.

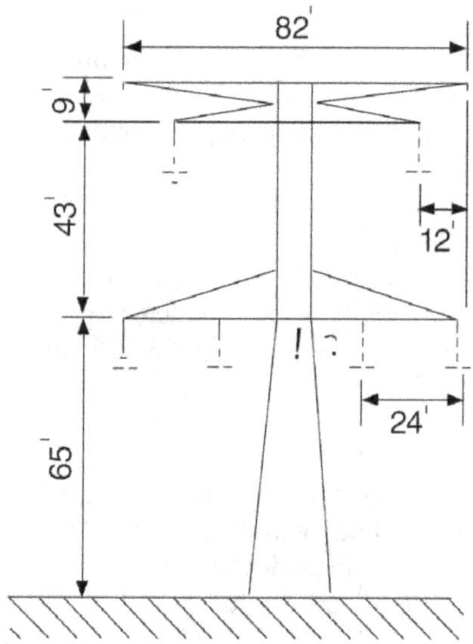


**Figure 5.30** Line for Problem 5.11.

**Problem 5.14**
Determine the capacitive reactance in ohm miles for the line of Problem 5.3. Assume the conductor's outside diameter is 1.602 in. Repeat by including earth effects given that the ground clearance is 136 ft.

**Problem 5.15**
Determine the capacitive reactance in ohm miles for the line of Problem 5.4. Assume the conductor's outside diameter is 1.7 in. Neglect earth effects.

**Problem 5.16**
Determine the capacitive reactance in ohm miles for the line of Problem 5.5. Assume the conductor's outside diameter is 1.762 in. Repeat by including earth effects given that the ground clearance is 63 ft.

**Problem 5.17**
Determine the capacitive reactance in ohm miles for the line of Problem 5.6. Assume the conductor's outside diameter is 1.165 in.

**Problem 5.18**
Determine the capacitive reactance in ohm miles for the line of Problem 5.7. Assume the conductor's outside diameter is 1.6 in. Repeat by including earth effects given that the ground clearance is 90 ft.

**Problem 5.19**
Calculate the capacitance in farads per meter per phase neglecting earth effect for the 1100-kV, bundle-conductor line of Problem 5.8. Assume the conductor diameter is 3.556 cm. Repeat including earth effects with $h_1 = 21.34$ m.

**Problem 5.20**

Determine the capacitive reactance in ohm mile for the line of Problem 5.9. Assume the conductor's outside diameter is 1.302 in. Neglect earth effect.

**Problem 5.21**

Determine the capacitive reactance in ohm mile for the line of Problem 5.10. Assume the conductor's outside diameter is 1.165 in.

**Problem 5.22**

Determine the capacitive reactance in ohm mile for the line of Problem 5.11. Assume the conductor's outside diameter is 1.302 in.

**Problem 5.23**

Assume that the 345-kV line of Problems 5.1 and 5.12 is 14 miles long and that the conductor's resistance is 0.0466 ohms/mile.

A. Calculate the exact *ABCD* parameters for the line.
B. Find the circuit elements of the equivalent $\pi$ model for the line. Neglect earth effects.

**Problem 5.24**

Assume that the 1100-kV line of Problems 5.8 and 5.19 is 400 km long and that the subconductor's resistance is 0.0435 ohms/km.

A. Calculate the exact *ABCD* parameters for the line.
B. Find the circuit elements of the equivalent $\pi$ model for the line. Neglect earth effects.

**Problem 5.25**

The following information is available for a single-circuit, three-phase, 345-kV, 360 mega volt amperes (MVA) transmission line:

Line length = 413 miles.
Number of conductors per phase = 2.
Bundle spacing = 18 in.
Outside conductor diameter = 1.165 in.
Conductor's GMR = 0.0374 ft.
Conductor's resistance = 0.1062 ohms/mile.
Phase separation = 30 ft.
Phase configuration is equilateral triangle.
Minimum ground clearance = 80 ft.

A. Calculate the line's inductive reactance in ohms per mile per phase.
B. Calculate the capacitive reactance including earth effects in ohm miles per phase.
C. Calculate the exact *A* and *B* parameters of the line.
D. Find the voltage at the sending end of the line if normal rating

power at 0.9 PF is delivered at 345-kV at the receiving end. Use the exact formulation.

E. Repeat (d) using the short-line approximation. Find the error involved in computing the magnitude of the sending-end voltage between this method and the exact one.

**Problem 5.26**

For the transmission line of Problem 5.24, calculate the sending-end voltage, sending-end current, power, and power factor when the line is delivering 4500 MVA at 0.9 PF lagging at rated voltage, using the following:

A. Exact formulation.
B. Nominal $\pi$ approximation.
C. Short-line approximation.

**Problem 5.27**

A 650-km, 765 kV, 60 Hz three phase transmission line has a series impedance $z = 0.02 + j0.32$ Ω/km and a shunt admittance $y = j\ 5\text{x}10^{-6}$ S/km. At the sending end of the line an apparent power of 2700 MVA is delivered at a lagging power factor of 0.9 at rated voltage. Determine:

A. The ABCD parameters of the line using the nominal $\pi$ approximation expressions.
B. The receiving end voltage, current, and power factor.

**Problem 5.28**

A 650-km, 765-kV, 60 Hz three phase transmission line has a series impedance $z = 0.0185 + j0.3170$ Ω/km and a shunt admittance $y = j\ 4.875\text{x}10^{-6}$ S/km. The full load at the receiving end is 2500 MVA at a lagging power factor of 0.85 and at 95% rated voltage. Determine:

A. The ABCD parameters of the line using exact long line hyperbolic expressions.
B. The sending end voltage, current, and power factor.

**Problem 5.29**

A. Results of tests conducted at the receiving end of a 500 kV, 60 Hz, three phase line involving open circuit ($I_r = 0$) and short circuit ($V_r = 0$) are given by:

$$Z_{oc} = \left.\frac{V_S}{I_S}\right|_{I_r=0} = 820\angle\underline{-88.8^\circ}$$

$$Z_{sc} = \left.\frac{V_S}{I_S}\right|_{V_r=0} = 200\angle\underline{78^\circ}$$

Find the line parameters A, B, and C.

B. Suppose that the load at the receiving end of the line of part A is 750 MVA at nominal voltage, and lagging power factor of 0.83 at rated voltage. Determine the sending end voltage, current, active and reactive power and power factor.

**Problem 5.30**

A. An open circuit test is conducted at the receiving end of a 500 kV, 60 Hz, three phase line, ($I_r = 0$), with the receiving end voltage maintained at 500 kV (line-to-line value). The sending end voltage is found to be 450 kV (line-to-line value) and leads the receiving end voltage by 1.3°. The sending end active power input to the line is 5.18 MW at 0.021-power factor leading. Find the line parameters A, B, and C.

B. Suppose that the load at the receiving end of the line of part A is 500 MVA at nominal voltage, and lagging power factor of 0.8 at rated voltage. Determine the sending end voltage, current, active and reactive power and power factor.

**Problem 5.31**

A 500-kV, 60 Hz three-phase transmission line is modeled using the exact A, B, C, and D parameters. It is known that:

$$A = 0.93\angle 0.2^{\circ} \qquad\qquad Z = 110\angle 87.3^{\circ}$$

$Z$ is the total series impedance of the line.

A. Find the values of the parameters B and C.

B. Suppose that the load at the receiving end of the line is 750 MVA at a lagging power factor of 0.9 at rated voltage. Determine the sending end voltage, current, active and reactive power and power factor.

# Chapter 6

# INDUCTION AND FRACTIONAL HORSEPOWER MOTORS

## 6.1 INTRODUCTION

In this chapter, we will discuss three-phase induction motors and their performance characteristics. We will then discuss motors of the fractional-horsepower class used for applications requiring low power output, small size, and reliability. Standard ratings for this class range from $\frac{1}{20}$ to 1 hp. Motors rated for less than $\frac{1}{20}$ hp are called subfractional-horsepower motors and are rated in millihorsepower and range from 1 to 35 mhp. These small motors provide power for all types of equipment in the home, office, and commercial installations. The majority are of the induction-motor type and operate from a single-phase supply.

## 6.2 THREE-PHASE INDUCTION MOTORS

The induction motor is characterized by simplicity, reliability, and low cost, combined with reasonable overload capacity, minimal service requirements, and good efficiency. An induction motor utilizes alternating current supplied to the stator directly. The rotor receives power by induction effects. The stator windings of an induction motor are similar to those of the synchronous machine. The rotor may be one of two types. In the *wound rotor motor*, windings similar to those of the stator are employed with terminals connected to insulated slip rings mounted on the shaft. The rotor terminals are made available through carbon brushes bearing on the slip rings. The second type is called the *squirrel-cage rotor*, where the windings are simply conducting bars embedded in the rotor and short-circuited at each end by conducting end rings.

When the stator of the motor is supplied by a balanced three-phase alternating current source, it will produce a magnetic field that rotates at synchronous speed as determined by the number of poles and applied frequency $f_s$.

$$n_s = \frac{120 f_s}{P} \text{ r/min} \qquad (6.1)$$

In steady state, the rotor runs at a steady speed $n_r$ r/min in the same direction as the rotating stator field. The speed $n_r$ is very close to $n_s$ when the motor is running slow, and runs slower as the mechanical load is increased. The speed difference $(n_s - n_r)$ is termed the *slip* and is commonly defined as a per unit value $s$.

$$s = \frac{n_s - n_r}{n_s} \tag{6.2}$$

Because of the relative motion between stator and rotor, induced voltages will appear in the rotor with a frequency $f_r$ called the *slip frequency*.

$$f_r = sf_s \tag{6.3}$$

From the above we observe that the induction motor is simply a transformer but that it has a secondary frequency $f_r$.

**Example 6.1**

Determine the number of poles, the slip, and the frequency of the rotor currents at rated load for three-phase, induction motors rated at:

A. 2200 V, 60 Hz, 588 r/min.
B. 120V, 600 Hz, 873 r/min.

*Solution*

We use $P = 120f/n$, to obtain $P$, using $n_r$, the rotor speed given to obtain the slip.

A.

$$P = \frac{120 \times 60}{588} = 12.245$$

But $P$ should be an even number. Therefore, take $P = 12$. Hence

$$n_s = \frac{120f}{P} = \frac{120 \times 60}{12} = 600 \text{ r/min}$$

The slip is thus given by

$$s = \frac{n_s - n_r}{n_s} = \frac{600 - 588}{600} = 0.02$$

The rotor frequency is

$$f_r = sf_s = 0.02 \times 60 = 1.2 \text{ Hz}$$

B.

$$P = \frac{120 \times 600}{873} = 82.47$$

Take $P = 82$.

$$n_s = \frac{120 \times 600}{82} = 878.05 \text{ r/min}$$

$$s = 0.006$$

$$f_r = 0.006 \times 600 = 3.6 \text{ Hz}$$

## Equivalent Circuits

An equivalent circuit of the three-phase induction motor can be developed on the basis of the above considerations and transformer models. Looking into the stator terminals, the applied voltage $V_s$ will supply the resistive drop $I_s R_1$ as well as the inductive voltage $jI_s X_1$ and the counter EMF $E_1$ where $I_s$ is the stator current and $R_1$ and $X_1$ are the stator effective resistance and inductive reactance respectively. In a manner similar to that employed for the analysis of the transformer, we model the magnetizing circuit by the shunt conductance $G_c$ and inductive susceptance $-jB_m$.

The rotor's induced voltage $E_{2s}$ is related to the stator EMF $E_1$ by

$$E_{2s} = sE_1 \tag{6.4}$$

This is due simply to the relative motion between stator and rotor. The rotor current $I_{rs}$ is equal to the current $I_r$ in the stator circuit. The induced EMF $E_{2s}$ supplies the resistive voltage component $I_r R_2$ and inductive component $jI_r(sX_2)$. $R_2$ is the rotor resistance, and $X_2$ is the rotor inductive reactance on the basis of the stator frequency.

$$E_{2s} = I_r R_2 + jI_r(sX_2)$$

or

$$sE_1 = I_r R_2 + jI_r(sX_2) \tag{6.5}$$

From the above we conclude that the equivalent rotor impedance seen from the stator is given by:

$$\frac{E_1}{I_r} = \frac{R_2}{s} + jX_2$$

The complete equivalent circuit of the induction motor is shown in Figure 6.1.

Considering the active power flow into the induction machine, we find that the input power $P_s$ supplies the stator $I^2R$ losses and the core losses. The remaining power denoted by the air-gap power $P_g$ is that transferred to the rotor circuit. Part of the air-gap power is expended as rotor $I^2R$ losses with the remainder being the mechanical power delivered to the motor shaft. We can express the air-gap power as

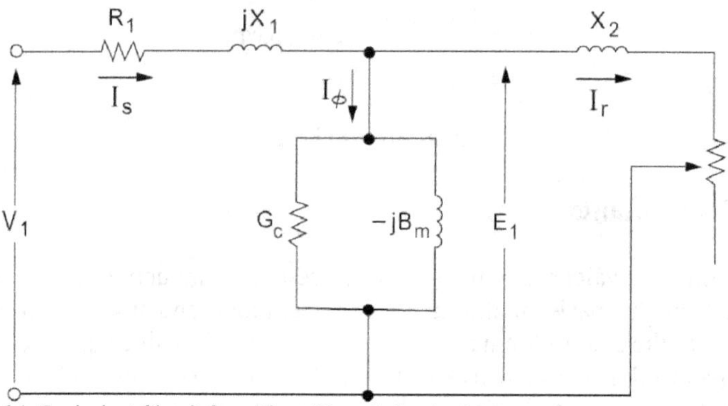


**Figure 6.1** Equivalent Circuit for a Three-Phase Induction Motor.

$$P_g = 3I_r^2\left(\frac{R_2}{s}\right) \tag{6.6}$$

The rotor $I^2R$ losses are given by

$$P_{lr} = 3I_r^2 R_2 \tag{6.7}$$

As a result, the mechanical power output (neglecting mechanical losses) is

$$\begin{aligned} P_r &= P_g - P_{lr} \\ &= 3I_r^2 \frac{(1-s)}{s} R_2 \end{aligned} \tag{6.8}$$

The last formula suggests a splitting of $R_2/s$ into the sum of $R_2$ representing the rotor resistance and a resistance

$$\frac{1-s}{s}(R_2)$$

which is the equivalent resistance of the mechanical load. As a result, it is customary to modify the equivalent circuit to the form shown in Figure 6.2.

## Motor Torque

The torque $T$ developed by the motor is related to $P_r$ by

$$T = \frac{P_r}{\omega_r} \tag{6.9}$$

with $\omega_r$ being the angular speed of the rotor. Thus,

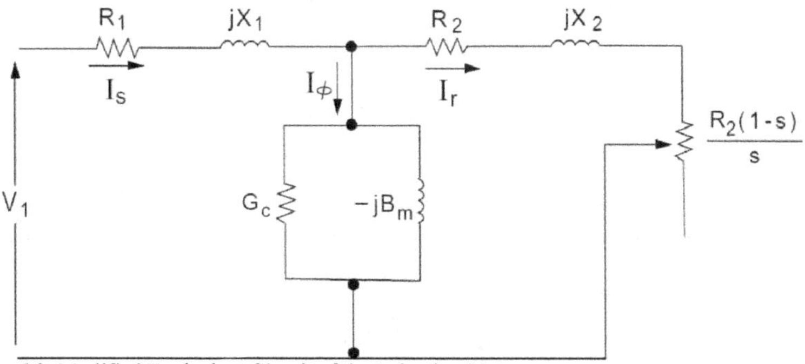


**Figure 6.2** Modified Equivalent Circuit of the Induction Motor.

$$\omega_r = \omega_s(1-s) \tag{6.10}$$

The angular synchronous speed $\omega_s$ is given by

$$\omega_s = \frac{2\pi n_s}{60} \tag{6.11}$$

As a result, the torque is given by

$$T = \frac{3I_r^2(R_2)}{s\omega_s} \tag{6.12}$$

The torque is slip-dependent. It is customary to utilize a simplified equivalent circuit for the induction motor in which the shunt branch is moved to the voltage source side. This situation is shown in Figure 6.3. The stator resistance and shunt branch can be neglected in many instances.

## Rotor Current

On the basis of the approximate equivalent circuit, we can find the rotor

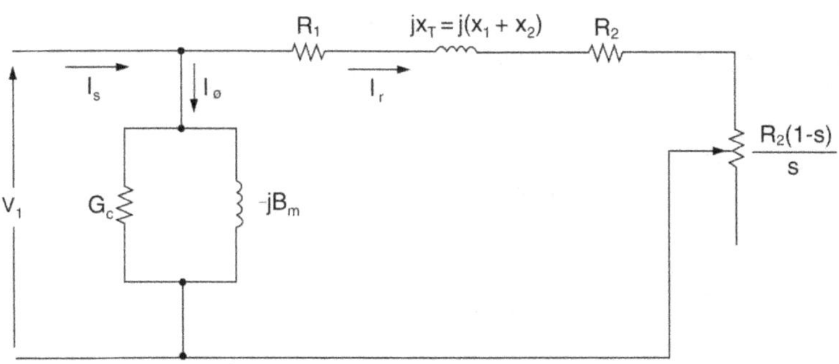


**Figure 6.3** Approximate Equivalent Circuit of the Induction Motor.

current as

$$I_r = \frac{V_1}{R_1 + \frac{R_2}{s} + jX_T} \tag{6.13}$$

At starting, we have $\omega_r = 0$; thus $s = 1$. The rotor starting current is hence given by

$$I_{r_{st}} = \frac{V_1}{(R_1 + R_2) + jX_T} \tag{6.14}$$

The starting current in much higher than the normal (or full-load) current. Depending on the motor type, the starting current can be as high as six to seven times the normal current. The power factor at starting is low compared to its value at full-load.

**Example 6.2**

A 15-hp, 220-V, three-phase, 60-Hz, six-pole, Y-connected induction motor has the following parameters per phase:

$$R_1 = 0.15 \text{ ohm}$$
$$R_2 = 0.1 \text{ ohm}$$
$$X_T = 0.5 \text{ ohm}$$
$$G_c = 6 \times 10^{-3}$$
$$B_m = 0.15 \text{ S}$$

The rotational losses are equal to the stator hysteresis and eddy-current losses. For a slip of 3 percent, find the following:

A. the line current and power factor;
B. the horsepower output;
C. the starting torque.

*Solution*

A. The voltage specified is line-to-line value as usual. Utilizing the approximate equivalent circuit of Figure 6.3, the rotor current can be seen to be given by

$$I_r = \frac{\frac{220}{\sqrt{3}}}{\left(0.15 + \frac{0.1}{0.03}\right) + j0.5}$$
$$= 36.09\angle -8.17^\circ \text{ A}$$

The no-load current $I_\phi$ is obtained as

$$I_\phi = \frac{220}{\sqrt{3}}\left(6\times10^{-3} - j0.15\right)$$
$$= 0.7621 - j19.05 \text{ A}$$

As a result, the line current (stator current) is

$$I_s = I_r + I_\phi$$
$$= 43.772\angle -33.535^\circ$$

Since $V_1$ is taken as reference, we conclude that

$$\phi_s = 33.535^\circ$$
$$\cos\phi_s = 0.8334$$

B. The air-gap power is given by

$$P_g = 3I_r^2\left(\frac{R_2}{s}\right) = 3(36.09)^2\left(\frac{0.1}{0.03}\right) = 13{,}024.881 \text{ W}$$

The mechanical power to the shaft is

$$P_m = (1-s)P_g = 12{,}634.135 \text{ W}$$

The core losses are

$$P_c = 3E_1^2(G_c) = 290.4 \text{ W}$$

The rotational losses are thus

$$P_{rl} = 290.4 \text{ W}$$

As a result, the net output mechanical power is

$$P_{\text{out}} = P_m - P_{rl}$$
$$= 12{,}343.735 \text{ W}$$

Therefore, in terms of horsepower, we get

$$\text{hp}_{\text{out}} = \frac{12{,}343.735}{746} = 16.547 \text{ hp}$$

C. At starting, $s = 1$:

$$|I_r| = \frac{\dfrac{220}{\sqrt{3}}}{|(0.15+0.1)+j0.5|} = 227.215 \text{ A}$$

$$P_g = 3(227.215)^2(0.1) = 15{,}487.997 \text{ W}$$

$$\omega_s = \frac{2\pi(60)}{3} = 40\pi$$

$$T = \frac{P_g}{\omega_s} = \frac{15{,}487.997}{40\pi} = 123.25 \text{ N.m.}$$

The following script implements Example 6.2 in MATLAB™:

```
% Example 6-2
%
V=220/3^.5;
s=0.03;
f=60;
R1=0.15;
R2=0.1;
Xt=0.5;
Gc=6*10^-3;
Bm=0.15;
Ir=V/((R1+R2/s)+i*Xt);
abs(Ir)
angle(Ir)*180/pi
Iphi=V*(Gc-i*Bm)
Is=Ir+Iphi;
abs(Is)
angle(Is)*180/pi
% V1 is taken as reference
phi_s=-angle(Is);
pf=cos(phi_s)
% B. The airgap power
Pg=3*(abs(Ir))^2*(R2/s)
% The mechanical power to the shaft
Pm=(1-s)*Pg
% The core loss
E1=V;
Pc=3*E1^2*Gc
% The rotational losses
Prl=Pc
% The net output mechanical power
Pout=Pm-Prl
hpout=Pout/746
```

MATLAB™ con't.

```
% At starting s=1
s=1;
Ir=V/((R1+R2/s)+i*Xt);
abs(Ir)
angle(Ir)*180/pi
Pg=3*(abs(Ir))^2*(R2/s)
omega_s=2*pi*f/3;
T=Pg/omega_s
```

The results obtained from MATLAB™ are as follows:

```
EDU»
ans =     36.0943
ans =     -8.1685
Iphi =     0.7621-19.0526i
ans =     43.7750
ans =    -33.5313
pf =     0.8336
Pg =    1.3028e+004
Pm =    1.2637e+004
Pc =     290.4000
Prl =    290.4000
Pout =   1.2347e+004
hpout =    16.5506
ans =    227.2150
ans =    -63.4349
Pg =          15488
T =   123.2496
```

## 6.3 TORQUE RELATIONS

The torque developed by the motor can be derived in terms of the motor parameters and slip using the expressions given before.

$$T = \frac{3|V_1|^2}{\omega_s} \frac{\frac{R_2}{s}}{\left(R_1 + \frac{R_2}{s}\right)^2 + X_T^2}$$

Neglecting stator resistance, we have

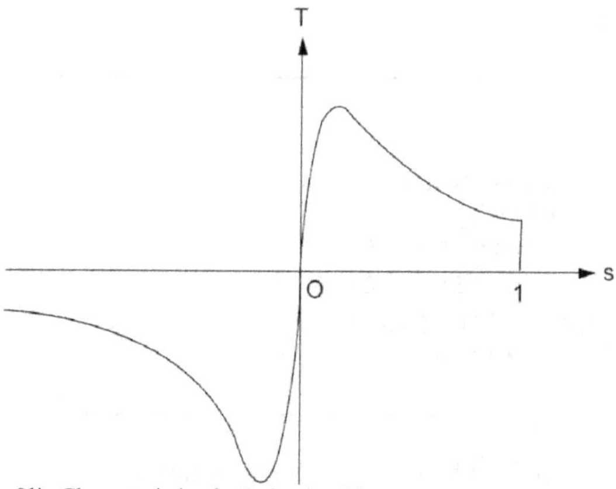


**Figure 6.4** Torque-Slip Characteristics for Induction Motor.

$$T = \frac{3|V_1|^2}{\omega_s} \frac{\frac{R_2}{s}}{\left(\frac{R_2}{s}\right)^2 + X_T^2}$$

The slip at which maximum torque occurs as

$$s_{\max_T} = \frac{R_2}{X_T} \quad \textbf{(6.15)}$$

The value of maximum torque is

$$T_{\max} = \frac{3|V_1|^2}{2\omega_s X_T} \quad \textbf{(6.16)}$$

The torque-slip variations are shown in Figure 6.4.

### Example 6.3

The resistance and reactance of a squirrel-cage induction motor rotor at standstill are 0.125 ohm per phase and 0.75 ohm per phase, respectively. Assuming a transformer ratio of unity, from the eight-pole stator having a phase voltage of 120 V at 60 Hz to the rotor secondary, calculate the following:

A. rotor starting current per phase, and
B. the value of slip producing maximum torque.

*Solution*

A. At starting, $s = 1$:

$$I_r = \frac{120}{0.125 + j0.75}$$
$$= 157.823\angle -80.538 \text{ A}$$

B.

$$s_{\max_T} = \frac{R_r}{X_T} = \frac{0.125}{0.75} = 0.1667$$

The following script implements Example 6.3 in MATLAB™:

```
% Example 6-3
% A squirrel cage induction motor
Rr=0.125; % ohm
XT=0.75; % ohm
V=120; % Volt
f=60; % Hz
% A. Rotor starting current per phase
% At starting s=1
Ir= V/(Rr+i*XT)
abs(Ir)
angle(Ir)*180/pi
% B. The value of slip producing
maximum torque
s_maxT=Rr/XT
```

The results obtained from MATLAB™ are as follows:

```
EDU»

Ir =  2.5946e+001 - 1.5568e+002I

ans =    157.8230

ans =    -80.5377

s_maxT =      0.1667
```

**Example 6.4**
The full-load slip of a squirrel-cage induction motor is 0.05, and the starting current is five times the full-load current. Neglecting the stator core and copper losses as well as the rotational losses, obtain:

A. the ratio of starting torque (st) to the full-load torque (fld), and

B. the ratio of maximum (max) to full-load torque and the corresponding slip.

*Solution*

$$s_{\text{fld}} = 0.05 \qquad \text{and} \qquad I_{\text{st}} = 5I_{\text{fld}}$$

$$\left(\frac{I_{\text{st}}}{I_{\text{fld}}}\right)^2 = \frac{\left(\frac{R_2}{0.05}\right)^2 + X_T^2}{R_2^2 + X_T^2} = (5)^2$$

This gives

$$\frac{R_2}{X_T} = \sqrt{\frac{24}{375}} \cong 0.25$$

A.

$$T = \frac{3I_r^2(R_2)}{s\omega_s}$$

$$\frac{T_{\text{st}}}{T_{\text{fld}}} = \frac{I_{\text{st}}^2}{I_{\text{fld}}^2}\left(\frac{s_{\text{fld}}}{s_{\text{st}}}\right) = (5)^2 \frac{0.05}{1} = 1.25$$

B.

$$s_{\max_T} = \frac{R_2}{X_T} = 0.25$$

$$\frac{T_{\max}}{T_{\text{fld}}} = \frac{I_{\max}^2}{I_{\text{fld}}^2}\left(\frac{s_{\text{fld}}}{s_{\max_T}}\right)$$

$$= \left(\frac{s_{\text{fld}}}{s_{\max_T}}\right)\frac{\left(\frac{R_2}{s_{\text{fld}}}\right)^2 + X_T^2}{\left(2X_T^2\right)}$$

$$= \frac{s_{\text{fld}}}{s_{\max_T}} \frac{\left(\frac{s_{\max_T}}{s_{\text{fld}}}\right)^2 + 1}{2}$$

$$= \frac{0.05}{0.25}\left[\frac{(5)^2 + 1}{2}\right]$$

Thus,

$$\frac{T_{\max}}{T_{\text{fld}}} = 2.6$$

The following script implements Example 6.4 in MATLAB™:

```
% Example 6-4
% A squirrel cage induction motor
sfld=0.05;
sst=1;
% Ist=5*Ifld;
% ratio1=Ist/Ifld=5
ratio1=5;
%
(ratio1)^2=((R2/sfld)^2+(XT)^2)/(R2^2+(
XT)^2)
% (R2/XT)^2*((1/sfld)^2-
ratio1^2)=ratio1^2-1
% ratio2=R2/XT
f=[((1/sfld)^2-ratio1^2) 0 -(ratio1^2-
1)]
ratio2=roots(f);
ratio2=ratio2(1)
% A. T=3*Ir^2*R2/(sfld*ws)
% ratio3=Tst/Tfld
ratio3=ratio1^2*(sfld/sst)
% B.
s_maxT=ratio2
%Tmax/Tfld=(Imax/Ifld)^2*(sfld/s_maxT)
%=(sfld/s_maxT)*((R2/sfld)^2+XT^2)/(2*X
T^2)
%
(Tmax/Tfld)=(sfld/s_maxT)*((s_maxT/sfld
)^2+1)/2
% ratio4=Tmax/Tfld
ratio4=(sfld/s_maxT)*((s_maxT/sfld)^2+1
)/2
```

The results obtained from MATLAB™ are as follows:

```
EDU»
f =     375      0    -24
ratio2 =       0.2530
ratio3 =       1.2500
s_maxT =       0.2530
ratio4 =       2.6286
```

## 6.4 CLASSIFICATION OF INDUCTION MOTORS

Integral-horsepower, three-phase, squirrel-cage motors are available from manufacturers' stock in a range of standard ratings up to 200 hp at standard frequencies, voltages, and speeds. (Larger motors are regarded as special-purposed.) Several standard designs are available to meet various starting and running requirements. Representative torque-speed characteristics of four designs are shown in Figure 6.5. These curves are typical of 1,800 r/min (synchronous-speed) motors in ratings from 7.5 to 200 hp.

The induction motor meets the requirements of substantially constant-speed drives. Many motor applications, however, require several speeds or a continuously adjustable range of speeds. The synchronous speed of an induction motor can be changed by (1) changing the number of poles, (2) varying the rotor resistance, or (3) inserting voltages of the appropriate frequency in the rotor circuits. A discussion of the details of speed control mechanisms is beyond the scope of this work. A common classification of induction motors is as follows.

### *Class A*

Normal starting torque, normal starting current, low slip. This design has a low-resistance, single-cage rotor. It provides good running performance at the expense of starting. The full-load slip is low and the full-load efficiency is high. The maximum torque usually is over 200 percent of full-load torque and occurs at a small slip (less than 20 percent). The starting torque at full voltage

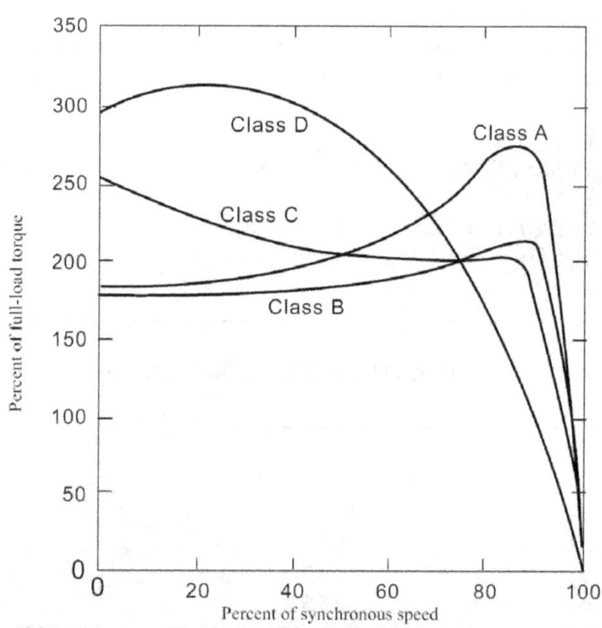


**Figure 6.5** Typical Torque-Speed Curves for 1,800 r/min General-Purpose Induction Motors.

varies from about 200 percent of full-load torque in small motors to about 100 percent in large motors. The high starting current (500 to 800 percent of full-load current when started at rated voltage) is the disadvantage of this design.

### *Class B*

Normal starting torque, low starting current, low slip. This design has approximately the same starting torque as the Class A with only 75 percent of the starting current. The full-load slip and efficiency are good (about the same as for the Class A). However, it has a slightly decreased power factor and a lower maximum torque (usually only slightly over 200 percent of full-load torque being obtainable). This is the commonest design in the 7.5 to 200-hp range of sizes used for constant-speed drives where starting-torque requirements are not severe.

### *Class C*

High starting torque, low starting current. This design has a higher starting torque with low starting current but somewhat lower running efficiency and higher slip than the Class A and Class B designs.

### *Class D*

High starting torque, high slip. This design produces very high starting torque at low starting current and high maximum torque at 50 to 100-percent slip, but runs at a high slip at full load (7 to 11 percent) and consequently has low running efficiency.

## 6.5 ROTATING MAGENTIC FIELDS IN SINGLE-PHASE INDUCTION MOTORS

To understand the operation of common single-phase induction motors, it is necessary to start by discussing two-phase induction machines. In a true two-phase machine two stator windings, labeled *AA′ and BB′*, are placed at 90° spatial displacement as shown in Figure 6.6. The voltages $\upsilon_A$ and $\upsilon_B$ form a set of balanced two-phase voltages with a 90° time (or phase) displacement. Assuming that the two windings are identical, then the resulting flux $\phi_A$ and $\phi_B$ are given by

$$\phi_A = \phi_M \cos \omega t \tag{6.17}$$

$$\phi_B = \phi_M \cos(\omega t - 90^\circ) = \phi_M \sin \omega t \tag{6.18}$$

where $\phi_M$ is the peak value of the flux. In Figure 6.6(B), the flux $\phi_A$ is shown to be at right angles to $\phi_B$ in space. It is clear that because of Eqs. (6.17) and (6.18), the phasor relation between $\phi_A$ and $\phi_B$ is shown in Figure 6.6(C) with $\phi_A$

taken as the reference phasor.

The resultant flux $\phi_P$ at a point $P$ displaced by a spatial angle $\theta$ from the reference is given by

$$\phi_P = \phi_{PA} + \phi_{PB}$$

where $\phi_{PA}$ is the component of $\phi_A$ along the $OP$ axis and $\phi_{PB}$ is the component of $\phi_B$ along the $OP$ axis, as shown in Figure 6.6(D). Here we have

$$\phi_{PA} = \phi_A \cos\theta$$
$$\phi_{PB} = \phi_B \sin\theta$$

As a result, we have

$$\phi_P = \phi_M (\cos\omega t \cos\theta + \sin\omega t \sin\theta)$$

The relationship above can be written alternatively as

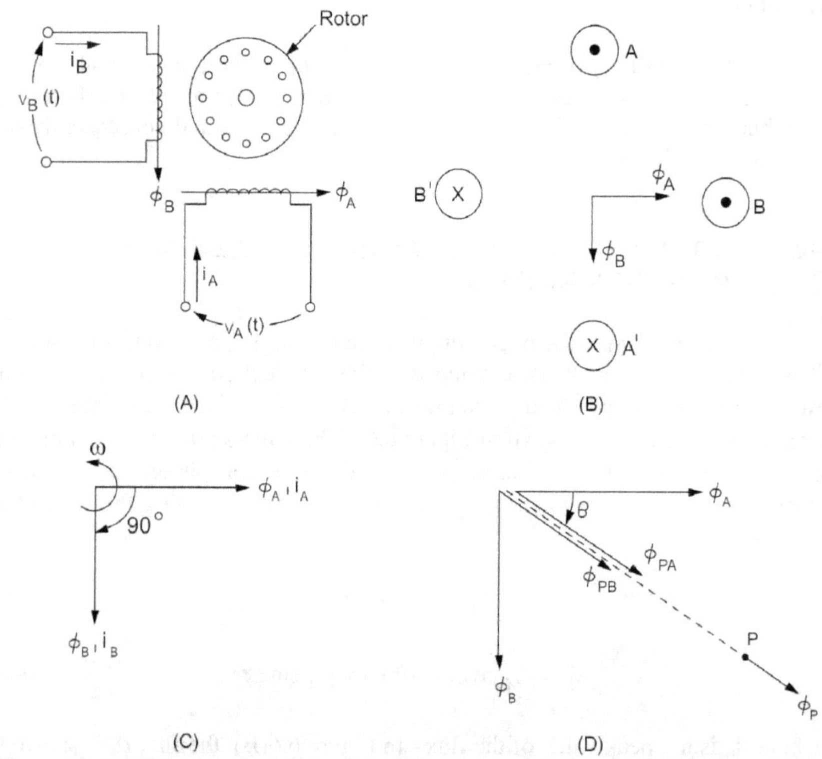


**Figure 6.6** Rotating magnetic field in a balanced two-phase stator: (A) winding schematic; (B) flux orientation; (C) phasor diagram; and (D) space phasor diagram.

$$\phi_P = \phi_M \cos(\theta - \omega t) \qquad \textbf{(6.19)}$$

The flux at point $P$ is a function of time and the spatial angle $\theta$, and has a constant amplitude $\phi_M$. this result is similar to that obtained earlier for the balanced three-phase induction motor.

The flux $\phi_P$ can be represented by a phasor $\phi_M$ that is coincident with the axis of phase $a$ at $t = 0$. The value of $\phi_P$ is $\phi_M \cos\theta$ at that instant as shown in Figure 6.7(A). At the instant $t = t_1$, the phasor $\phi_M$ has rotated an angle of $\omega t_1$ in the positive direction of $\theta$, as shown in Figure 6.7(B). The value of $\phi_P$ is seen to be $\phi_M \cos(\theta - \omega t_1)$ at that instant. It is thus clear that the flux waveform is a rotating field that travels at an angular velocity $\omega$ in the forward direction of increase in $\theta$.

The result obtained here for a two-phase stator winding set and for a three-phase stator winding set can be extended to an $N$-phase system. In this case the $N$ windings are placed at spatial angles of $2\pi/N$ and excited by sinusoidal voltages of time displacement $2\pi/N$. Our analysis proceeds as follows. The flux waveforms are given by

$$\phi_1 = \phi_M \cos \omega t$$

$$\phi_2 = \phi_M \cos\left(\omega t - \frac{2\pi}{N}\right)$$

$$\vdots$$

$$\phi_i = \phi_M \cos\left[\omega t - (i-1)\frac{2\pi}{N}\right]$$

The resultant flux at a point $P$ can be shown to be given by:

$$\phi_P = \sum_{i=1}^{N} \phi_{Pi}$$

$$\phi_P = \frac{N\phi_M}{2}\cos(\theta - \omega t) \qquad \textbf{(6.20)}$$

From (6.20), it is evident that a rotating magnetic field of constant magnitude will be produced by an $N$-phase winding excited by balanced $N$-phase currents when each phase is displaced $2\pi/N$ electrical degrees from the next phase in space.

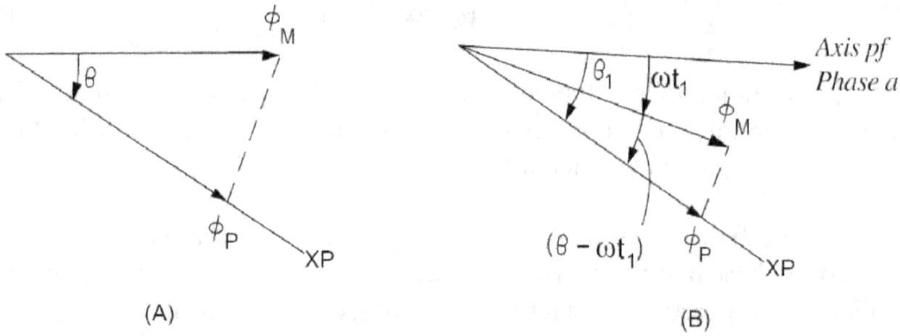


**Figure 6.7** Illustrating forward rotating magnetic field: (A) $t = 0$; (B) $t = t_1$.

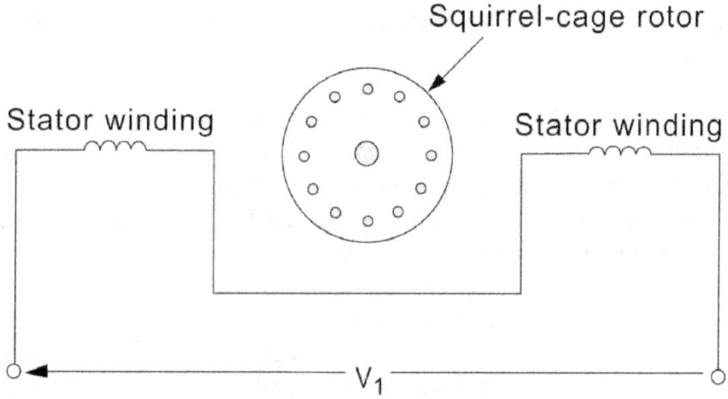


**Figure 6.8** Schematic of a single-phase induction motor.

In order to understand the operation of a single-phase induction motor, we consider the configuration shown in Figure 6.8. The stator carries a single-phase winding and the rotor is of the squirrel-cage type. This configuration corresponds to a motor that has been brought up to speed, as is discussed presently.

Let us now consider a single-phase stator winding as shown in Figure 6.9(A). The flux $\phi_A$ is given by

$$\phi_A = \phi_M \cos \omega t \tag{6.21}$$

The flux at point $P$ displaced by angle $\theta$ from the axis of phase $a$ is clearly given by

$$\phi_P = \phi_A \cos\theta$$

Using Eq. (6.21), we obtain

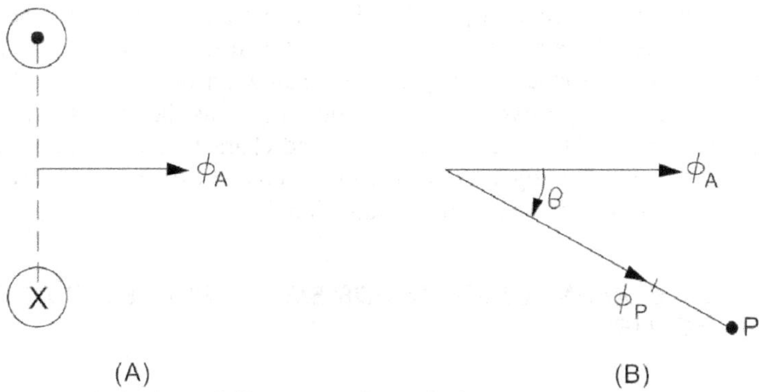


**Figure 6.9** (A) Single-phase winding; (B) the flux at a point P.

$$\phi_P = \frac{\phi_M}{2}[\cos(\theta - \omega t) + \cos(\theta + \omega t)] \tag{6.22}$$

The flux at point $P$ can therefore be seen to be the sum of two waveforms $\phi_f$ and $\phi_b$ given by

$$\phi_f = \frac{\phi_M}{2}\cos(\theta - \omega t) \tag{6.23}$$

$$\phi_b = \frac{\phi_M}{2}\cos(\theta + \omega t) \tag{6.24}$$

The waveform $\phi_f$ is of the same form as that obtained in Eq. (6.19), which was shown to be rotating in the forward direction (increase in $\theta$ from the axis of phase $a$). The only difference between Eqs. (6.23) and (6.21) is that the amplitude of $\phi_f$ is half of that of $\phi_P$ in Eq. (6.21). The subscript $f$ in Eq. (6.23) signifies the fact that cos ($\theta$ - $\omega t$) is a forward rotating wave.

Consider now the waveform $\phi_b$ of Eq. (6.24). At $t = 0$, the value of $\phi_b$ is ($\phi_M/2$) cos$\theta$ and is represented by the phasor ($\phi_M/2$), which is coincident with the axis of phase $a$ as shown in Figure 6.10(a). Note that at $t = 0$, both $\phi_f$ and $\phi_b$ are equal in value. At a time instant $t = t_1$, the phasor ($\phi_M/2$) is seen to be at angle $\omega t_1$ with the axis of phase $a$, as shown in Figure 6.9(B). The waveform $\phi_b$ can therefore be seen to be rotating at an angular velocity $\omega$ in a direction opposite to that of $\phi_f$ and we refer to $\phi_b$ as a backward-rotating magnetic field. The subscript ($b$) in Eq. (6.24) signifies the fact that cos ($\theta$ + $\omega t$) is a backward-rotating wave.

The preceding discussion reveals that in a single-phase induction machine there are two magnetic fields rotating in opposite directions. Each field produces an induction-motor torque in a direction opposite to the other. When the rotor is at rest, the forward torque is equal and opposite to the backward torque and the resulting torque is zero. A single-phase induction motor therefore cannot produce a torque when it is at rest and therefore it is not a self-starting machine. If the rotor is made to rotate by an external means, each of the two fields would produce a torque-speed characteristic similar to that of a balanced three-phase (or two-phase) induction motor, as shown in Figure 6.11 in the dashed curves. The resultant torque-speed characteristic is shown using a solid line. The foregoing argument will be confirmed once we develop an equivalent circuit for the single-phase induction motor.

## 6.6 EQUIVALENT CIRCUITS FOR SINGLE-PHASE INDUCTION MOTORS

In a single-phase induction motor, the pulsating flux wave resulting from a single winding stator MMF is equal to the sum of two rotating flux components. The first component is referred to as the forward field and has a constant amplitude equal to half of that of the stator waveform. The forward field rotates at synchronous speed. The second component, referred to as the backward field, is of the same constant amplitude but rotates in the opposite (or backward) direction at synchronous speed. Each component induces its own rotor current and creates induction motor action in the same manner as in a balanced three-phase induction motor. It is on this basis that we conceive of the circuit model of Figure 6.12(A). Note that $R_1$ and $X_1$ are the stator resistance and leakage reactance, respectively, and $V_1$ is the stator input voltage. The EMF $E_1$ is assumed to be the sum of two components, $E_{1_f}$ and $E_{1_b}$, corresponding to the forward and backward field waves, respectively. Note that since the two waves have the same amplitude, we have

$$E_{1_f} = E_{1_b} = \frac{E_1}{2} \tag{6.25}$$

The rotor circuit is modeled as the two blocks shown in Figure 6.12(A), representing the rotor forward circuit model $Z_f$ and the rotor backward circuit model $Z_b$, respectively.

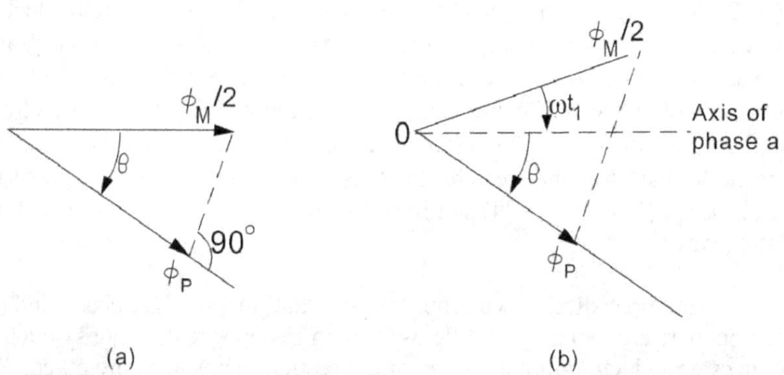


**Figure 6.10** Showing that $\phi_b$ is a backward-rotating wave: (A) $t = 0$; (B) $t = t_1$.

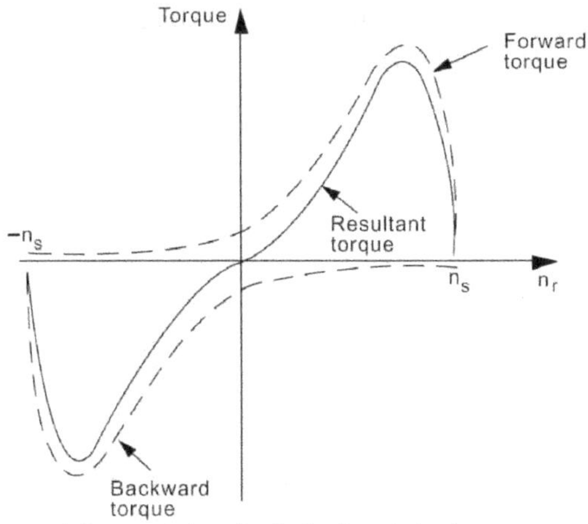


**Figure 6.11** Torque-speed characteristics of a single-phase induction motor.

The model of the rotor circuit for the forward rotating wave $Z_f$ is simple since we are essentially dealing with induction-motor action and the rotor is set in motion in the same direction as the stator synchronous speed. The model of $Z_f$ is shown in Figure 6.12(B) and is similar to that of the rotor of a balanced three-phase induction motor. The impedances dealt with are half of the actual values to account for the division of $E_1$ into two equal voltages. In this model, $X_m$ is the magnetizing reactance, and $R_2'$ and $X_2'$ are rotor resistance and leakage reactance, both referring to the stator side. The slip $s_f$ is given by

$$s_f = \frac{n_s - n_r}{n_s} \tag{6.26}$$

This is the standard definition of slip as the rotor is revolving in the same direction as that of the forward flux wave.

The model of the rotor circuit for the backward-rotating wave $Z_b$ is shown in Figure 6.12(C) and is similar to that of $Z_f$, with the exception of the backward slip, denoted by $s_b$. The backward wave is rotating at a speed of $-n_s$, and the rotor is rotating at $n_r$. We thus have

$$s_b = \frac{(-n_s) - n_r}{-n_s} = 1 + \frac{n_r}{n_s} \tag{6.27}$$

Using Eq. (6.26), we have

$$s_f = 1 - \frac{n_r}{n_s} \tag{6.28}$$

As a result, we conclude that the slip of the rotor with respect to the backward wave is related to its slip with respect to the forward wave by

$$s_b = 2 - s_f \tag{6.29}$$

We now let $s$ be the forward slip,

$$s_f = s \tag{6.30}$$

and thus

$$s_b = 2 - s \tag{6.31}$$

On the basis of Eqs. (6.30) and (6.31), a complete equivalent circuit as shown in Figure 6.12(D) is now available. The core losses in the present model are treated separately in the same manner as the rotational losses.

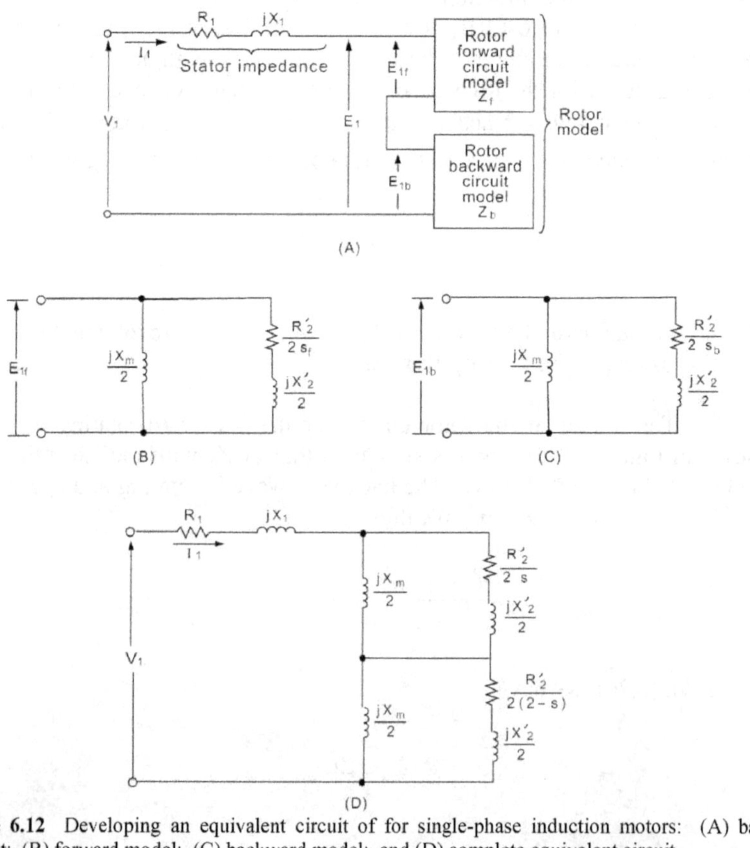


**Figure 6.12** Developing an equivalent circuit of for single-phase induction motors: (A) basic concept; (B) forward model; (C) backward model; and (D) complete equivalent circuit.

The forward impedance $Z_f$ is obtained as the parallel combination of $(jX_m/2)$ and $[(R_2'/2s)+j(X_2'/2)]$, given by

$$Z_f = \frac{j(X_m/2)[(R_2'/2s)+j(X_2'/2)]}{(R_2'/2s)+j[(X_m+X_2')/2]} \tag{6.32}$$

Similarly, for the backward impedance, we get

$$Z_b = \frac{j(X_m/2)\{[R_2'/2(2-s)]+j(X_2'/2)]\}}{[R_2'/2(2-s)+j[(X_m+X_2')/2]} \tag{6.33}$$

Note that with the rotor at rest, $n_r = 0$, and thus with $s = 1$, we get $Z_f = Z_b$.

**Example 6.5**

The following parameters are available for a 60-Hz four-pole single-phase 110-V ½-hp induction motor:

$$R_1 = 1.5\ \Omega \qquad R_2' = 3\ \Omega$$
$$X_1 = 2.4\ \Omega \qquad X_2' = 2.4\ \Omega$$
$$X_m = 73.4\ \Omega$$

Calculate $Z_f$, $Z_b$, and the input impedance of the motor at a slip of 0.05.

*Solution*

$$Z_f = \frac{j36.7(30+j1.2)}{30+j37.9} = 22.796\angle 40.654^\circ$$
$$= 17.294 + j14.851\,\Omega$$

The result above is a direct application of Eq. (6.32). Similarly, using Eq. (6.33), we get

$$Z_b = \frac{j36.7[(1.5/1.95)+j1.2]}{(1.5/1.95)+j37.9} = 1.38\angle 58.502^\circ\ \Omega$$
$$= 0.721 + j1.766\,\Omega$$

We observe here that $|Z_f|$ is much larger than $|Z_b|$ at this slip, in contrast to the situation at starting ($s = 1$), for which $Z_f = Z_b$.

The input impedance $Z_i$ is obtained as

$$Z_i = Z_1 + Z_f + Z_b = 19.515 + j18.428$$
$$= 26.841\angle 43.36^\circ\ \Omega$$

Equations (6.32) and (6.33) yield the forward and backward

impedances on the basis of complex number arithmetic. The results can be written in the rectangular forms

$$Z_f = R_f + jX_f \tag{6.34}$$

and

$$Z_b = R_b + jX_b \tag{6.35}$$

Using Eq. (6.32), we can write

$$2R_f = \frac{a_f X_m^2}{a_f^2 + X_t^2} \tag{6.36}$$

and

$$X_f = \frac{R_f}{a_f X_m}(a_f^2 + X_t X_2') \tag{6.37}$$

where

$$a_f = \frac{R_2'}{s} \tag{6.38}$$

$$X_t = X_2' + X_m \tag{6.39}$$

In a similar manner we have, using Eq. (6.33),

$$2R_b = \frac{a_b X_m^2}{a_b^2 + X_t^2} \tag{6.40}$$

$$X_b = \frac{R_b}{a_b X_m}(a_b^2 + X_t X_2') \tag{6.41}$$

where

$$a_b = \frac{R_2'}{2-s} \tag{6.42}$$

It is often desirable to introduce some approximations in the formulas just derived. As is the usual case, for $X_t > 10\ a_b$, we can write an approximation to Eq. (6.40) as

$$2R_b \cong a_b \left( \frac{X_m}{X_t} \right)^2 \tag{6.43}$$

As a result, by substitution in Eq. (6.41), we get

$$X_b \cong \frac{X_2' X_m}{2X_t} + \frac{a_b R_b}{X_m} \tag{6.44}$$

We can introduce further simplifications by assuming that $X_m/X_t \cong 1$, to obtain from Eq. (6.43)

$$2R_b \cong a_b = \frac{R_2'}{2-s} \tag{6.45}$$

Equation (6.44) reduces to the approximate form

$$X_b \cong \frac{X_2'}{2} + \frac{a_b^2}{2X_m} \tag{6.46}$$

Neglecting the second term in Eq. (6.46), we obtain the most simplified representation of the backward impedance as

$$R_b = \frac{R_2'}{2(2-s)} \tag{6.47}$$

$$X_b = \frac{X_2'}{2} \tag{6.48}$$

Equations (6.47) and (6.48) imply that $X_m/2$ is considered an open circuit in the backward field circuit, as shown in Figure 6.13.

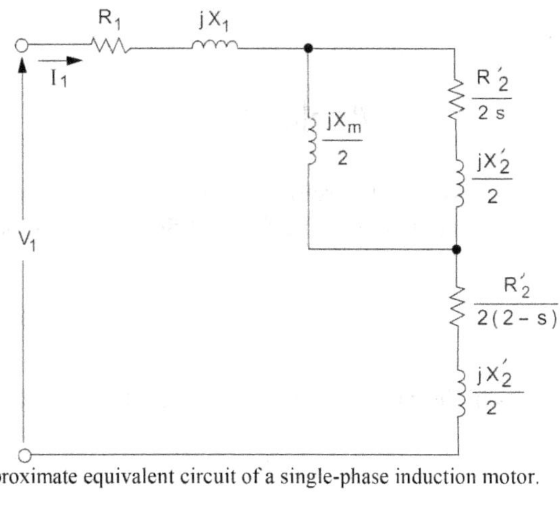


**Figure 6.13** Approximate equivalent circuit of a single-phase induction motor.

## 6.7 POWER AND TORQUE RELATIONS

The development of an equivalent-circuit model of a running single-phase induction motor enables us to quantify power and torque relations in a simple way. The power input to the stator $P_i$ is given by

$$P_i = V_1 I_1 \cos\phi_1 \tag{6.49}$$

where $\phi_1$ is the phase angle between $V_1$ and $I_1$. Part of this power will be dissipated in stator ohmic losses, $P_{\ell_s}$, given by

$$P_{\ell_s} = I_1^2 R_1 \tag{6.50}$$

The core losses will be accounted for as a fixed loss and is treated in the same manner as the rotational losses at the end of the analysis. The air-gap power $P_g$ is thus given by

$$P_g = P_i - P_{\ell_s} \tag{6.51}$$

The air-gap power is the power input to the rotor circuit and can be visualized to be made up of two components. The first component is the power taken up by the forward field and is denoted by $P_{gf}$, and the second is the backward field power denoted by $P_{gb}$. Thus we have

$$P_g = P_{gf} + P_{gb} \tag{6.52}$$

As we have modeled the forward field circuit by an impedance $Z_f$, it is natural to write

$$P_{gf} = |I_1|^2 R_f \tag{6.53}$$

Similarly, we write

$$P_{gb} = |I_1|^2 R_b \tag{6.54}$$

The ohmic losses in the rotor circuit are treated in a similar manner. The losses in the rotor circuit due to the forward field $P_{\ell_{rf}}$ can be written as

$$P_{\ell_{rf}} = s_f P_{gf} \tag{6.55}$$

Similarly, the losses in the rotor circuit due to the backward field are written as

$$P_{\ell_{rb}} = s_b P_{gb} \tag{6.56}$$

Equations (6.55) and (6.56) are based on arguments similar to those used with the balanced three-phase induction motor. Specifically, the total rotor equivalent resistance in the forward circuit is given by

$$R_{rf} = \frac{R_2'}{2s_f} \tag{6.57}$$

This is written as

$$R_{rf} = \frac{R_2'}{2} + \frac{R_2'(1-s_f)}{2s_f} \tag{6.58}$$

The first term corresponds to the rotor ohmic loss due to the forward field and the second represents the power to mechanical load and fixed losses. It is clear from Figure 6.14 that

$$P_{\ell_{rf}} = \left|I_{rf}\right|^2 \frac{R_2'}{2} \tag{6.59}$$

and

$$P_{gf} = \left|I_{rf}\right|^2 \frac{R_2'}{2s_f} \tag{6.60}$$

Combining Eqs. (6.59) and (6.60), we get Eq. (6.55). A similar argument leads to Eq. (6.56). It is noted here that Eqs. (6.53) and (6.60) are equivalent, since the active power to the rotor circuit is consumed only in the right-hand branch, with $jX_m/2$ being a reactive element.

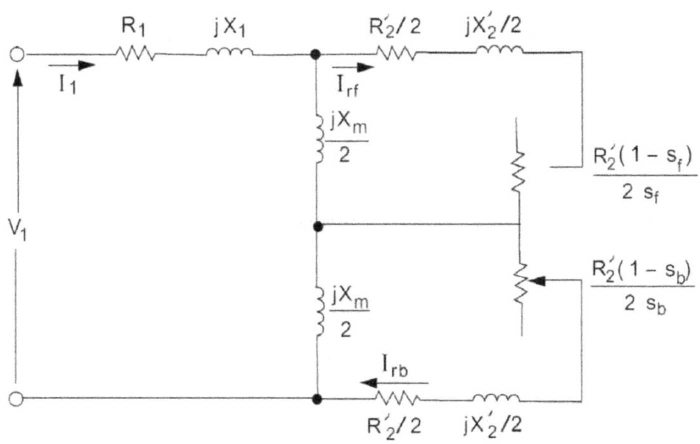


**Figure 6.14** Equivalent circuit of single-phase induction motor showing rotor loss components in the forward and backward circuits.

The net power form the rotor circuit is denoted by $P_m$ and is given by

$$P_m = P_{mf} + P_{mb} \tag{6.61}$$

The component $P_{mf}$ is due to the forward circuit and is given by

$$P_{mf} = P_{gf} - P_{\ell_{rf}} \tag{6.62}$$

Using Eq. (6.55), we get

$$P_{mf} = (1 - s_f)P_{gf} \tag{6.63}$$

Similarly, $P_{mb}$ is due to the backward circuit and is given by

$$P_{mb} = P_{gb} - P_{\ell_{rb}} \tag{6.64}$$

Using Eq. (6.56), we get

$$P_{mb} = (1 - s_b)P_{gb} \tag{6.65}$$

Recall that

$$s_f = s$$
$$s_b = 2 - s$$

As a result,

$$P_{mf} = (1 - s)P_{gf} \tag{6.66}$$

$$P_{mb} = (s - 1)P_{gb} \tag{6.67}$$

We now substitute Eqs. (6.66) and (6.67) into Eq. (6.61), to obtain

$$P_m = (1 - s)(P_{gf} - P_{gb}) \tag{6.68}$$

The shaft power output $P_o$ can now be written as

$$P_o = P_m - P_{\text{rot}} - P_{\text{core}} \tag{6.69}$$

The rotational losses are denoted by $P_{\text{rot}}$ and the core losses are denoted by $P_{\text{core}}$.

The output torque $T_o$ is obtained as

$$T_o = \frac{P_o}{\omega_r} \tag{6.70}$$

If fixed losses are neglected, then

$$T_m = \frac{P_m}{\omega_s(1-s)} \tag{6.71}$$

As a result, using Eq. (6.68), we get

$$T_m = \frac{1}{\omega_s}(P_{gf} - P_{gb}) \tag{6.72}$$

The torque due to the forward field is

$$T_{mf} = \frac{P_{mf}}{\omega_r} = \frac{P_{gf}}{\omega_s} \tag{6.73}$$

The torque due to the backward field is

$$T_{mb} = \frac{P_{mb}}{\omega_r} = -\frac{P_{gb}}{\omega_s} \tag{6.74}$$

It is thus clear that the net mechanical torque is the algebraic sum of a forward torque $T_{mf}$ (positive) and a backward torque $T_{mb}$ (negative). Note that at starting, $s = 1$ and $R_f = R_b$, and as a result $P_{gf} = P_{gb}$, giving zero output torque. This confirms our earlier statements about the need for starting mechanisms for a single-phase induction motor. This is discussed in the next section.

**Example 6.6**

For the single-phase induction motor of Example 6.5, it is necessary to find the power and torque output and the efficiency when running at a slip of 5 percent. Neglect core and rotational losses.

*Solution*

In Example 6.5 we obtained

$$Z_i = 26.841\angle 43.36^\circ$$

As a result, with $V_1 = 110\angle 0$, we obtain

$$I_1 = \frac{110\angle 0}{26.841\angle 43.36^\circ} = 4.098\angle -43.36^\circ \text{ A}$$

The power factor is thus

$$\cos\phi_1 = \cos 43.36^\circ = 0.727$$

The power input is

$$P_1 = V_1 I_1 \cos\phi_1 = 327.76 \text{ W}$$

We have from Example 6.5 for $s = 0.05$,

$$R_f = 17.294\ \Omega \qquad R_b = 0.721\ \Omega$$

Thus we have

$$P_{gf} = |I_1|^2 R_f = (4.098)^2(17.294) = 290.46 \text{ W}$$
$$P_{gb} = |I_1|^2 R_b = (4.098)^2(0.721) = 12.109 \text{ W}$$

The output power is thus obtained as

$$\begin{aligned} P_m &= (1-s)(P_{gf} - P_{gb}) \\ &= 0.95(290.46 - 12.109) = 264.43 \text{ W} \end{aligned}$$

As we have a four-pole machine, we get

$$\begin{aligned} n_s &= \frac{120(60)}{4} = 1800 \text{ r/min} \\ \omega_s &= \frac{2\pi n_s}{60} = 188.5 \text{ rad/s} \end{aligned}$$

The output torque is therefore obtained as

$$\begin{aligned} T_m &= \frac{1}{\omega_s}(P_{gf} - P_{gb}) \\ &= \frac{290.46 - 12.109}{188.5} = 1.4767 \text{ N}\cdot\text{m} \end{aligned}$$

The efficiency is now calculated as

$$\eta = \frac{P_m}{P_1} = \frac{264.43}{327.76} = 0.8068$$

It is instructive to account for the losses in the motor. Here we have the stator ohmic losses obtained as

$$P_{\ell_s} = |I_1|^2 R_1 = (4.098)^2 (1.5) = 25.193 \text{ W}$$

The forward rotor losses are

$$P_{\ell_{rf}} = sP_{gf} = 0.05(290.46) = 14.523 \text{ W}$$

The backward rotor losses are

$$P_{\ell_{rb}} = (2-s)P_{gb} = 1.95(12.109) = 23.613 \text{ W}$$

The sum of the losses is

$$P_\ell = 25.193 + 14.523 + 23.613 = 63.329 \text{ W}$$

The power output and losses should match the power input

$$P_m + P_\ell = 264.43 + 63.33 = 327.76 \text{ W}$$

which is indeed the case.

**Example 6.7**

A single-phase induction motor takes an input power of 490 W at a power factor of 0.57 lagging from a 110-V supply when running at a slip of 5 percent. Assume that the rotor resistance and reactance are 1.78 Ω and 1.28 Ω, respectively, and that the magnetizing reactance is 25 Ω. Find the resistance and reactance of the stator.

*Solution*

The equivalent circuit of the motor yields

$$Z_f = \frac{\{[1.78/2(0.5)] + j0.64\}(j12.5)}{[1.78/2(0.05)] + j(0.64 + 12.5)} = 5.6818 + j8.3057$$

$$Z_b = \frac{\{[1.78/2(1.95)] + j0.64\}(j12.5)}{[1.78/2(1.95)] + j(0.64 + 12.5)} = 0.4125 + j0.6232$$

As a result of the problem specifications

$$P_i = 490 \text{ W} \qquad \cos\phi = 0.57 \qquad V = 110$$

$$I_i = \frac{P_i}{V\cos\phi} = \frac{590}{110(0.57)} = 7.815\angle -55.2488^\circ$$

Thus the input impedance is

$$Z_i = \frac{V}{I_i} = \frac{110}{7.815}\angle 55.2498^\circ = 8.023 + j11.5651$$

The stator impedance is obtained as

$$Z_1 = Z_i - (Z_f + Z_b) = 1.9287 + j2.636\,\Omega$$

## 6.8 STARTING SINGLE-PHASE INDUCTION MOTORS

We have shown earlier that a single-phase induction motor with one stator winding is not capable of producing a torque at starting [see, for example, Eq. (6.68) with $s = 1$]. Once the motor is running, it will continue to do so, since the forward field torque dominates that of the backward field component. We have also seen that with two stator windings that are displaced by 90° in space and with two-phase excitation a purely forward rotating field is produced, and this form of a motor (like the balanced three-phase motor) is self-starting.

Methods of starting a single-phase induction motor rely on the fact that given two stator windings displaced by 90° in space, a starting torque will result if the flux in one of the windings lags that of the other by a certain phase angle $\psi$. To verify this, we consider the situation shown in Figure 6.15. Assume that

$$\phi_A = \phi_M \cos\omega t \tag{6.75}$$

$$\phi_B = \phi_M \cos(\omega t - \psi) \tag{6.76}$$

Clearly, the flux at $P$ is given by the sum of $\phi_{PA}$ and $\phi_{PB}$

$$\phi_{P_A} = \frac{\phi_M}{2}[\cos(\theta - \omega t) + \cos(\theta + \omega t)] \tag{6.77}$$

$$\begin{aligned}\phi_{P_B} = \frac{\phi_M}{2}\{&\cos\psi[\sin(\theta + \omega t) + \sin(\theta - \omega t)] \\ &+ \sin\psi[\cos(\theta - \omega t) - \cos(\theta + \omega t)]\}\end{aligned} \tag{6.78}$$

The flux at $P$ is therefore obtained as

$$\begin{aligned}\phi_P = \frac{\phi_M}{2}[&a_{f_r}\cos(\theta - \omega t) + a_{f_i}\sin(\theta - \omega t) \\ &+ a_{b_r}\cos(\theta + \omega t) + a_{b_r}\sin(\theta + \omega t)]\end{aligned} \tag{6.79}$$

where we have

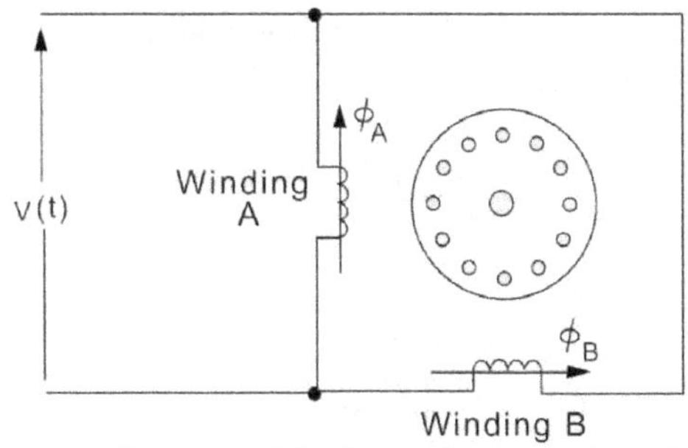


**Figure 6.15** Two stator windings to explain the starting mechanism of single-phase induction motors.

$$a_{f_r} = 1 + \sin\psi$$
$$a_{f_i} = \cos\psi$$
$$a_{b_r} = 1 - \sin\psi$$
$$a_{b_i} = \cos\psi$$

Note that we can also define the magnitudes $a_f$ and $a_b$ by

$$a_f^2 = a_{f_r}^2 + a_{f_r}^2 = 2(1 + \sin\psi) \tag{6.80}$$

$$a_b^2 = a_{b_r}^2 + a_{b_r}^2 = 2(1 - \sin\psi) \tag{6.81}$$

The angles $\alpha_f$ and $\alpha_b$ are defined next by

$$\cos\alpha_f = \frac{a_{f_r}}{a_f} = \sqrt{\frac{1 + \sin\psi}{2}}$$

$$\cos\alpha_b = \frac{a_{b_r}}{a_b} = \sqrt{\frac{1 - \sin\psi}{2}}$$

$$\sin\alpha_f = \frac{a_{f_i}}{a_f} = \frac{\cos\psi}{\sqrt{2(1 + \sin\psi)}}$$

$$\sin\alpha_b = \frac{a_{b_i}}{a_b} = \frac{\cos\psi}{\sqrt{2(1 - \sin\psi)}}$$

We can now write the flux $\phi_P$ as

$$\phi_P = \frac{\phi_M}{2}[a_f \cos(\theta - \omega t + \alpha_f) + a_b \cos(\theta + \omega t - \alpha_b)] \tag{6.82}$$

It is clear that $\phi_P$ is the sum of a forward rotating component $\phi_f$ and a backward rotating component $\phi_b$ given by

$$\phi_P = \phi_f(t) + \phi_b(t) \tag{6.83}$$

where

$$\phi_f(t) = \frac{a_f \phi_M}{2} \cos(\theta - \omega t + \alpha_f) \tag{6.84}$$

$$\phi_b(t) = \frac{a_b \phi_M}{2} \cos(\theta + \omega t - \alpha_b) \tag{6.85}$$

Let us note here that from Eqs. (6.80) and (6.81), we can see that

$$a_f > a_b \tag{6.86}$$

As a result, the magnitude of the forward rotating wave is larger than that of the backward rotating wave. It is clear that for the arrangement of Figure 6.15, a starting torque should result. This is the basis of the starting mechanisms for single-phase induction motors.

## 6.9 SINGLE-PHASE INDUCTION MOTOR TYPES

Single-phase induction motors are referred to by names that describe the method of starting. A number of types of single-phase induction motors are now discussed.

### Split-Phase Motors

A single-phase induction motor with two distinct windings on the stator that are displaced in space by 90 electrical degrees is called a split-phase motor. The main (or running) winding has a lower *R*/*X* ratio than the auxiliary (or starting) winding. A starting switch disconnects the auxiliary windings when the motor is running at approximately 75 to 80 percent of synchronous speed. The switch is centrifugally operated. The rotor of a split-phase motor is of the squirrel-cage type. At starting, the two windings are connected in parallel across the line as shown in Figure 6.16.

The split-phase design is one of the oldest single-phase motors and is most widely used in the ratings of 0.05 to 0.33 hp. A split-phase motor is used in machine tools, washing machines, oil burners, and blowers, to name just a few of its applications.

The torque-speed characteristic of a typical split-phase induction motor

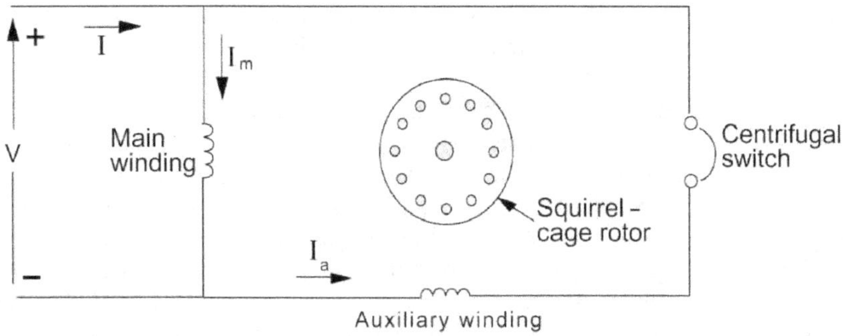


**Figure 6.16** Schematic diagram of a split-phase induction motor.

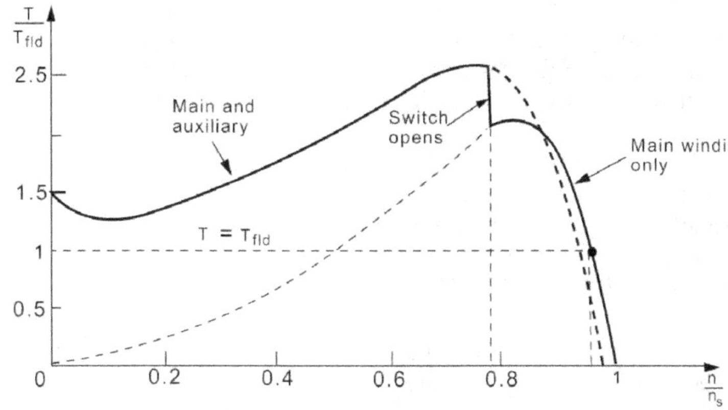


**Figure 6.17** Torque-speed characteristic of a split-phase induction motor.

is shown in Figure 6.17. At starting the torque is about 150 percent of its full-load value. As the motor speed picks up, the torque is increased (except for a slight decrease at low speed) and may reach higher than 250 percent of full-load value. The switch is opened and the motor runs on its main winding alone and the motor reaches its equilibrium speed when the torque developed is matched by the load.

## Capacitor-Start Motors

The class of single-phase induction motors in which the auxiliary winding is connected in series with a capacitor is referred to as that of capacitor motors. The auxiliary winding is placed 90 electrical degrees form the main winding. There are three distinct types of capacitor motors in common practice. The first type, which we discuss presently, employs the auxiliary winding and capacitor only during starting and is thus called a capacitor-start motor. It is thus clear that a centrifugal switch that opens at 75 to 80 percent of synchronous speed is used in the auxiliary winding circuit (sometimes called the capacitor phase). A sketch of the capacitor-start motor connection is shown in Figure 6.18. A commercial capacitor-start motor is not simply a split-phase motor with a capacitor inserted in the auxiliary circuit but is a specially designed motor that produces higher torque than the corresponding split-phase version.

Capacitor-start motors are extremely popular and are available in all ratings from 0.125 hp up. For ratings at 1/3 hp and above, capacitor-start motors are wound as dual-voltage so that they can be operated on either a 115- or a 230-V supply. In this case, the main winding is made of two sections that are connected in series for 230-V operation or in parallel for 115-V operation. The auxiliary winding in a dual-voltage motor is made of one section which is connected in parallel with one section of the main winding for 230-V operation. The auxiliary winding in a dual-voltage motor is made of one section which is connected in parallel with one section of the main winding for 230-V operation.

It is important to realize that the capacitor voltage increases rapidly above the switch-open speed and the capacitor can be damaged if the centrifugal switch fails to open at the designed speed. It is also important that switches not flutter, as this causes a dangerous rise in the voltage across the capacitor.

A typical torque-speed characteristic for a capacitor-start single-phase induction motor is shown in Figure 6.19. The starting torque is very high, which is a desirable feature of this type of motor.

## Permanent-Split Capacitor Motors

The second type of capacitor motors is referred to as the permanent-split capacitor motor, where the auxiliary winding and the capacitor are retained at normal running speed. This motor is used for special-purpose applications requiring high torque and is available in ratings from $10^{-3}$ to $\frac{1}{3}$ - $\frac{3}{4}$ hp. A schematic of the permanent-split capacitor motor is shown in Figure 6.20.

A typical torque-speed characteristic for a permanent-split capacitor motor is shown in Figure 6.21. The starting torque is noticeably low since the capacitance is a compromise between best running and starting conditions. The next type of motor overcomes this difficulty.

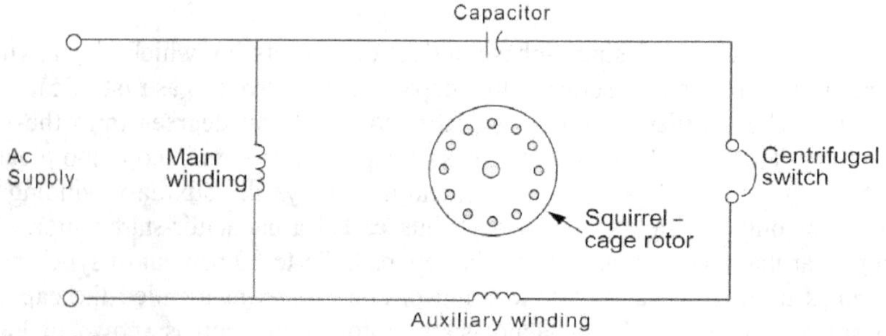


**Figure 6.18** Capacitor-start motor.

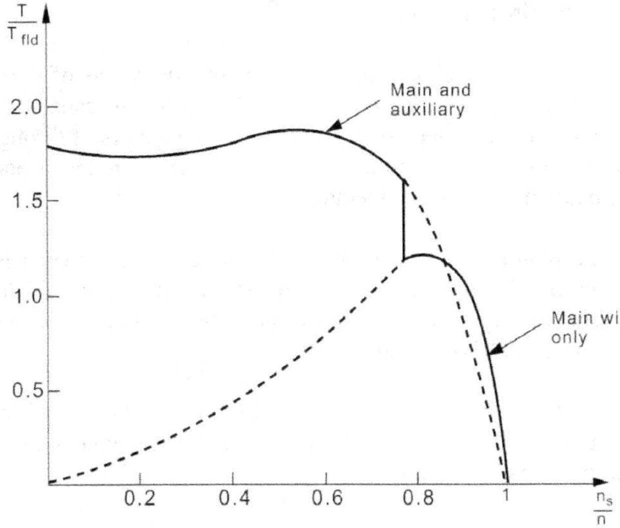


**Figure 6.19** Torque-speed characteristic of a capacitor-start motor.

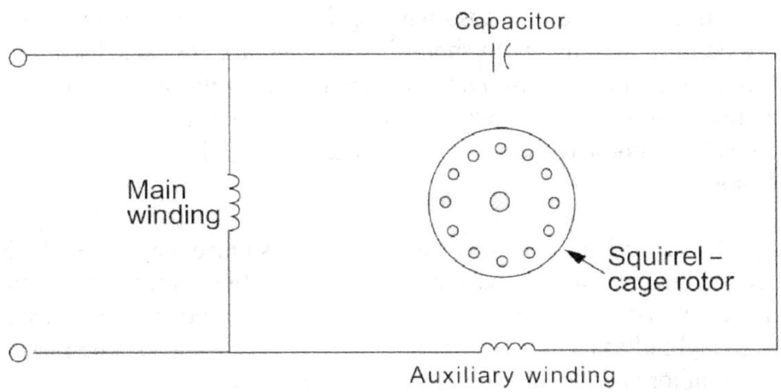


**Figure 6.20** Permanent-split capacitor motor.

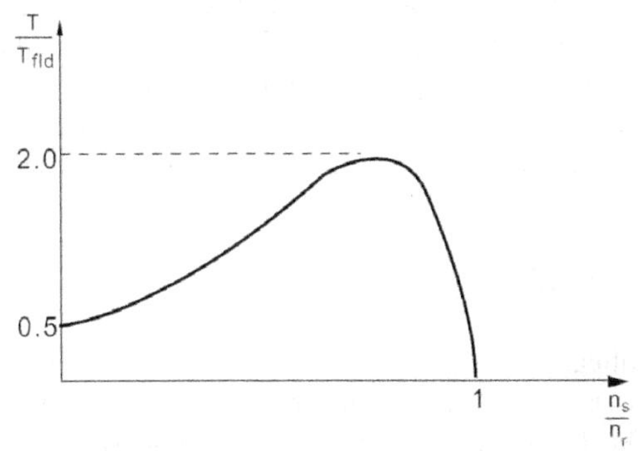


**Figure 6.21** Torque-speed characteristic of a permanent-split single-phase induction motor.

## Two-Value Capacitor Motors

A two-value capacitor motor starts with one value of capacitors in series with the auxiliary winding and runs with a different capacitance value. This change can be done either using two separate capacitors or through the use of an autotransformer. This motor has been replaced by the capacitor-start motor for applications such as refrigerators and compressors.

For the motor using an autotransformer, a transfer switch is used to change the tap on the autotransformer, as shown in Figure 6.22(A). This arrangement appears to be obsolete now and the two-capacitor mechanism illustrated in Figure 6.22(B) is used.

A typical torque-speed characteristic for a two-value capacitor motor is shown in Figure 6.23. Note that optimum starting and running conditions can be accomplished in this type of motor.

## Repulsion-Type Motors

A repulsion motor is a single-phase motor with power connected to the stator winding and a rotor whose winding is connected to a commutator. The brushes on the commutator are short-circuited and are positioned such that there is an angle of 20 to 30° between the magnetic axis of the stator winding and the magnetic axis of the rotor winding. A representative torque-speed characteristic for a repulsion motor is shown in Figure 6.24. A repulsion motor is a variable-speed motor.

If in addition to the repulsion winding, a squirrel-cage type of winding is embedded in the rotor, we have a repulsion-induction motor. The torque-speed characteristic for a repulsion-induction motor is shown in Figure 6.25 and can be thought of as a combination of the characteristics of a single-phase induction motor and that of a straight repulsion motor.

A repulsion-start induction motor is a single-phase motor with the same windings as a repulsion motor, but at a certain speed the rotor winding is short circuited to give the equivalent of a squirrel-cage winding. The repulsion-start motor is the first type of single-phase motors that gained wide acceptance. In recent years, however, it has been replaced by capacitor-type motors. A typical torque-speed characteristic of a repulsion-start induction motor is shown in Figure 6.26.

## Shaded-Pole Induction Motors

For applications requiring low power of ¼ hp or less, a shaded-pole induction motor is the standard general-purpose device for constant-speed applications. The torque characteristics of a shaded-pole motor are similar to those of a permanent-split capacitor motor as shown in Figure 6.27.

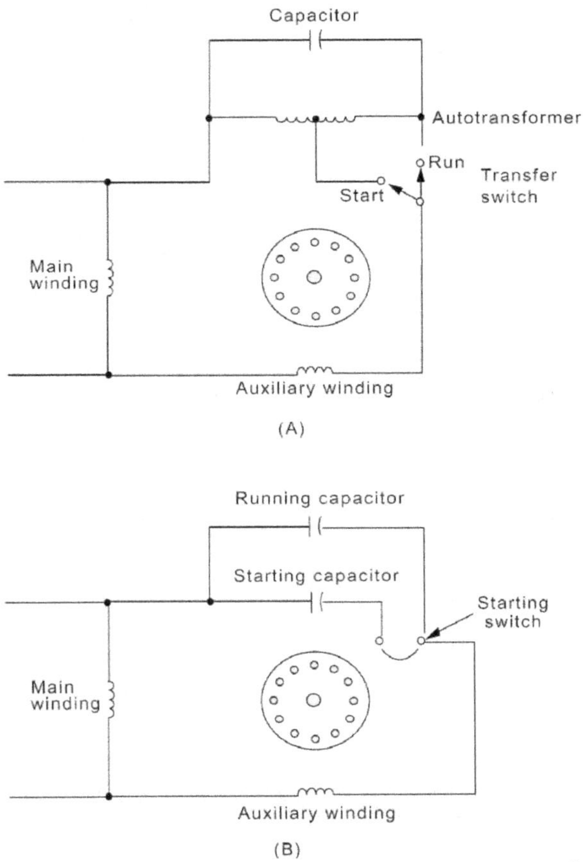


**Figure 6.22** Two-value capacitor motor: (A) autotransformer type; (B) two-capacitor type.

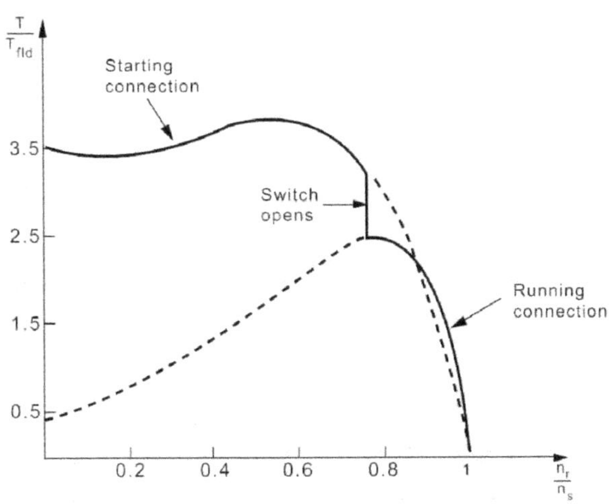


**Figure 6.23** Torque-speed characteristic of a two-value capacitor motor.

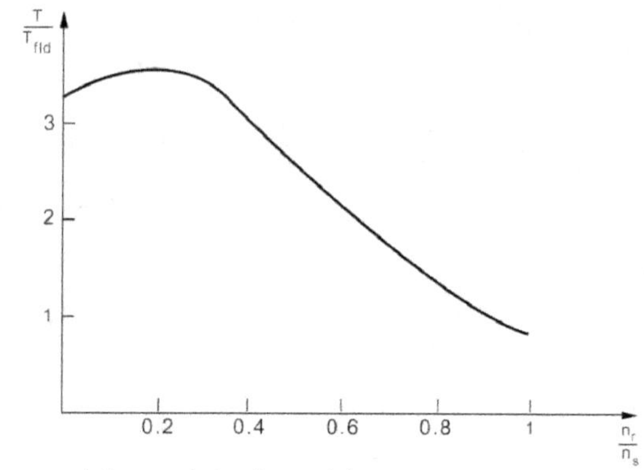


**Figure 6.24** Torque-speed characteristic of a repulsion motor.

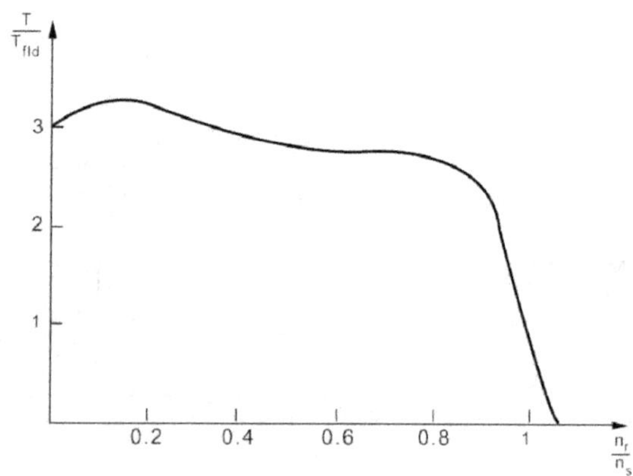


**Figure 6.25** Torque-speed characteristic of a repulsion-induction motor.

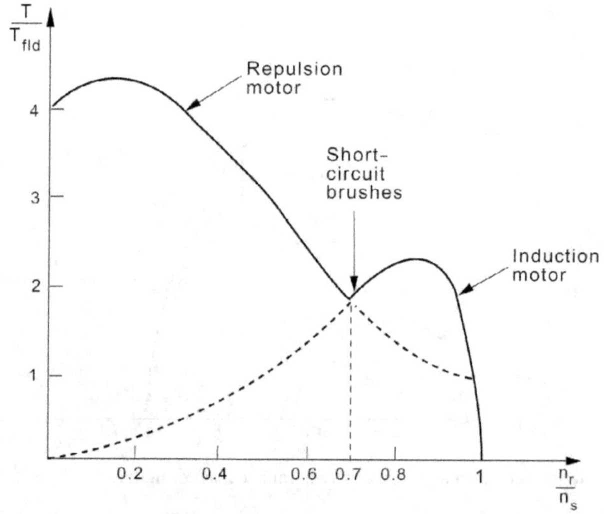


**Figure 6.26** Torque-speed characteristic of a repulsion-start single-phase induction motor.

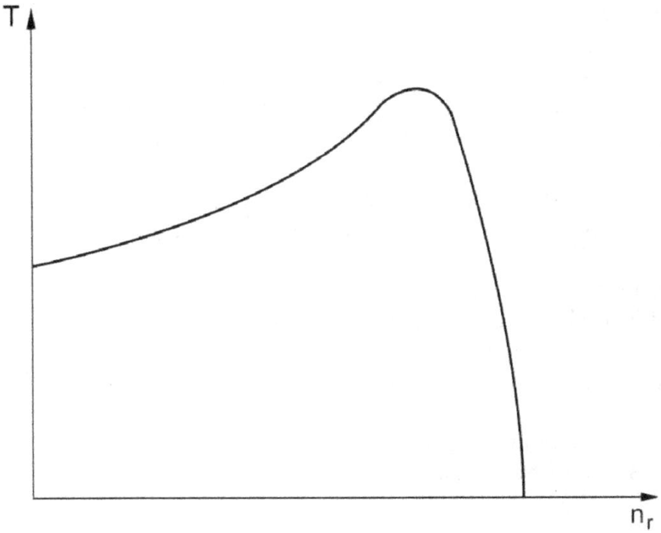


**Figure 6.27** Torque-speed characteristic of a shaded-pole induction motor.

## *PROBLEMS*

### Problem 6.1

Determine the number of poles, the slip, and the frequency of the rotor currents at rated load for three-phase, induction motors rated at:

A. 220 V, 50 Hz, 1440 r/min.
B. 120 V, 400 Hz, 3800 r/min.

### Problem 6.2

A 50-HP, 440-V, three-phase, 60-Hz, six-pole, *Y*-connected induction motor has the following parameters per phase:

$$R_2 = 0.15 \text{ ohm}$$
$$R_1 = 0.12 \text{ ohm}$$
$$G_c = 6 \times 10^{-3} \text{ siemens}$$
$$X_T = 0.75 \text{ ohm}$$
$$B_m = 0.07 \text{ siemens}$$

The rotational losses are equal to the stator hysteresis and eddy-current losses. For a slip of 4 percent, find the following

A. the line current and power factor.
B. the horsepower output.
C. the starting torque.

### Problem 6.3

Use MATLAB™ to verify the results of Problem 6.2.

**Problem 6.4**

The rotor resistance and reactance of a squirrel-cage induction motor rotor at standstill are 0.14 ohm per phase and 0.8 ohm per phase respectively. Assuming a transformer ratio of unity, from the eight-pole stator having a phase voltage of 254 at 60 Hz to the rotor secondary, calculate the following

A. rotor starting current per phase.
B. the value of slip producing maximum torque.

**Problem 6.5**

The full-load slip of a squirrel-cage induction motor is 0.06, and the starting current is five times the full-load current. Neglecting the stator core and copper losses as well as the rotational losses, obtain:

A. the ratio of starting torque to the full-load torque.
B. the ratio of maximum to full-load torque and the corresponding slip.

**Problem 6.6**

The rotor resistance and reactance of a wound-rotor induction motor at standstill are 0.14 ohm per phase and 0.8 ohm per phase, respectively. Assuming a transformer ratio of unity, from the eight-pole stator having a phase voltage of 254 V at 60 Hz to the rotor secondary, find the additional rotor resistance required to produce maximum torque at:

A. starting $s = 1$.
B. a speed of 450 r/min.

**Problem 6.7**

A two-pole 60-Hz induction motor develops a maximum torque of twice the full-load torque. The starting torque is equal to the full load torque. Determine the full load speed.

**Problem 6.8**

The starting torque of a three-phase induction motor is 165 percent and its maximum torque is 215 percent of full-load torque. Determine the slips at full load and at maximum torque. Find the rotor current at starting in per unit of full-load rotor current.

**Problem 6.9**

Consider a 25-hp, 230-V three-phase, 60-Hz squirrel cage induction motor operating at rated voltage and frequency. The rotor $I^2R$ loss at maximum torque is 9.0 times that at full-load torque, and the slip at full load torque is 0.028. Neglect stator resistance and rotational losses. Find the maximum torque in per unit of full load torque and the slip at which it takes place. Find the starting torque in per unit of full load torque.

**Problem 6.10**

The slip at full load for a three-phase induction motor is 0.04 and the rotor current at starting is 5 times its value at full load. Find the starting torque in per unit of full-load torque and the ratio of the maximum torque to full load torque and the slip at which it takes place.

**Problem 6.11**

A 220-V three phase four-pole 60 Hz squirrel-cage induction motor develops a maximum torque of 250 percent at a slip of 14 percent when operating at rated voltage and frequency. Now, assume that the motor is operated at 180 V and 50 Hz. Determine the maximum torque and the speed at which it takes place.

**Problem 6.12**

A six-pole, 60-Hz three-phase wound rotor induction motor has a rotor resistance of 0.8 Ω and runs at 1150 rpm at a given load. The motor drives a constant torque load. Suppose that we need the motor to run at 950 rpm while driving the same load. Find the additional resistance required to be inserted in the rotor circuit to fulfill this requirement.

**Problem 6.13**

Assume for a 3-phase induction motor that for a certain operating condition the stator $I^2R$ = rotor $I^2R$ = core loss = rotational loss and that the output is 30 KW at 86% efficiency. Determine the slip under this operating condition.

**Problem 6.14**

Find the required additional rotor resistance to limit starting current to 45 A for a 3-phase 600-V induction motor with $R_T = 1.66\ \Omega$ and $X_T = 4.1\ \Omega$.

**Problem 6.15**

The rotor $I^2R$ at starting are 6.25 times that at full load with slip of 0.035 for a three-phase induction motor. Find the slip at maximum torque and the ratio of starting to full-load torques.

**Problem 6.16**

The following parameters are available for a single-phase induction motor

$$R_1 = 1.5\,\Omega \qquad R_2' = 3.4\,\Omega$$
$$X_1 = X_2' = 3\,\Omega \qquad X_m = 100\,\Omega$$

Calculate $Z_f$, $Z_b$, and the input impedance of the motor for a slip of 0.06.

**Problem 6.17**

The induction motor of Problem 6.16 is a 60-Hz 110-V four-pole machine. Find the output power and torque under the conditions of Problem 6.16 assuming that the core losses are 66 W. Neglect rotational losses.

**Problem 6.18**

A four-pole 110-V 60-Hz single-phase induction motor has the following parameters:

$$R_1 = 0.8\,\Omega \qquad R_2' = 1\,\Omega$$
$$X_1 = X_2' = 1.92\,\Omega \qquad X_m = 42\,\Omega$$

The core losses are equal to the rotational losses, which are given by 40 W. Find the output power and efficiency at a slip of 0.05.

**Problem 6.19**

The following parameters are available for a single-phase 110-V induction motor:

$$R_1 = R_2' = 2.7\,\Omega$$
$$X_1 = X_2' = 2.7\,\Omega$$
$$X_m = 72\,\Omega$$

The core losses are 18.5 W and rotational losses are 17 W. Assume that the machine has four poles and operates on a 60-Hz supply. Find the rotor ohmic losses, output power, and torque for a slip of 5%.

**Problem 6.20**

The stator resistance of a single-phase induction motor is 1.96 Ω and the rotor resistance referred to the stator is 3.6 Ω. The motor takes a current of 4.2 A from the 110-V supply at a power factor of 0.624 when running at slip of 0.05. Assume that the core loss is 36 W and that the approximation of Eq. (6.47) is applicable. Find the motor's output power and efficiency neglecting rotational losses.

**Problem 6.21**

A single-phase induction motor takes an input power of 280 W at a power factor of 0.6 lagging from a 110-V supply when running at a slip of 5 percent. Assume that the rotor resistance and reactance are 3.38 and 2.6 Ω, respectively, and that the magnetizing reactance is 60 Ω. Find the resistance and the reactance of the motor.

**Problem 6.22**

For the motor of Problem 6.21, assume that the core losses are 35 W and the rotational losses are 14 W. Find the output power and efficiency when running at a slip of 5 percent.

**Problem 6.23**

The output torque of a single-phase induction motor is 0.82 N · m at a speed of 1710 rpm. The efficiency is 60 percent and the fixed losses are 37 W. Assume that the motor operates on a 110-V supply and that the stator resistance is 2 Ω.

Find the input power factor and input impedance. Assume that the rotor ohmic losses are 35.26 W. Find the forward and backward gap power and the values of $R_f$ and $R_b$. Assume a four-pole machine.

### Problem 6.24

The forward field impedance of a $\frac{1}{4}$–hp four-pole 110-V 60-Hz single-phase induction motor for a slip of 0.05 is given by

$$Z_f = 12.4 + j16.98\,\Omega$$

Assume that

$$X_m = 53.5\,\Omega$$

Find the values of the rotor resistance and reactance.

### Problem 6.25

For the motor of Problem 6.24, assume that the stator impedance is given by

$$Z_1 = 1.86 + j2.56\,\Omega$$

Find the internal mechanical power, output power, power factor, input power, developed torque, and efficiency, assuming that friction losses are 15 W.

# Chapter 7

# FAULTS AND PROTECTION OF ELECTRIC ENERGY SYSTEMS

## 7.1 INTRODUCTION

A short-circuit fault takes place when two or more conductors come in contact with each other when normally they operate with a potential difference between them. The contact may be a physical metallic one, or it may occur through an arc. In the metal-to-metal contact case, the voltage between the two parts is reduced to zero. On the other hand, the voltage through an arc will be of a very small value. Short-circuit faults in three-phase systems are classified as:

1. Balanced or symmetrical three-phase faults.
2. Single line-to-ground faults.
3. Line-to-line faults.
4. Double line-to-ground faults.

Generator failure is caused by insulation breakdown between turns in the same slot or between the winding and the steel structure of the machine. The same can take place in transformers. The breakdown is due to insulation deterioration combined with switching and/or lightning overvoltages. Overhead lines are constructed of bare conductors. Wind, sleet, trees, cranes, kites, airplanes, birds, or damage to supporting structure are causes for accidental faults on overhead lines. Contamination of insulators and lightning overvoltages will in general result in short-circuit faults. Deterioration of insulation in underground cables results in short circuit faults. This is mainly attributed to aging combined with overloading. About 75 percent of the energy system's faults are due to single-line-to-ground faults and result from insulator flashover during electrical storms. Only one in twenty faults is due to the balanced category.

A fault will cause currents of high value to flow through the network to the faulted point. The amount of current may be much greater than the designed thermal ability of the conductors in the power lines or machines feeding the fault. As a result, temperature rise may cause damage by annealing of conductors and insulation charring. In addition, the low voltage in the neighborhood of the fault will cause equipment malfunction.

Short-circuit and protection studies are an essential tool for the electric energy systems engineer. The task is to calculate the fault conditions and to provide protective equipment designed to isolate the faulted zone from the remainder of the system in the appropriate time. The least complex fault category computationally is the balanced fault. It is possible that a balanced fault could (in some locations) result in currents smaller than that due to some other type of fault. The interrupting capacity of breakers should be chosen to accommodate the largest of fault currents, and hence, care must be taken not to base protection decisions on the results of a balanced three phase fault.

## 7.2 TRANSIENTS DURING A BALANCED FAULT

The value and severity of short-circuit current in the electric power system depends on the instant in the cycle at which the short circuit occurs. This can be verified using a simple model, consisting of a generator with series resistance $R$ and inductance $L$ as shown in Figure 7.1. The voltage of the generator is assumed to vary as

$$e(t) = E_m \sin(\omega t + \alpha) \tag{7.1}$$

A dc term will in general exist when a balanced fault is placed on the generator terminals at $t = 0$. The initial magnitude may be equal to the magnitude of the steady-state current term.

The worst possible case of transient current occurs for the value of short circuit time placement corresponding to $\alpha$ given by

$$\tan\alpha = -\frac{R}{\omega L}$$

Here, the current magnitude will approach twice the steady-state maximum value immediately after the short circuit. The transient current is given in this case by the small $t$ approximation

$$i(t) = \frac{E_m}{Z}(1 - \cos\omega t) \tag{7.2}$$

It is clear that

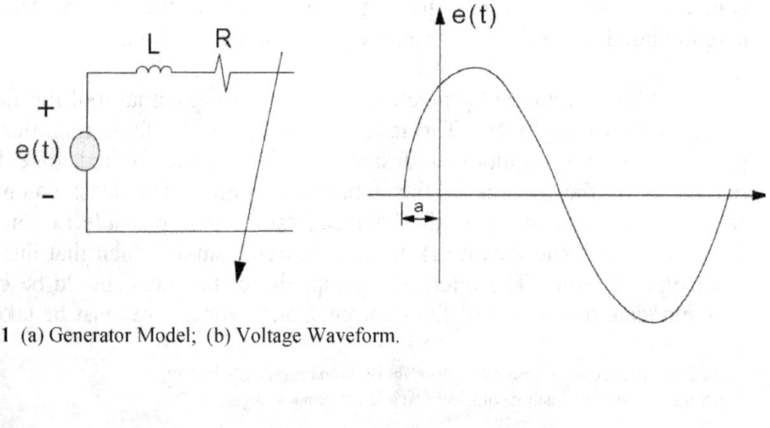


**Figure 7.1** (a) Generator Model; (b) Voltage Waveform.

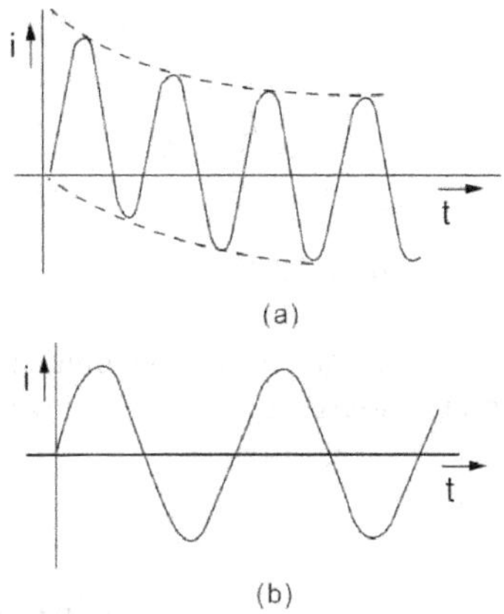


**Figure 7.2** (a) Short-Circuit Current Wave Shape for tan $\alpha = -(R/\omega L)$; (b) Short-Circuit Current Wave Shape for tan $\alpha = (\omega L/R)$.

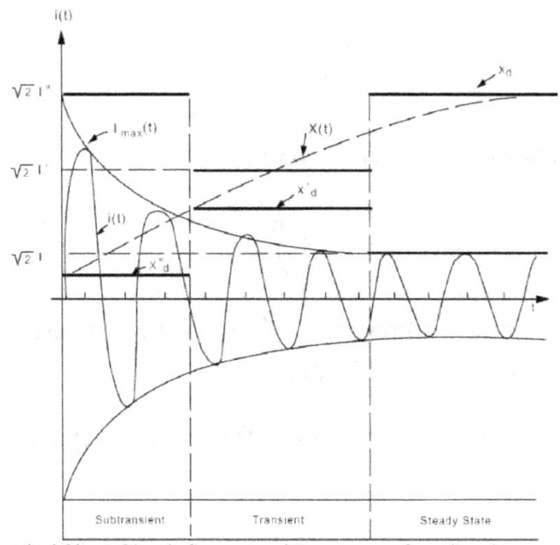


**Figure 7.3** Symmetrical Short-Circuit Current and Reactances for a Synchronous Machine.

$$i_{max} = \frac{2E_m}{Z}$$

This waveform is shown in Figure 7.2(a).

For the case of short circuit application corresponding to

$$\tan\alpha = \frac{\omega L}{R}$$

we have

$$i(t) = \frac{E_m}{Z} \sin \omega t \tag{7.3}$$

This waveform is shown in Figure 7.2(b).

It is clear that the reactance of the machine appears to be time-varying, if we assume a fixed voltage source $E$. For our power system purposes, we let the reactance vary in a stepwise fashion $X''_d$, $X'_d$, and $X_d$ as shown in Figure 7.3.

The current history $i(t)$ can be approximated considering three time segments by three different expressions. The first is called the subtransient interval and lasts up to two cycles, the current is $I''$. This defines the direct-axis subtransient reactance:

$$X''_d = \frac{E}{I''} \tag{7.4}$$

The second, denoted the transient interval, gives rise to

$$X'_d = \frac{E}{I'} \tag{7.5}$$

where $I'$ is the transient current and $X'_d$ is direct-axis transient reactance. The transient interval lasts for about 30 cycles.

The steady-state condition gives the direct-axis synchronous reactance:

$$X_d = \frac{E}{I} \tag{7.6}$$

Table 7.1 list typical values of the reactances defined in Eqs. (7.4), (7.5), and (7.6). Note that the subtransient reactance can be as low as 7 percent of the synchronous reactance.

## 7.3 THE METHOD OF SYMMETRICAL COMPONENTS

The method of symmetrical components is used to transform an unbalanced three-phase system into three sets of balanced three-phase phasors. The basic idea of the transformations is simple. Given three voltage phasors $V_A$, $V_B$, and $V_C$, it is possible to express each as the sum of three phasors as follows:

$$V_A = V_{A+} + V_{A-} + V_{A0} \tag{7.7}$$

**Table 7.1**

Typical Average Reactance Values for Synchronous Machines

| | **Two-Pole Turbine Generator** | **Four-Pole Turbine Generator** | **Salient-Pole Machine with Dampers** | **Salient-Pole Generator without Dampers** | **Synchronous Condensers** |
|---|---|---|---|---|---|
| $X_d$ | 1.2 | 1.2 | 1.25 | 1.25 | 2.2 |
| $X'_d$ | 0.15 | 0.23 | 0.30 | 0.30 | 0.48 |
| $X''_d$ | 0.09 | 0.14 | 0.2 | 0.30 | 0.32 |
| $X_-$ | 0.09 | 0.14 | 0.2 | 0.48 | 0.31 |
| $X_0$ | 0.03 | 0.08 | 0.18 | 0.19 | 0.14 |

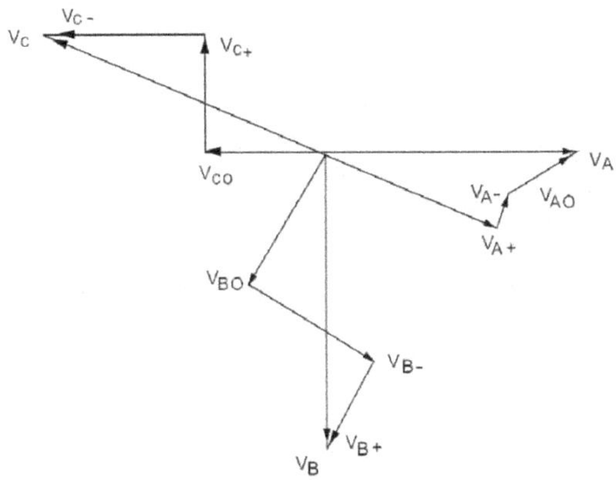


**Figure 7.4** An Unbalanced Set of Voltage Phasors and a Possible Decomposition.

$$V_B = V_{B+} + V_{B-} + V_{B0} \tag{7.8}$$

$$V_C = V_{C+} + V_{C-} + V_{C0} \tag{7.9}$$

Figure 7.4 shows the phasors $V_A$, $V_B$, and $V_C$ as well as a particular possible choice of the decompositions.

Obviously there are many possible decompositions. For notational simplicity, we introduce the complex operator $\alpha$ defined by

$$\alpha = e^{j120^\circ} \tag{7.10}$$

We require that the sequence voltages $V_{A+}$, $V_{B+}$, and $V_{C+}$ form a balanced positively rotating system. Thus the phasor magnitudes are equal, and the phasors are 120° apart in a sequence *A-B-C*.

$$V_{B+} = \alpha^2 V_{A+} \tag{7.11}$$

$$V_{C+} = \alpha V_{A+} \tag{7.12}$$

Similarly, we require that the sequence voltages $V_{A-}$, $V_{B-}$, and $V_{C-}$ form a balanced negatively rotating system. This requires that the sequence is *C-B-A*

$$V_{B-} = \alpha V_{A-} \tag{7.13}$$

$$V_{C-} = \alpha^2 V_{A-} \tag{7.14}$$

The sequence voltages $V_{A0}$, $V_{B0}$, $V_{C0}$ are required to be equal in magnitude and phase. Thus,

$$V_{B0} = V_{A0} \tag{7.15}$$

$$V_{C0} = V_{A0} \tag{7.16}$$

The original phasor voltages $V_A$, $V_B$, and $V_C$ are expressed in terms of the sequence voltages as

$$V_A = V_{A+} + V_{A-} + V_{A0} \tag{7.17}$$

$$V_B = \alpha^2 V_{A+} + \alpha V_{A-} + V_{A0} \tag{7.18}$$

$$V_C = \alpha V_{A+} + \alpha^2 V_{A-} + V_{A0} \tag{7.19}$$

The inverse relation giving the positive sequence voltage $V_{A+}$, the negative sequence voltage $V_{A-}$, and the zero sequence voltage $V_{A0}$ is obtained by solving the above three simultaneous equations to give

$$V_{A+} = \frac{1}{3}\left(V_A + \alpha V_B + \alpha^2 V_C\right) \tag{7.20}$$

$$V_{A-} = \frac{1}{3}\left(V_A + \alpha^2 V_B + \alpha V_C\right) \tag{7.21}$$

$$V_{A0} = \frac{1}{3}\left(V_A + V_B + V_C\right) \tag{7.22}$$

Some of the properties of the operator $\alpha$ are as follows:

$$\alpha^2 = \alpha^{-1}$$
$$\alpha^3 = 1$$
$$1 + \alpha + \alpha^2 = 0$$

For clarity, we will drop the suffix $A$ from the sequence voltage symbols, and we have

$$V_A = V_+ + V_- + V_0 \quad \textbf{(7.23)}$$

$$V_B = \alpha^2 V_+ + \alpha V_- + V_0 \quad \textbf{(7.24)}$$

$$V_C = \alpha V_+ + \alpha^2 V_- + V_0 \quad \textbf{(7.25)}$$

$$\alpha^2 = \alpha^{-1}$$
$$\alpha^3 = 1$$
$$1 + \alpha + \alpha^2 = 0$$

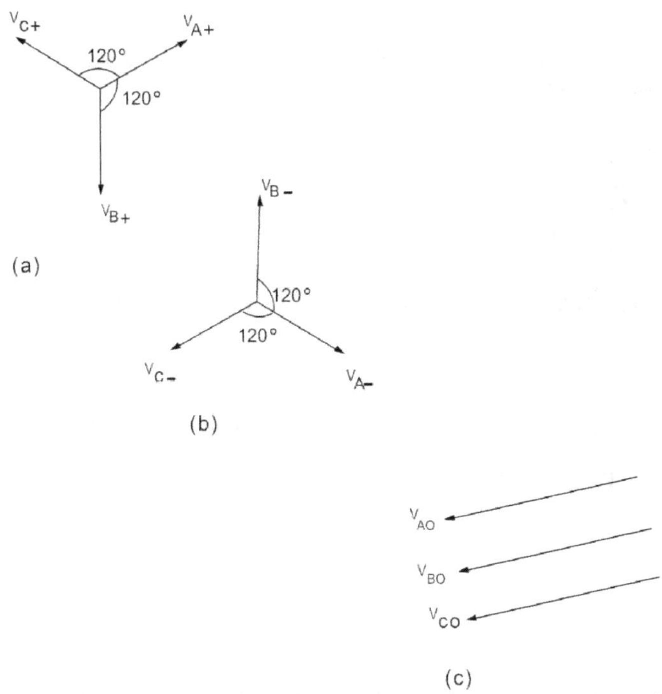


**Figure 7.5** (a) Positive Sequence Voltage Phasors; (b) Negative Sequence Voltage Phasors; and (c) Zero Sequence Voltage Phasors.

and

$$V_+ = \frac{1}{3}\left(V_A + \alpha V_B + \alpha^2 V_C\right) \tag{7.26}$$

$$V_- = \frac{1}{3}\left(V_A + \alpha^2 V_B + \alpha V_C\right) \tag{7.27}$$

$$V_0 = \frac{1}{3}\left(V_A + V_B + V_C\right) \tag{7.28}$$

The ideas of symmetrical components apply to currents in the same manner.

We have the following two examples:

**Example 7.1**

The following currents were recorded under fault conditions in a three-phase system:

$$I_A = 150\angle 45^\circ \text{ A}$$
$$I_B = 250\angle 150^\circ \text{ A}$$
$$I_C = 100\angle 300^\circ \text{ A}$$

Calculate the values of the positive, negative, and zero phase sequence components for each line.

*Solution*

$$I_0 = \frac{1}{3}(I_A + I_B + I_C)$$
$$= \frac{1}{3}\left(106.04 + j106.07 + j106.07 - 216.51 + j125.00 + 50 - j86.6\right)$$
$$= 52.2\angle 112.7^\circ$$
$$I_+ = \frac{1}{3}\left(I_A + \alpha I_B + \alpha^2 I_C\right) = \frac{1}{3}\left(150\angle 45^\circ + 250\angle 270^\circ + 100\angle 180^\circ\right)$$
$$= 48.02\angle -87.6^\circ$$
$$I_- = \frac{1}{3}\left(I_A + \alpha^2 I_B + \alpha I_C\right)$$
$$= 163.21\angle 40.45^\circ$$

**Example 7.2**

Given that

$$V_0 = 100$$
$$V_+ = 200\angle 60^\circ$$
$$V_- = 100\angle 120^\circ$$

find the phase voltage $V_A$, $V_B$, and $V_C$.

*Solution*

$$\begin{aligned}
V_A &= V_+ + V_- + V_0 \\
&= 200\angle -120^\circ + 100\angle -60^\circ + 100 = 300\angle 60^\circ \\
V_B &= \alpha^2 V_+ + \alpha V_- + V_0 \\
&= \left(1\angle 240^\circ\right)\left(200\angle 60^\circ\right) + \left(1\angle 120^\circ\right)\left(100\angle 120^\circ\right) + 100 \\
&= 300\angle -60^\circ \\
V_C &= \alpha V_+ + \alpha^2 V_- + V_0 \\
&= \left(1\angle 120^\circ\right)\left(200\angle 60^\circ\right) + \left(1\angle 240^\circ\right)\left(100\angle 120^\circ\right) + 100 \\
&= 0
\end{aligned}$$

## Power in Symmetrical Components

The total power in a three-phase network is given in terms of phase variables by

$$S = V_A I_A^* + V_B I_B^* + V_C I_C^* \tag{7.29}$$

where the asterisk denotes complex conjugation. We can show that the corresponding expression in terms of sequence variables is given by

$$S = 3\left(V_+ I_+^* + V_- I_-^* + V_0 I_0^*\right) \tag{7.30}$$

The total power is three times the sum of powers in individual sequence networks.

## *7.4 SEQUENCE NETWORKS*

## Positive Sequence Networks

For a given power system the positive sequence network shows all the paths for the flow of positive sequence currents in the system. The one-line diagram of the system is converted to an impedance diagram that shows the equivalent circuit of each component under balanced operating conditions.

Each generator in the system is represented by a source voltage in series with the appropriate reactance and resistance. To simplify the calculations, all resistance and the magnetizing current for each transformer are neglected. For transmission lines, the line's shunt capacitance and resistance are neglected. Motor loads, whether synchronous or induction, are included in the network as generated EMF's in series with the appropriate reactance. Static loads are mostly neglected in fault studies.

## Negative Sequence Networks

Balanced three-phase generators and motors have only positive sequence-generated voltages. Thus, the negative sequence network model will not contain voltage sources associated with rotating machinery. Note that the negative sequence impedance will in general be different from the positive sequence values. For static devices such as transmission lines and transformers, the negative sequence impedances have the same values as the corresponding positive sequence impedances.

The current-limiting impedances between the generator's neutral and ground will not appear in either the positive or negative sequence network. This arises simply because positive and negative sequence currents are balanced.

## Zero Sequence Networks

The zero sequence network of a system depends on the nature of the connections of the three-phase windings for each of the system's components.

### *Delta-Connected Winding*

Zero sequence currents can exist in the phase windings of the delta connection. However, since we have the requirement

$$I_{A0} = I_{B0} = I_{C0} = I_0$$

we conclude that the line currents coming out of a delta winding are zero. For example,

$$I_{AB} = I_{A0} - I_{B0} = 0$$

This situation is shown in Figure 7.6.

The single-phase equivalent zero sequence network for a delta-connected load with zero sequence impedance $Z_0$ is shown in Figure 7.7.

### *Wye-Connected Winding*

When a neutral return wire is present, zero sequence currents will pass both in the phase windings as well as on the lines. The neutral current $I_N$ will be

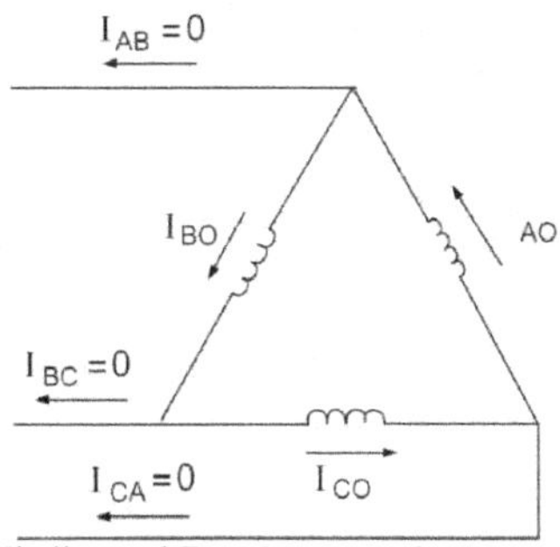


**Figure 7.6** Delta-Connected Winding and Zero Sequence Currents.

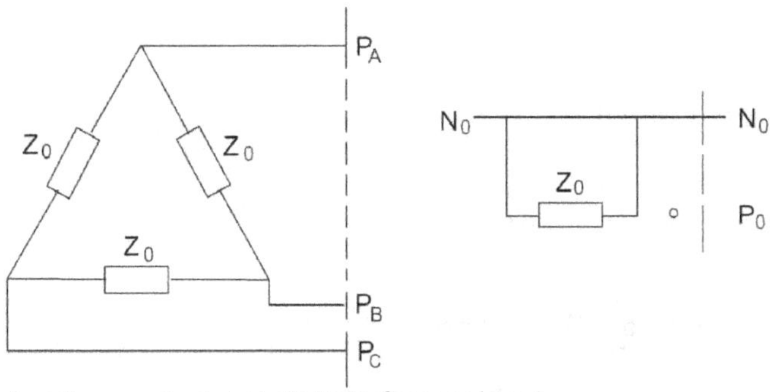


**Figure 7.7** Zero Sequence Equivalent of a Delta-Connected Load.

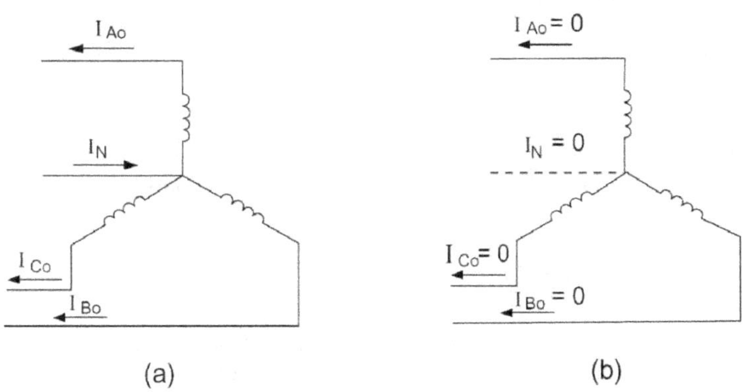


**Figure 7.8** Wye-Connected Winding with and without Neutral Return.

$$
\begin{aligned}
I_N &= I_{A0} + I_{B0} + I_{C0} \\
&= 3I_0
\end{aligned}
$$

This is shown in Figure 7.8(a). In the case of a system with no neutral return, $I_N = 0$ shows that no zero sequence currents can exist. This is shown in Figure 7.8(b). Zero sequence equivalents are shown in Figure 7.9.

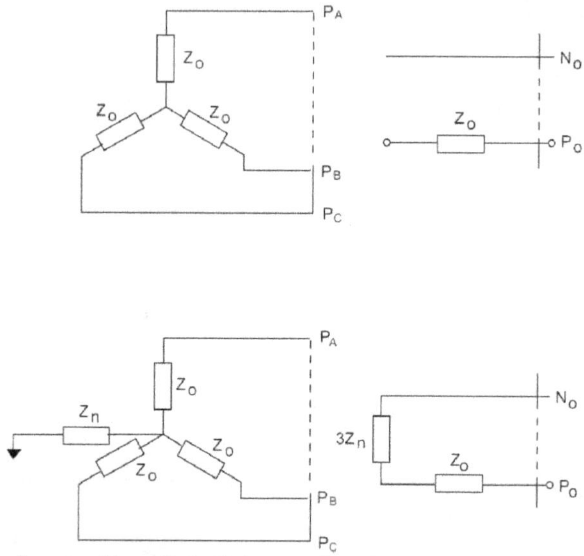


**Figure 7.9** Zero Sequence Networks for Y-Connected Loads.

## Transformer's Zero Sequence Equivalents

There are various possible combinations of the primary and secondary connections for three-phase transformers. These change the corresponding zero sequence network.

### *Delta-delta Bank*

In a delta circuit no return path for zero sequence current exists, and thus no zero sequence current can flow into a delta-delta bank, although it can circulate within the delta windings. The equivalent circuit connections are shown in Figure 7.10.

### *Wye-delta Bank, Ungrounded Wye*

For an ungrounded wye connection, no path exists for zero sequence current to the neutral. The equivalent circuit is shown in Figure 7.11.

### *Wye-delta Bank, Grounded Wye*

Zero sequence currents will pass through the wye winding to ground. As a result, secondary zero sequence currents will circulate through the delta winding. No zero sequence current will exist on the lines of the delta secondary. The equivalent circuit is shown in Figure 7.12.

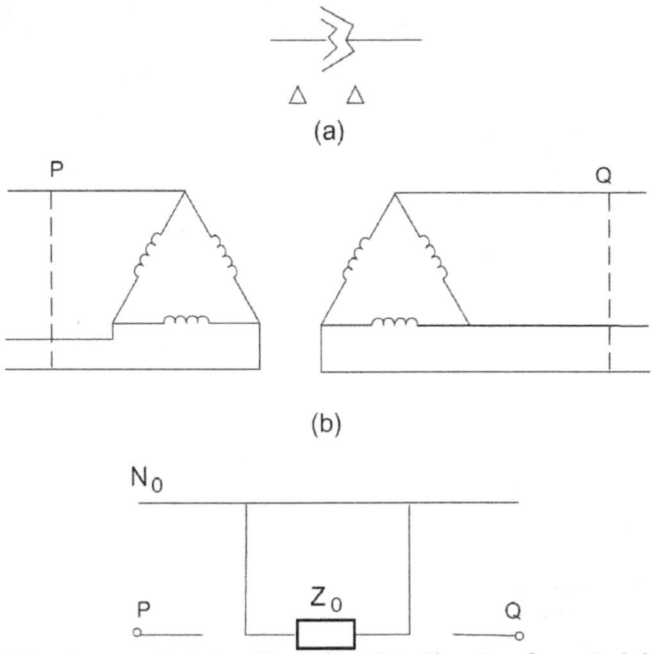


**Figure 7.10** Zero Sequence Equivalent Circuits for a Three-Phase Transformer Bank Connected in delta-delta.

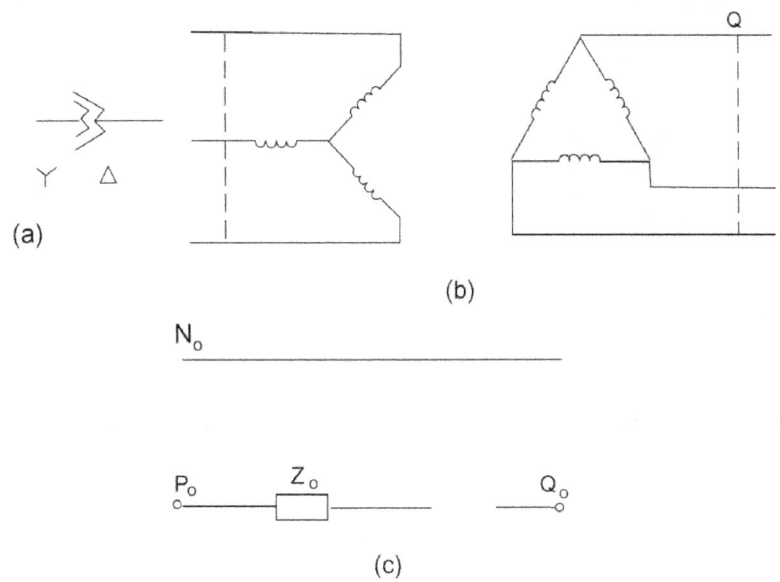


**Figure 7.11** Zero Sequence Equivalent Circuits for a Three-Phase Transformer Bank Connected in Wye-delta.

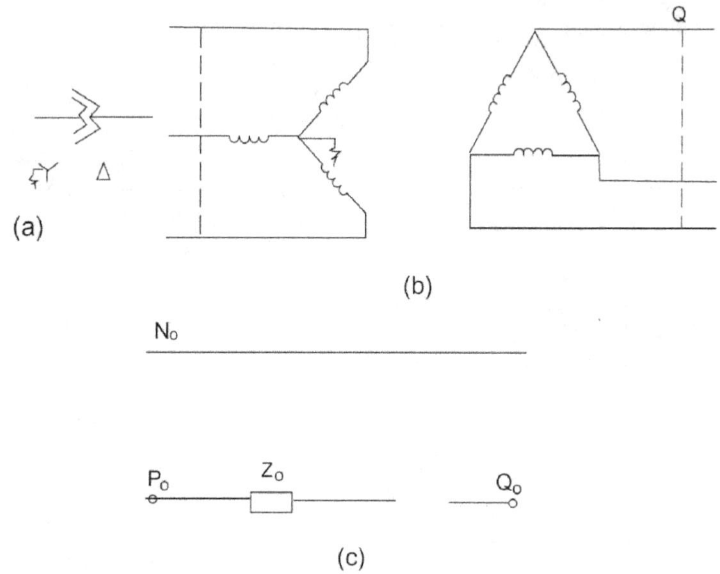


**Figure 7.12** Zero Sequence Equivalent Circuit for a Three-Phase Transformer Bank Connected in Wye-Delta Bank with Grounded Y.

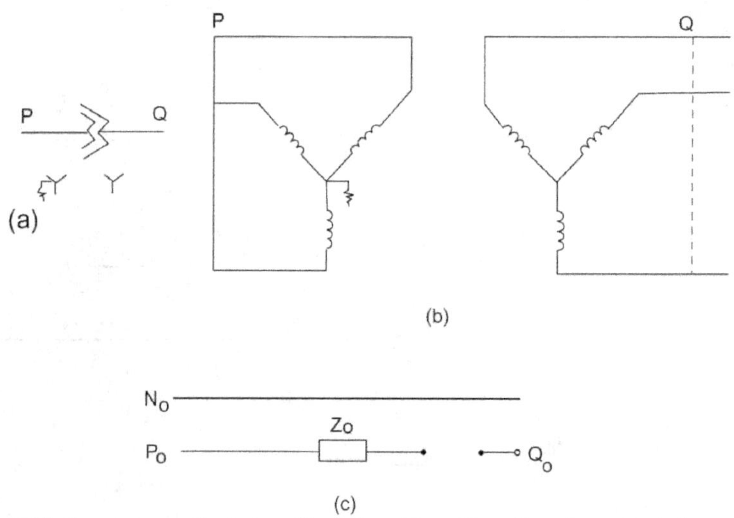


**Figure 7.13** Zero Sequence Equivalent Circuit for a Three-Phase Transformer Bank Connected in Wye-Wye with One Grounded Neutral.

## *Wye-Wye Bank, One Neutral Grounded*

With an ungrounded wye, no zero sequence current can flow. No current in one winding means that no current exists in the other. Figure 7.13 illustrates the situation.

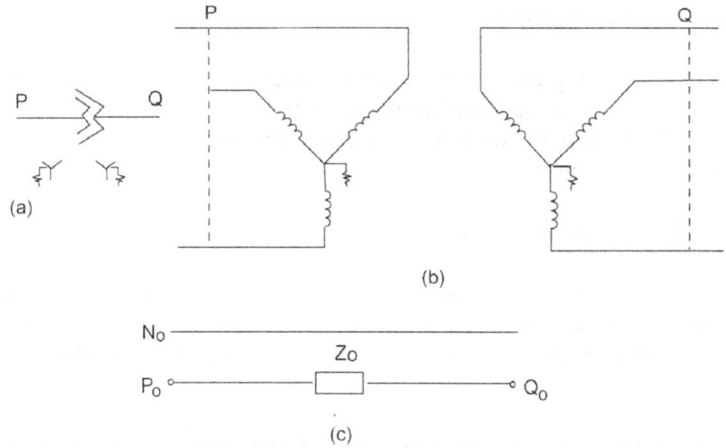


**Figure 7.14** Zero Sequence Equivalent Circuit for a Three-Phase Transformer Bank Connected in Wye-Wye with Neutrals Grounded.

### *Wye-Wye Bank, Both Neutrals Grounded*

With both wyes grounded, zero sequence current can flow. The presence of the current in one winding means that secondary current exists in the other. Figure 7.14 illustrates the situation.

## Sequence Impedances for Synchronous Machines

For a synchronous machine, sequence impedances are essentially reactive. The positive, negative, and zero sequence impedances have in general different values.

### *Positive Sequence Impedance*

Depending on the time interval of interest, one of three reactances may be used:

1. For the subtransient interval, we use the subtransient reactance:

$$Z_+ = jX_d''$$

2. For the transient interval, we use the corresponding reactance:

$$Z_+ = jX_d'$$

3. In the steady state, we have

$$Z_+ = jX_d$$

### Negative Sequence Impedance

The MMF produced by negative sequence armature three-phase current rotates in a direction opposite to the rotor and hence opposite to the dc field winding. Therefore the reactance of the machine will be different from that for the positively rotating sequence.

### Zero Sequence Impedance

The zero sequence impedance of the synchronous machine is quite variable and depends on the nature of the stator windings. In general, these will be much smaller than the corresponding positive and negative sequence reactance.

## Sequence Impedances for a Transmission Link

Consider a three-phase transmission link of impedance $Z_L$ per phase. The return (or neutral) impedance is $Z_N$. If the system voltages are unbalanced, we have a neutral current $I_N$. Thus,

$$I_N = I_A + I_B + I_C$$

The voltage drops $\Delta V_A$, $\Delta V_B$, and $\Delta V_C$ across the link are as shown below:

$$\begin{aligned}\Delta V_A &= I_A Z_L + I_N Z_N \\ \Delta V_B &= I_B Z_L + I_N Z_N \\ \Delta V_C &= I_C Z_L + I_N Z_N\end{aligned}$$

In terms of sequence voltages and currents, we have

$$\begin{aligned}\Delta V_+ &= I_+ Z_L \\ \Delta V_- &= I_- Z_L \\ \Delta V_0 &= I_0 (Z_L + 3Z_N)\end{aligned}$$

Therefore the sequence impedances are given by:

$$\begin{aligned}Z_0 &= Z_L + 3Z_N \\ Z_- &= Z_L \\ Z_+ &= Z_L\end{aligned}$$

The impedance of the neutral path enters into the calculation of zero sequence impedance in addition to the link's impedance $Z_L$. However, for the positive and negative sequence impedances, only the link's impedance appears.

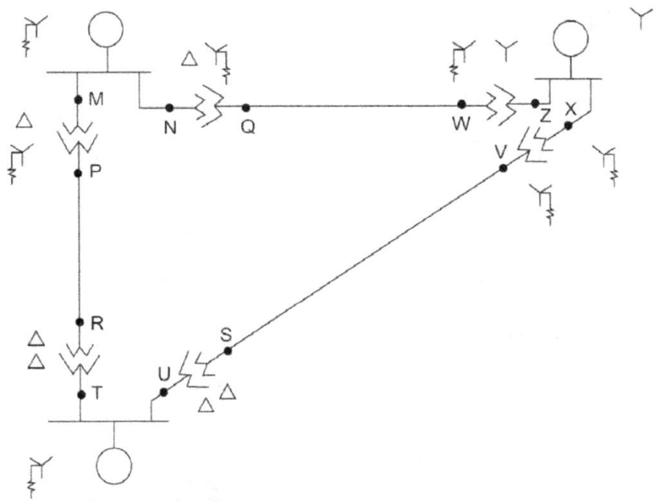


Figure 7.15 System for Example 7.3.

## Example 7.3

Draw the zero sequence network for the system shown in Figure 7.15.

### *Solution*

The zero sequence network is shown in Figure 7.16.

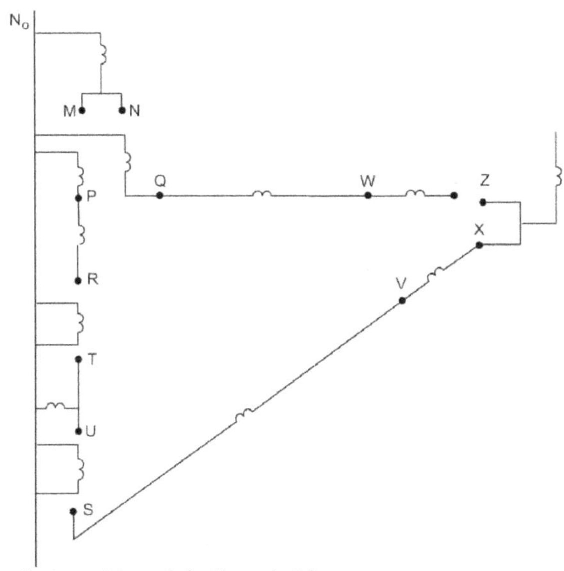


Figure 7.16 Zero Sequence Network for Example 7.3.

### Example 7.4

Obtain the sequence networks for the system shown in Figure 7.17. Assume the following data in p.u. on the same base.

| | |
|---|---|
| Generator $G_1$: | $X_+ = 0.2$ p.u. |
| | $X_- = 0.12$ p.u. |
| | $X_0 = 0.06$ p.u. |
| Generator $G_2$: | $X_+ = 0.33$ p.u. |
| | $X_- = 0.22$ p.u. |
| | $X_0 = 0.066$ p.u. |
| Transformer $T_1$: | $X_+ = X_- = X_0 = 0.2$ p.u. |
| Transformer $T_2$: | $X_+ = X_- = X_0 = 0.225$ p.u. |
| Transformer $T_3$: | $X_+ = X_- = X_0 = 0.27$ p.u. |
| Transformer $T_4$: | $X_+ = X_- = X_0 = 0.16$ p.u. |
| Line $L_1$: | $X_+ = X_- = 0.14$ p.u. |
| | $X_0 = 0.3$ p.u. |
| Line $L_2$: | $X_+ = X_- = 0.20$ p.u. |
| | $X_0 = 0.4$ p.u. |
| Line $L_3$: | $X_+ = X_- = 0.15$ p.u. |
| | $X_0 = 0.2$ p.u. |
| Load: | $X_+ = X_- = 0.9$ p.u. |
| | $X_0 = 1.2$ p.u. |

Assume an unbalanced fault occurs at $F$. Find the equivalent sequence networks for this condition.

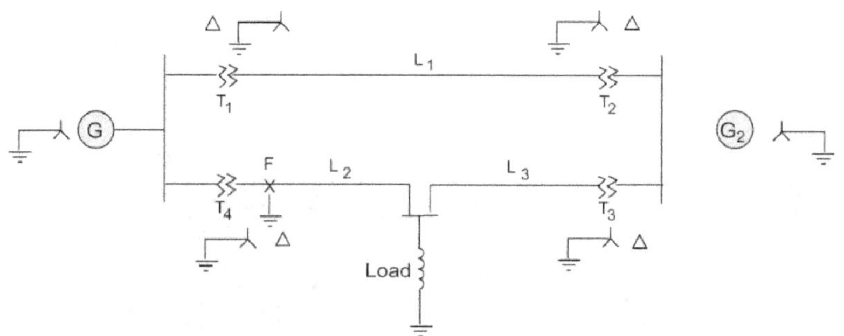


**Figure 7.17** Network for Example 7.4.

*Solution*

The positive sequence network is as shown in Figure 7.18(A). One step in the reduction can be made, the result of which is shown in Figure 7.18(B). To avoid tedious work we utilize Thévenin's theorem to obtain the positive sequence network in reduced form. We assign currents $I_1$, $I_2$, and $I_3$ as shown in Figure 7.18(B) and proceed to solve for the open-circuit voltage between $F_+$ and $N_+$.

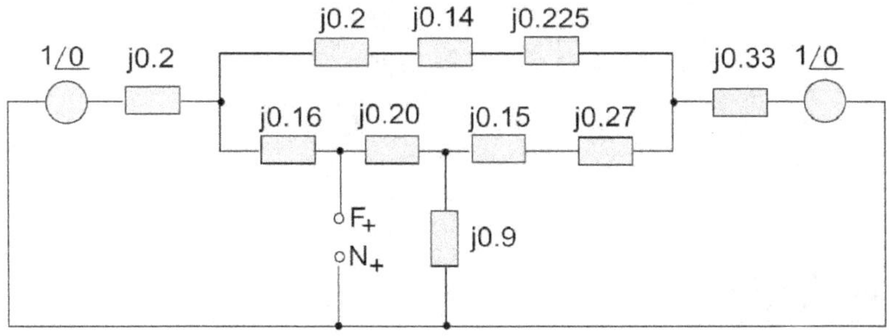


(A)

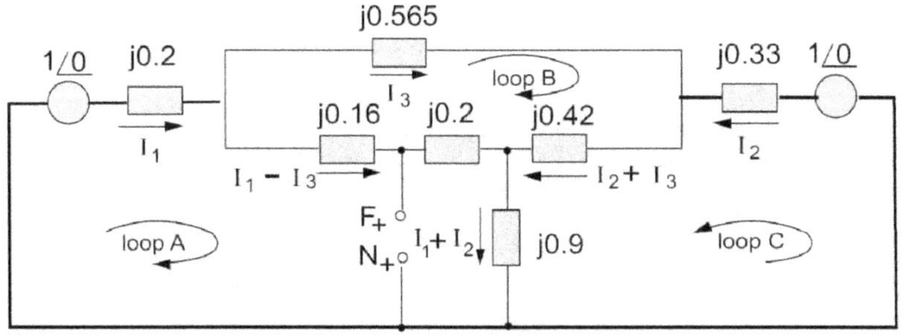


(B)

**Figure 7.18** Positive Sequence Network for Example 7.4.

Consider loop *A*. We can write

$$1\angle 0 = j\left[0.2I_1 + 0.36(I_1 - I_3) + 0.9(I_1 + I_2)\right]$$

For loop *B*, we have

$$0 = j\left[0.565I_3 + 0.42(I_2 + I_3) - 0.36(I_1 - I_3)\right]$$

For loop *C*, we have

$$1\angle 0 = j\left[0.33I_2 + 0.42(I_2 + I_3) + 0.9(I_1 + I_2)\right]$$

The above three equations are rearranged to give

$$\begin{aligned} 1\angle 0 &= j(1.46I_1 + 0.9I_2 - 0.36I_3) \\ 0 &= 0.36I_1 - 0.42I_2 - 1.345I_3 \\ 1\angle 0 &= j(0.9I_1 + 1.65I_2 + 0.42I_3) \end{aligned}$$

Solving the three equations we obtain the current values as follows:

$$I_1 = -j0.4839$$
$$I_2 = -j0.3357$$
$$I_3 = -j0.0247$$

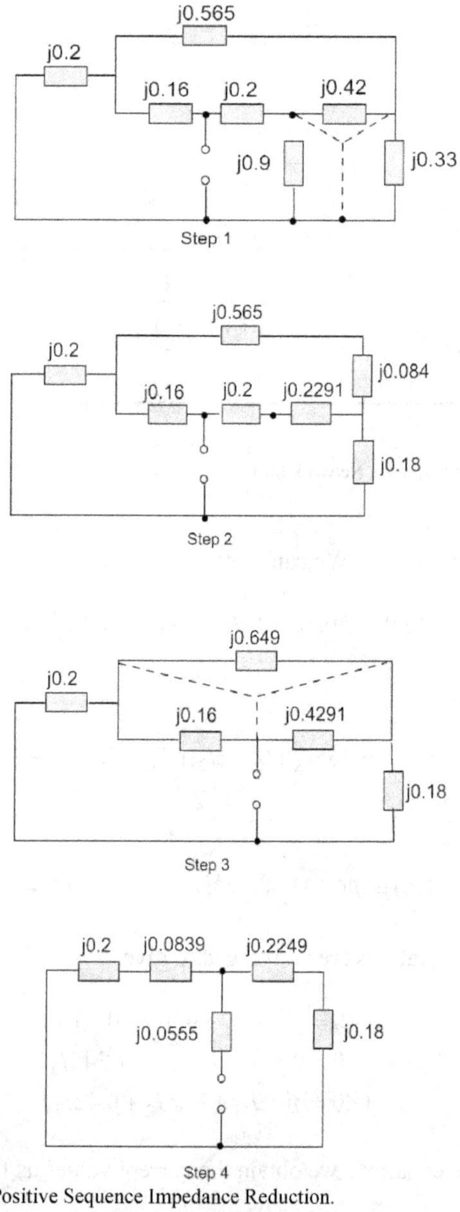


**Figure 7.19** Steps in Positive Sequence Impedance Reduction.

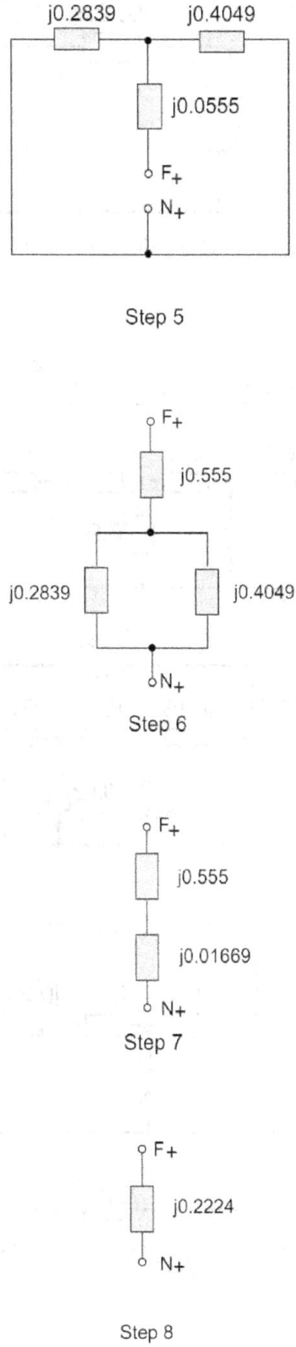


**Figure 7.19** (*Cont.*)

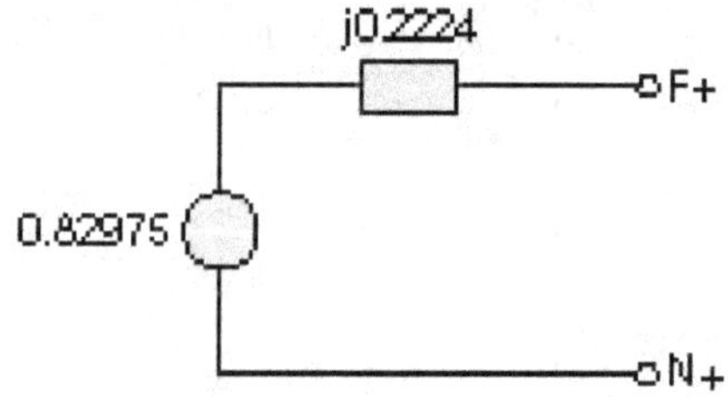


**Figure 7.20** Positive Sequence Network Equivalent for Example 7.4.

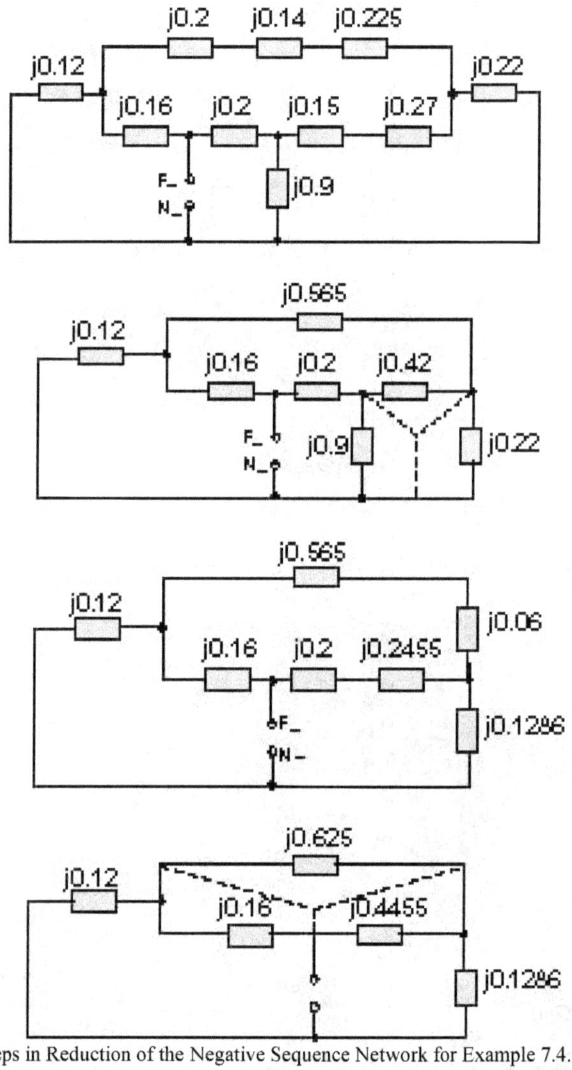


**Figure 7.21** Steps in Reduction of the Negative Sequence Network for Example 7.4.

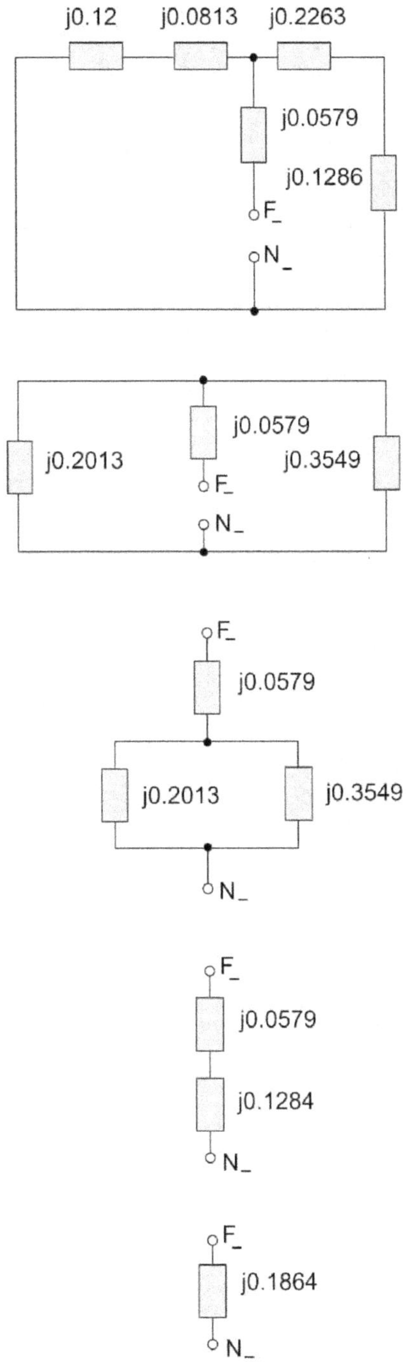


**Figure 7.21 (*Cont.*)**

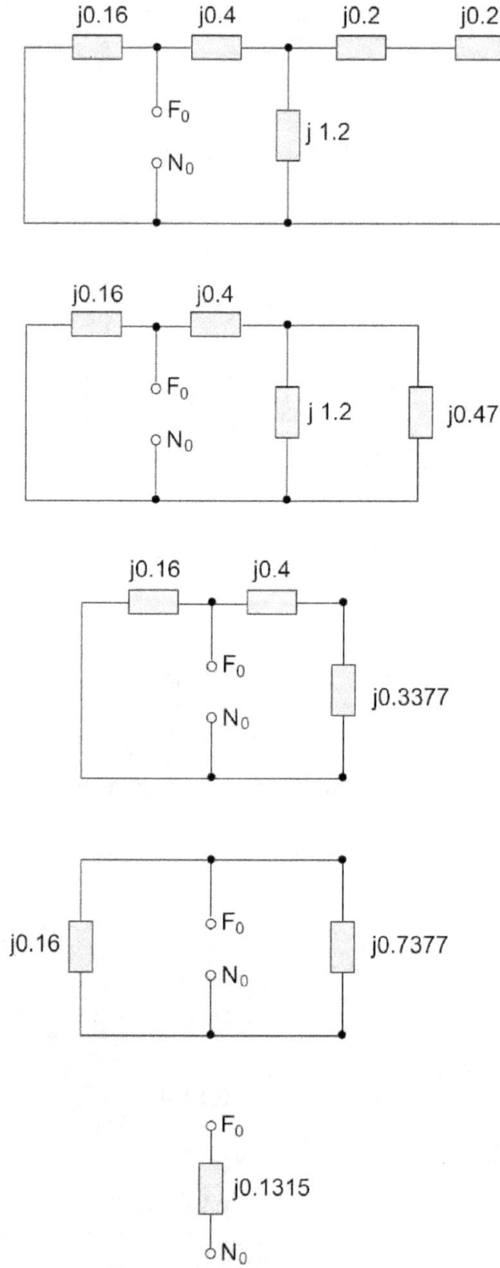


**Figure 7.22** Steps in Reducing the Zero Sequence Network for Example 7.4.

As a result, we get

$$V_{F_+N_+} = V_{TH} = 1 - j0.2I_1 - j0.16(I_1 - I_3)$$
$$= 1 - (02)(0.4839) - (0.16)(0.4839 - 0.0247)$$
$$= 0.82975$$

We now turn our attention to the Thévenin's equivalent impedance, which is obtained by shorting out the sources and using network reduction. The steps are shown in Figure 7.19. As a result, we get

$$Z_+ = j0.224$$

The positive sequence equivalent is shown in Figure 7.20.

The negative sequence and zero sequence impedance networks and steps in their reduction are shown in Figure 7.21 and Figure 7.22. As a result, we get

$$Z_- = j0.1864$$
$$Z_0 = j0.1315$$

## 7.5 LINE-TO-GROUND FAULT

Assume that phase $A$ is shorted to ground at the fault point $F$ as shown in Figure 7.23. The phase $B$ and $C$ currents into the fault are zero, and we can thus write $I_B = 0$, $I_C = 0$. The sequence currents are obtained as:

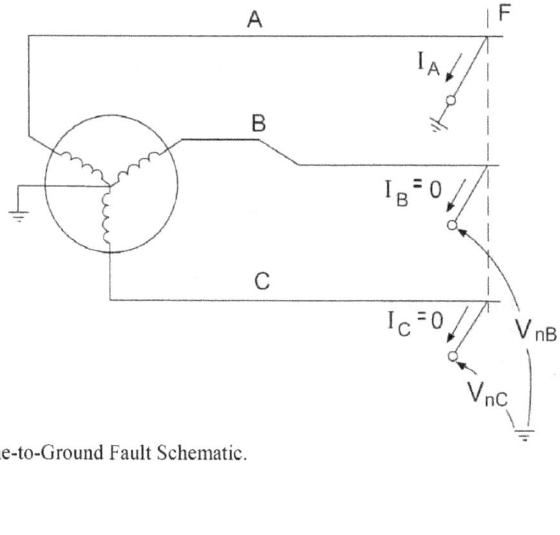


**Figure 7.23** Line-to-Ground Fault Schematic.

$$I_+ = I_- = I_0 = \frac{I_A}{3} \tag{7.31}$$

With the generators normally producing balanced three-phase voltages, which are positive sequence only, we can write

$$E_+ = E_A \tag{7.32}$$

$$E_- = 0 \tag{7.33}$$

$$E_0 = 0 \tag{7.34}$$

Let us assume that the sequence impedances from the source to the fault are given by $Z_+$, $Z_-$, $Z_0$. We can write the following expressions for sequence voltages at the fault:

$$V_+ = E_+ - I_+ Z_+ \tag{7.35}$$

$$V_- = 0 - I_- Z_- \tag{7.36}$$

$$V_0 = 0 - I_0 Z_0 \tag{7.37}$$

The fact that phase *A* is shorted to ground is used. Thus,

$$V_A = 0$$

This leads to

$$0 = E_+ - I_0 (Z_+ + Z_- + Z_0)$$

or

$$I_0 = \frac{E_+}{Z_+ + Z_- + Z_0} \tag{7.38}$$

The resulting equivalent circuit is shown in Figure 7.24.

We can now state the solution in terms of phase currents:

$$\begin{aligned} I_A &= \frac{3E_+}{Z_+ + Z_- + Z_0} \\ I_B &= 0 \\ I_C &= 0 \end{aligned} \tag{7.39}$$

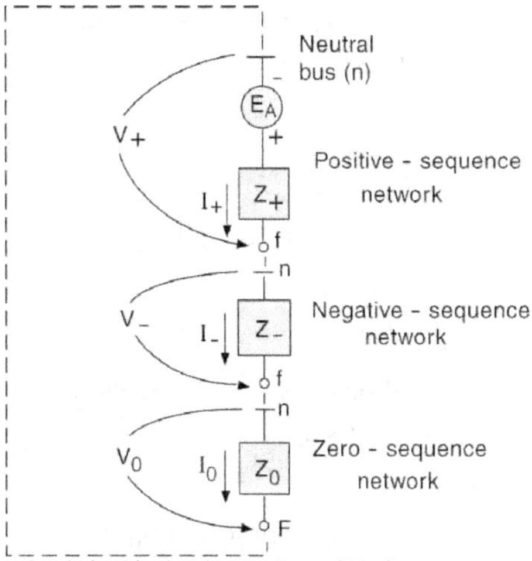


**Figure 7.24** Equivalent circuit for Single Line-to-Ground Fault.

For phase voltages we have

$$
\begin{aligned}
V_A &= 0 \\
V_B &= \frac{E_B(1-\alpha)[Z_0+(1+\alpha)Z_-]}{Z_0+Z_-+Z_+} \\
V_C &= \frac{E_C(1-\alpha)[(1+\alpha)Z_0+Z_-]}{Z_0+Z_-+Z_+}
\end{aligned}
\tag{7.40}
$$

## Example 7.5

Consider a system with sequence impedances given by $Z_+ = j0.2577$, $Z_- = j0.2085$, and $Z_0 = j0.14$; find the voltages and currents at the fault point for a single line-to-ground fault.

*Solution*

The sequence networks are connected in series for a single line-to-ground fault.

The sequence currents are given by

$$
\begin{aligned}
I_+ = I_- = I_0 &= \frac{1}{j(0.2577+0.2085+0.14)} \\
&= 1.65\angle -90^\circ \text{ p.u.}
\end{aligned}
$$

Therefore,

$$I_A = 3I_+ = 4.95\angle -90^\circ \text{ p.u.}$$
$$I_B = I_C = 0$$

The sequence voltages are as follows:

$$\begin{aligned}
V_+ &= E_+ - I_+ Z_+ \\
&= 1\angle 0 - (1.65\angle -90^\circ)(0.2577\angle 90^\circ) \\
&= 0.57 \text{ p.u.} \\
V_- &= -I_- Z_- \\
&= -(1.65\angle -90^\circ)(0.2085\angle 90^\circ) \\
&= -0.34 \text{ p.u.} \\
V_0 &= -I_0 Z_0 \\
&= -(1.65\angle -90^\circ)(0.14\angle 90^\circ) \\
&= -0.23 \text{ p.u.}
\end{aligned}$$

The phase voltages are thus

$$\begin{aligned}
V_A &= V_+ + V_- + V_0 = 0 \\
V_B &= \alpha^2 V_+ + \alpha V_- + V_0 \\
&= (1\angle 240^\circ)(0.57) + (1\angle 120^\circ)(-0.34) + (-0.23) \\
&= 0.86\angle -113.64^\circ \text{ p.u.} \\
V_C &= \alpha V_+ + \alpha^2 V_- + V_0 \\
&= (1\angle 120^\circ)(0.57) + (1\angle 240^\circ)(-0.34) + (-0.23) \\
&= 0.86\angle 113.64^\circ \text{ p.u.}
\end{aligned}$$

## 7.6 DOUBLE LINE-TO-GROUND FAULT

We will consider a general fault condition. In this case we assume that phase *B* has fault impedance of $Z_f$; phase *C* has a fault impedance of $Z_f$; and the common line-to-ground fault impedance is $Z_g$. This is shown in Figure 7.25.

The boundary conditions are as follows:

$$\begin{aligned}
I_A &= 0 \\
V_{Bn} &= I_B(Z_f + Z_g) + I_C Z_g \\
V_{Cn} &= I_B Z_g + (Z_f + Z_g) I_C
\end{aligned}$$

We can demonstrate that

$$E_+ - I_+(Z_+ + Z_f) = -I_-(Z_- + Z_f) = -I_0(Z_0 + Z_f + 3Z_g) \tag{7.41}$$

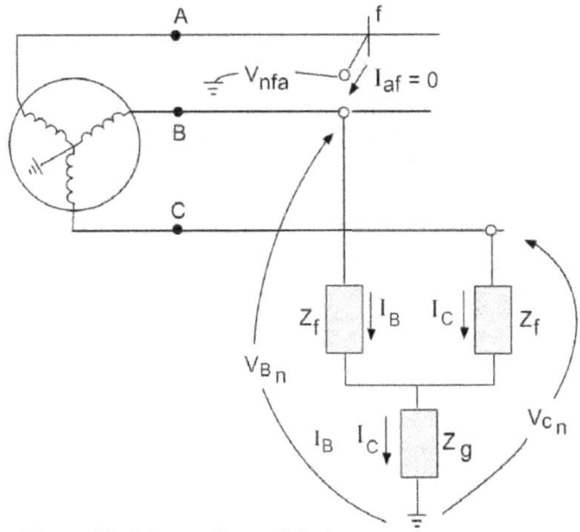


**Figure 7.25** Circuit with Double Line-to-Ground fault.

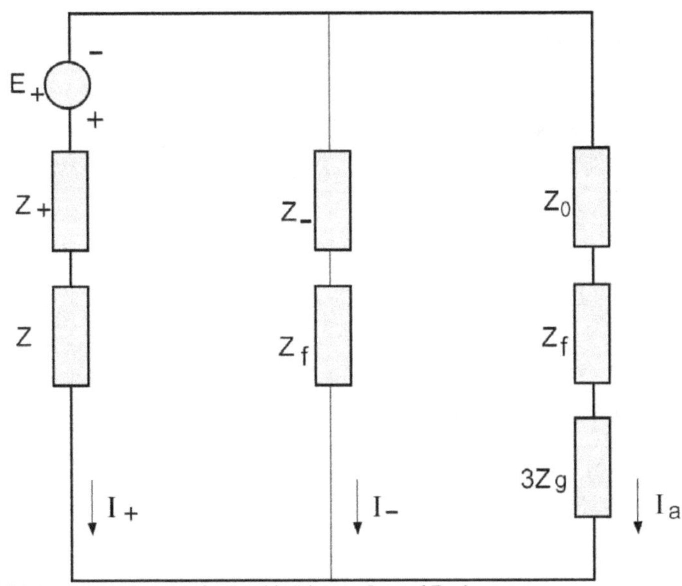


**Figure 7.26** Sequence Network for Double Line-to-Ground Fault.

The equivalent circuit is shown in Figure 7.26. It is clear from Eq. (7.41) that the sequence networks are connected in parallel. From the equivalent circuit we can obtain the positive, negative, and zero sequence currents easily.

**Example 7.6**

For the system of Example 7.5 find the voltages and currents at the fault point for a double line-to-ground fault. Assume

$$
\begin{aligned}
Z_f &= j0.05 \text{ p.u.} \\
Z_g &= j0.033 \text{ p.u.}
\end{aligned}
$$

*Solution*

The sequence network connection is as shown in Figure 7.27. Steps of the network reduction are also shown. From the figure, sequence currents are as follows:

$$
\begin{aligned}
I_+ &= \frac{1\angle 0}{0.45\angle 90^\circ} = 2.24\angle -90^\circ \\
I_- &= -I_+\left(\frac{0.29}{0.29+0.2585}\right) \\
&= -1.18\angle -90^\circ \\
I_0 &= -1.06\angle -90^\circ
\end{aligned}
$$

The sequence voltages are calculated as follows.

$$
\begin{aligned}
V_+ &= E_+ - I_+ Z_+ \\
&= 1\angle 0 - (2.24\angle -90^\circ)(0.26\angle -90^\circ) \\
&= 0.42 \\
V_- &= -I_- Z_- \\
&= +(1.18)(0.2085) = 0.25 \\
V_0 &= -I_0 Z_0 \\
&= (1.06)(0.14) = 0.15
\end{aligned}
$$

The phase currents are obtained as

$$I_A = 0$$

$$\begin{aligned} I_B &= \alpha^2 I_+ + \alpha I_- + I_0 \\ &= (1\angle 240)(2.24\angle -90^\circ) + (1\angle 120)(-1.18\angle -90^\circ) \\ &\qquad + (-1.06\angle -90^\circ) \\ &= 3.36\angle 151.77^\circ \end{aligned}$$

$$\begin{aligned} I_C &= \alpha I_+ + \alpha^2 I_- + I_0 \\ &= (1\angle 120)(2.24\angle -90^\circ) + (1\angle 240)(-1.18\angle -90^\circ) \\ &\qquad + (-1.06\angle -90^\circ) \\ &= 3.36\angle 28.23^\circ \end{aligned}$$

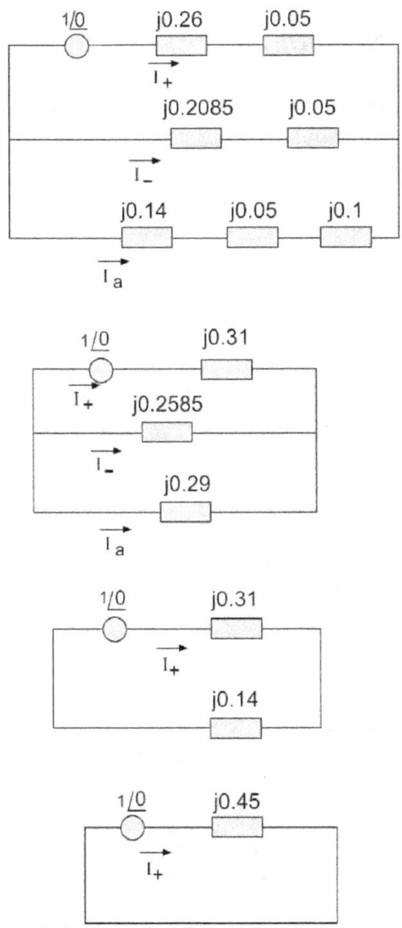


**Figure 7.27** Sequence Network for Example 7.6.

The phase voltages are found as

$$
\begin{aligned}
V_A &= V_+ + V_- + V_0 \\
&= 0.42 + 0.25 + 0.15 \\
&= 0.82 \\
V_B &= \alpha^2 V_+ + \alpha V_- + V_0 \\
&= (1\angle 240^\circ)(0.42) + (1\angle 120^\circ)(0.25) + (0.15) \\
&= 0.24\angle -141.49^\circ \\
V_C &= \alpha V_+ + \alpha^2 V_- + V_0 \\
&= (1\angle 120^\circ)(0.42) + (1\angle 240^\circ)(0.25) + 0.15 \\
&= 0.24\angle 141.49^\circ
\end{aligned}
$$

## 7.7 LINE-TO-LINE FAULT

Let phase $A$ be the unfaulted phase. Figure 7.28 shows a three-phase system with a line-to-line short circuit between phases $B$ and $C$. The boundary conditions in this case are

$$
\begin{aligned}
I_A &= 0 \\
I_B &= -I_C \\
V_B - V_C &= I_B Z_f
\end{aligned}
$$

The first two conditions yield

$$
\begin{aligned}
I_0 &= 0 \\
I_+ &= -I_- = \frac{1}{3}(\alpha - \alpha^2) I_B
\end{aligned}
$$

The voltage conditions give

$$V_+ - V_- = Z_f I_+ \tag{7.42}$$

The equivalent circuit will take on the form shown in Figure 7.29. Note that the zero sequence network is not included since $I_0 = 0$.

**Example 7.7**

For the system of Example 7.5, find the voltages and currents at the fault point for a line-to-line fault through an impedance $Z_f = j0.05$ p.u.

*Solution*

The sequence network connection is as shown in Figure 7.30. From the diagram,

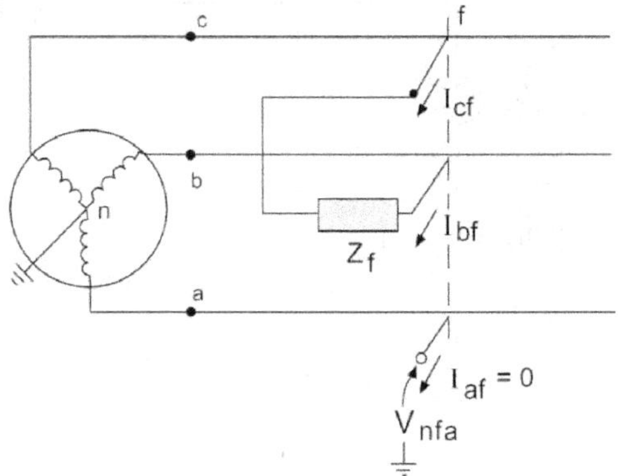


**Figure 7.28** Example of a Line-to-Line Fault.

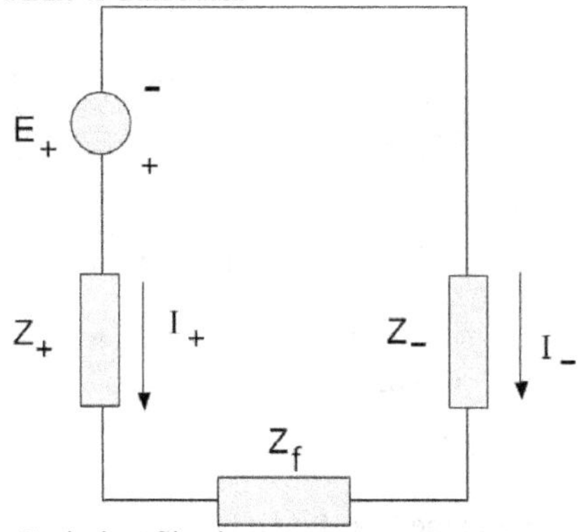


**Figure 7.29** Line-to-Line Equivalent Circuit.

$$
\begin{aligned}
I_+ &= -I_- = \frac{1\angle 0}{0.5185\angle 90^\circ} \\
&= 1.93\angle -90^\circ \text{ p.u.} \\
I_0 &= 0
\end{aligned}
$$

The phase currents are thus

$$
\begin{aligned}
I_A &= 0 \\
I_B &= -I_C \\
&= \left(\alpha^2 - \alpha\right)I_+ \\
&= \left(1\angle 240^\circ - 1\angle 120^\circ\right)\left(1.93\angle -90^\circ\right) \\
&= 3.34\angle -180^\circ \text{ p.u.}
\end{aligned}
$$

The sequence voltages are

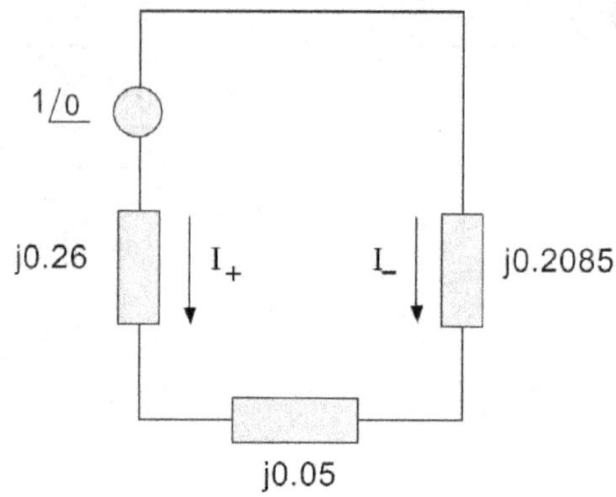


**Figure 7.30** Sequence Network Connection for Example 7.7.

$$
\begin{aligned}
V_+ &= E_+ - I_+ Z_+ \\
&= 1\angle 0 - (1.93\angle -90^\circ)(0.26\angle 90^\circ) \\
&= 0.5 \text{ p.u.} \\
V_- &= -I_- Z_- \\
&= -(1.93\angle -90^\circ)(0.2085\angle 90^\circ) \\
&= 0.4 \text{ p.u.} \\
V_0 &= -I_0 Z_0 \\
&= 0
\end{aligned}
$$

The phase voltages are obtained as shown below:

$$
\begin{aligned}
V_A &= V_+ + V_- + V_0 \\
&= 0.9 \text{ p.u.} \\
V_B &= \alpha^2 V_+ + \alpha V_- + V_0 \\
&= (1\angle 240^\circ)(0.5) + (1\angle 120^\circ)(0.4) \\
&= 0.46\angle -169.11^\circ \\
V_C &= \alpha V_+ + \alpha^2 V_- + V_0 \\
&= (1\angle 120^\circ)(0.5) + (1\angle 240^\circ)(0.4) \\
&= 0.46\angle 169.11^\circ
\end{aligned}
$$

As a check, we calculate

$$
\begin{aligned}
V_B - V_C &= 0.17\angle -90^\circ \\
I_B Z_f &= (3.34\angle -180^\circ)(0.05\angle 90^\circ) \\
&= 0.17\angle -90^\circ
\end{aligned}
$$

Hence,

$$V_B - V_C = I_B Z_f$$

## 7.8 THE BALANCED THREE-PHASE FAULT

Let us now consider the situation with a balanced three-phase fault on phases $A$, $B$, and $C$, all through the same fault impedance $Z_f$. This fault condition is shown in Figure 7.31. It is clear from inspection in Figure 7.31 that the phase voltage at the faults are given by

$$V_A = I_A Z_f \tag{7.43}$$

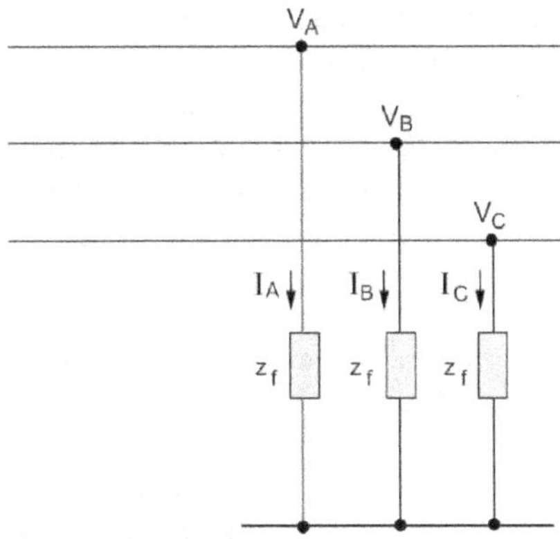


**Figure 7.31** A Balanced Three-Phase Fault.

$$V_B = I_B Z_f \tag{7.44}$$

$$V_C = I_C Z_f \tag{7.45}$$

We can show that

$$I_+ = \frac{E}{Z_+ + Z_f} \tag{7.46}$$

$$I_- = 0 \tag{7.47}$$

$$I_0 = 0 \tag{7.48}$$

The implications of Eqs. (7.47) and (7.48) are obvious. No zero sequence nor negative sequence components of the current exist. Instead, only positive sequence quantities are obtained in the case of a balanced three-phase fault.

### Example 7.8

For the system of Example 7.5, find the short-circuit currents at the fault point for a balanced three-phase fault through three impedances each having a value of $Z_f = j0.05$ p.u.

*Solution*

$$I_{A_{sc}} = I_+ = \frac{1\angle 0}{j(0.26+0.05)} = 3.23\angle -90^\circ$$

## 7.9 SYSTEM PROTECTION, AN INTRODUCTION

The result of the preceding section provides a basis to determine the conditions that exist in the system under fault conditions. It is important to take the necessary action to prevent the faults, and if they do occur, to minimize possible damage or possible power disruption. A protection system continuously monitors the power system to ensure maximum continuity of electrical supply with minimum damage to life, equipment, and property.

The following are some consequences of faults:

1. Abnormally large currents will flow in parts of system with associated overheating of components.
2. System voltages will be off their normal acceptable levels, resulting in possible equipment damage.
3. Parts of the system will be caused to operate as unbalanced three-phase systems, which will mean improper operation of some or all of the equipment.

A number of requirements for protective systems provide the basis for design criteria.

1. *Reliability*: Provide both dependability (guaranteed correct operation in response to faults) and security (avoiding unnecessary operation). Reliability requires that relay systems perform correctly under adverse system and environmental conditions.
2. *Speed*: Relays should respond to abnormal conditions in the least possible time. This usually means that the operation time should not exceed three cycles on a 60-Hz base.
3. *Selectivity*: A relay system should provide maximum possible service continuity with minimum system disconnection.
4. *Simplicity and economy*: The requirements of simplicity and economy are common in any engineering design, and relay systems are no exception.

A protective system detects fault conditions by continuously monitoring variables such as current, voltage, power, frequency, and impedance. Measuring currents and voltages is performed by instrument transformers of the potential type (P.T.) or current type (C.T.). Instrument transformers feed the measured variables to the relay system, which in turn, upon detecting a fault, commands a circuit-interrupting device known as the circuit breaker (C.B.) to disconnect the faulted section of the system.

An electric power system is divided into protective zones for each apparatus in the system. The division is such that zones are given adequate protection while keeping service interruption to a minimum. A single-line diagram of a part of a power system with its zones of protection is given in Figure 7.32. It is to be noted that each zone is overlapped to avoid unprotected (blind) areas.

## 7.10 PROTECTIVE RELAYS

A relay is a device that opens and closes electrical contacts to cause the operation of other devices under electric control. The relay detects intolerable or undesirable conditions within an assigned area. The relay acts to operate the appropriate circuit breakers to disconnect the area affected to prevent damage to personnel and property.

We classify relays according to their function, that is, as measuring or on-off relays. The latter class is also known as all-or-nothing and includes relays such as time-lag relays, auxiliary relays, and tripping relays. Here the relay does not have a specified setting and is energized by a quantity that is

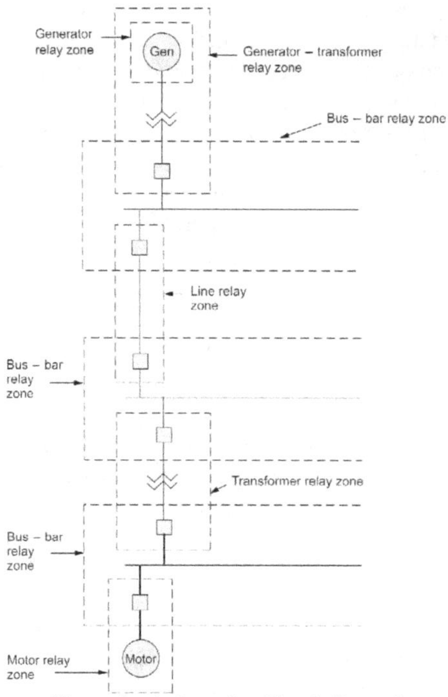


**Figure 7.32** Typical Zones of Protection in Part of an Electric Power System.

either higher than that at which it operates or lower than that at which it resets.

The class of measuring relays includes a number of types with the common feature that they operate at a predetermined setting. Examples are as follows:

- *Current relays*: Operate at a predetermined threshold value of current. These include overcurrent and undercurrent relays.
- *Voltage relays*: Operate at a predetermined value of voltage. These include overvoltage and undervoltage relays.
- *Power relays*: Operate at a predetermined value of power. These include overpower and underpower relays.
- *Directional relays*:
  (i) Alternating current: Operate according to the phase relationship between alternating quantities.
  (ii) Direct current: Operate according to the direction of the current and are usually of the permanent-magnetic, moving-coil pattern.
- *Differential relays*: Operate according to the scalar or vectorial difference between two quantities such as current, voltage, etc.
- *Distance relays*: Operate according to the "distance" between the relay's current transformer and the fault. The "distance" is measured in terms of resistance, reactance, or impedance.

Relays are made up of one or more fault-detecting units along with the necessary auxiliary units. Basic units for relay systems can be classified as being electromechanical units, sequence networks, solid-state units, or computer-based units. The electromechanical types include those based on magnetic attraction, magnetic induction, D'Arsonval, and thermal principles. Static networks with three-phase inputs can provide a single-phase output proportional to positive, negative, or zero sequence quantities. These are used as fault sensors and are known as sequence filters. Solid-state relays use low power components, which are designed into logic units used in many relays.

## Electromechanical Relays

We consider some electromechanical type relays such as the plunger unit, the clapper unit, the polar unit, and the induction disc types.

The *plunger type* has cylindrical coils with an external magnetic structure and a center plunger. The plunger moves upward to operate a set of contacts when the current or voltage applied to the coil exceeds a certain value. The moving force is proportional to the square of the current in the coil. These units operate instantaneously since no delay is intentionally introduced.

*Clapper units* have a U-shaped magnetic frame with a movable armature across the open end. The armature is hinged at one side and spring-restrained at the other. When the coil is energized, the armature moves toward the magnetic core, opening or closing a set of contacts with a torque proportional to the square of the coil current. Clapper units are less accurate than plunger units and are primarily applied as auxiliary or "go/no go" units.

*Polar units* use direct current applied to a coil wound around the hinged armature in the center of the magnetic structure. A permanent magnet across the structure polarizes the armature-gap poles. Two nonmagnetic spacers, located at the rear of the magnetic frames, are bridged by two adjustable magnetic shunts. This arrangement enables the magnetic flux paths to be adjusted for pickup and contact action. With balanced air gaps the armature will float in the center with the coil deenergized. With the gaps unbalanced, polarization holds the armature against one pole with the coil deenergized. The coil is arranged so that its magnetic axis is in line with the armature and at a right angle to the permanent magnet axis. Current in the coil magnetizes the armature either north or south, increasing or decreasing any prior polarization of the armature. If the magnetic shunt adjustment normally makes the armature a north pole, it will move to the right. Direct current in the operating coil, which tends to make the contact end a south pole, will overcome this tendency, and the armature will move to the left to close the contacts.

*Induction disc units* employ the watt hour meter design and use the same operating principles. They operate by torque resulting from the interaction of fluxes produced by an electromagnet with those from induced currents in the plane of a rotatable aluminum disc. The unit shown in Figure 7.33 has three poles on one side of the disc and a common magnetic keeper on the opposite side. The main coil is on the center leg. Current ($I$) in the main coil produces flux ($\phi$), which passes through the air gap and disc to the keeper. The flux $\phi$ is divided into $\phi_L$ through the left-hand leg and $\phi_R$ through the right-hand leg. A short-circuited lagging coil on the left leg causes $\phi_L$ to lag both $\phi_R$ and $\phi$, producing a split-phase motor action. The flux $\phi_L$ induces a voltage $V_s$, and current $I_s$ flows, in phase, in the shorted lag coil. The flux $\phi_T$ is the total flux produced by the main coil current ($I$). The three fluxes cross the disc air gap and produce eddy currents in the disc. As a result, the eddy currents set up counter fluxes, and the interaction of the two sets of fluxes produces the torque that rotates the disc.

A spiral spring on the disc shaft conducts current to the moving contact. This spring, together with the shape of the disc and the design of electromagnet, provides a constant minimum operating current over the contact's travel range. A permanent magnet with adjustable keeper (shunt) damps the disc, and the magnetic plugs in the electromagnet control the degree of saturation. The spring tension, the damping magnet, and the magnetic plugs allow separate and relatively independent adjustment of the unit's inverse time overcurrent characteristics.

## Solid-State Units

Solid-state, linear, and digital-integrated circuit logic units are combined in a variety of ways to provide modules for relays and relay systems.

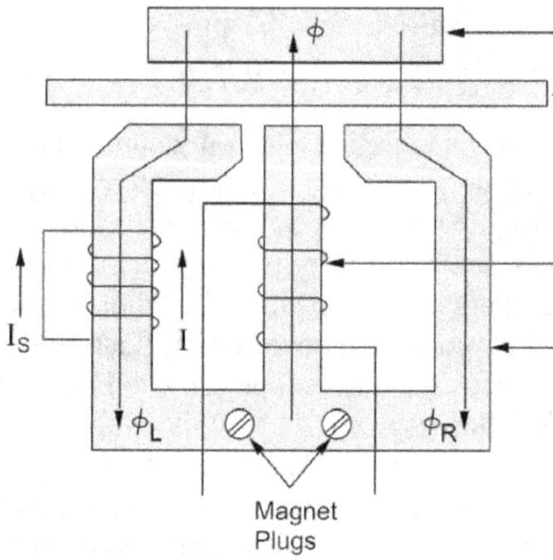


**Figure 7.33** Induction Disc-Type Relay Unit.

Three major categories of circuits can be identified: (1) fault-sensing and data-processing logic units, (2) amplification logic units, and (3) auxiliary logic units.

Logic circuits in the fault-sensing and data-processing category employ comparison units to perform conventional fault-detection duties. Magnitude comparison logic units are used for overcurrent detection both of instantaneous and time overcurrent categories. For instantaneous overcurrent protection, a dc level detector, or a fixed reference magnitude comparator, is used. A variable reference magnitude comparator circuit is used for ground-distance protection. Phase-angle comparison logic circuits produce an output when the phase angle between two quantities is in the critical range. These circuits are useful for phase, distance, and directional relays.

## 7.11 TRANSFORMER PROTECTION

A number of fault conditions can arise within a power transformer. These include:

1. *Earth faults*: A fault on a transformer winding will result in currents that depend on the source, neutral grounding impedance, leakage reactance of the transformer, and the position of the fault in the windings. The winding connections also influence the magnitude of fault current. In the case of a Y-connected winding with neutral point connected to ground through an impedance $Z_g$, the fault current depends on $Z_g$ and is proportional to the distance of the fault from the neutral point. If the neutral is solidly grounded, the fault current is controlled by the leakage reactance, which depends on fault location. The reactance decreases as the fault becomes closer to the neutral point. As a result, the fault current is highest for a fault close to the neutral point. In the case

of a fault in a Δ-connected winding, the range of fault current is less than that for a Y-connected winding, with the actual value being controlled by the method of grounding used in the system. Phase fault currents may be low for a Δ-connected winding due to the high impedance to fault of the Δ winding. This factor should be considered in designing the protection scheme for such a winding.

2. *Core faults* due to insulation breakdown can permit sufficient eddy-current to flow to cause overheating, which may reach a magnitude sufficient to damage the winding.
3. *Interturn faults* occur due to winding flashovers caused by line surges. A short circuit of a few turns of the winding will give rise to high currents in the short-circuited loops, but the terminal currents will be low.
1. *Phase-to-phase* faults are rare in occurrence but will result in substantial currents of magnitudes similar to those caused by earth faults.
5. *Tank faults* resulting in loss of oil reduce winding insulation as well as producing abnormal temperature rises.

In addition to fault conditions within the transformer, abnormal conditions due to external factors result in stresses on the transformer. These conditions include: *overloading*, *system faults*, *overvoltages*, and *underfrequency operation*.

When a transformer is switched into the circuit at any point of the supply voltage wave, the peak values of the core flux wave will depend on the residual flux as well as on the time of switching. The peak value of the flux will be higher than the corresponding steady-state value and will be limited by core saturation. The magnetizing current necessary to produce the core flux can have a peak of eight to ten times the normal full-load peak and has no equivalent on the secondary side. This phenomenon is called *magnetizing inrush current* and appears to the system as an internal fault. Maximum inrush occurs if the transformer is switched in when the supply voltage is zero. Realizing this, is important for the design of differential relays for transformer protection so that no tripping takes place due to the magnetizing inrush current. A number of schemes based on the harmonic properties of the inrush current are used to prevent tripping due to large inrush currents.

Overheating protection is provided for transformers by placing a thermal-sensing element in the transformer tank. Overcurrent relays are used as a backup protection with time delay higher than that for the main protection. Restricted earth fault protection is utilized for Y-connected windings. This scheme is shown in Figure 7.34. The sum of the phase currents is balanced against the neutral current, and hence the relay will not respond to faults outside the winding.

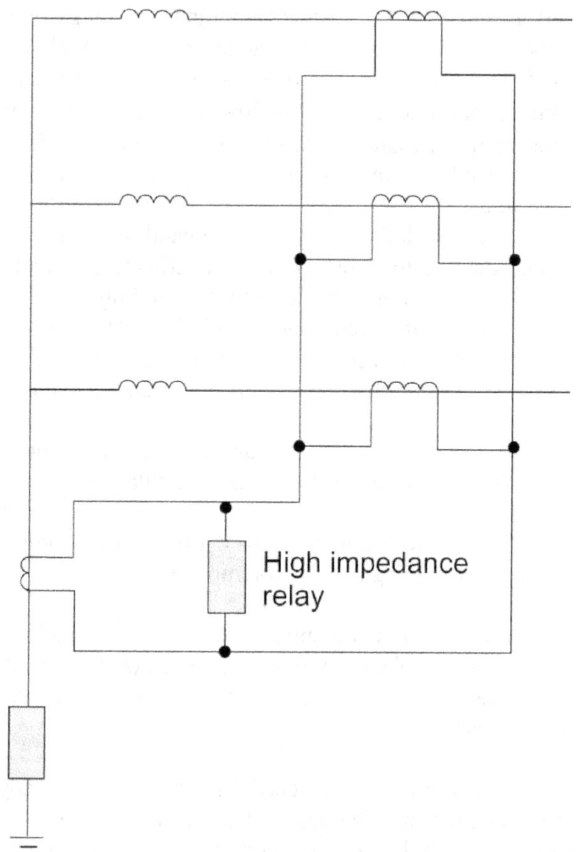


**Figure 7.34** Restricted Ground Fault Protection for a Y Winding.

Differential protection is the main scheme used for transformers. The principle of a differential protection system is simple. Here the currents on each side of the protected apparatus for each phase are compared in a differential circuit. Any difference in currents will operate a relay. Figure 7.35 shows the relay circuit for one phase only. On normal operation, only the difference between the current transformer magnetizing currents $i_{m_1}$ and $i_{m_2}$ passes through the relay. This is due to the fact that with no faults within the protected apparatus, the currents entering and leaving are equal to $i$. If a fault occurs between the two sets of current transformers, one or more of the currents (in a three-phase system) on the left-hand side will suddenly increase, while that on the right-hand side may decrease or increase with a direction reversal. In both instances, the total fault current will flow through the relay, causing it to operate. In units where the neutral ends are inaccessible, differential relays are not used, but reverse power relays are employed instead.

A number of considerations should be dealt with in applying differential protection, including:

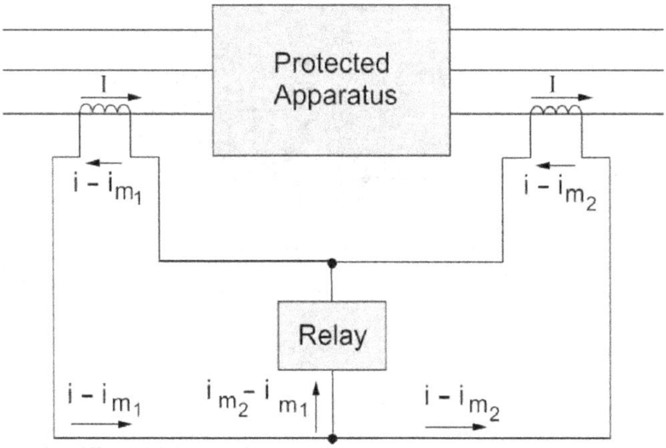


**Figure 7.35** Basic Differential Connection.

1. Transformer ratio: The current transformers should have ratings to match the rated currents of the transformer winding to which they are applied.
2. Due to the 30°-phase change between Y- and Δ-connected windings and the fact that zero sequence quantities on the Y side do not appear on the terminals of the Δ side, the current transformers should be connected in Y for a Δ winding and in Δ for a Y winding. Figure 7.36 shows the differential protection scheme applied to a Δ/Y transformer. When current transformers are connected in Δ, their secondary ratings must be reduced to $\left(1/\sqrt{3}\right)$ times the secondary rating of Y-connected transformers.
3. Allowance should be made for tap changing by providing restraining coils (bias). The bias should exceed the effect of the maximum ratio deviation.

**Example 7.9**

Consider a Δ/Y-connected, 20-MVA, 33/11-kV transformer with differential protection applied, for the current transformers ratios shown in Figure 7.37. Calculate the relay currents on full load. Find the minimum relay current setting to allow 125 percent overload.

*Solution*

The primary line current is given by

$$I_p = \frac{20\times10^6}{\left(\sqrt{3}\right)\left(33\times10^3\right)} = 349.91\text{ A}$$

The secondary line current is

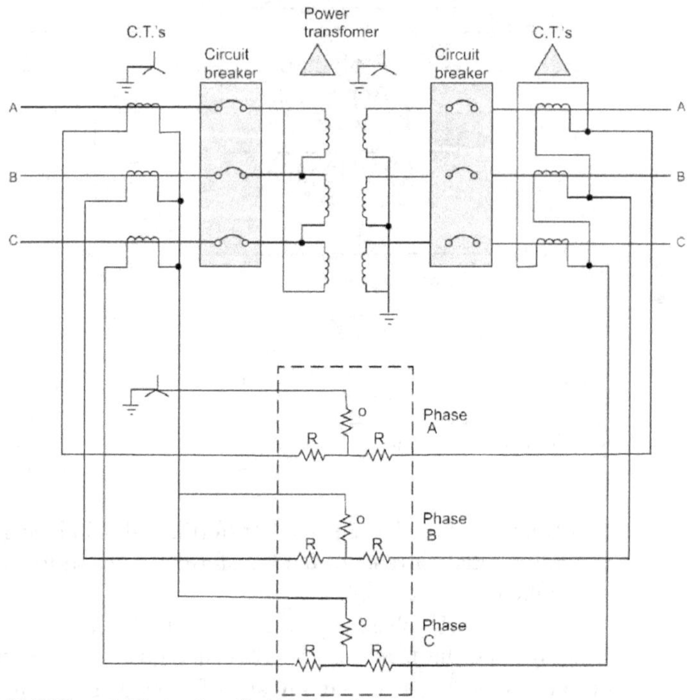


**Figure 7.36** Differential Protection of a Δ/Y Transformer.

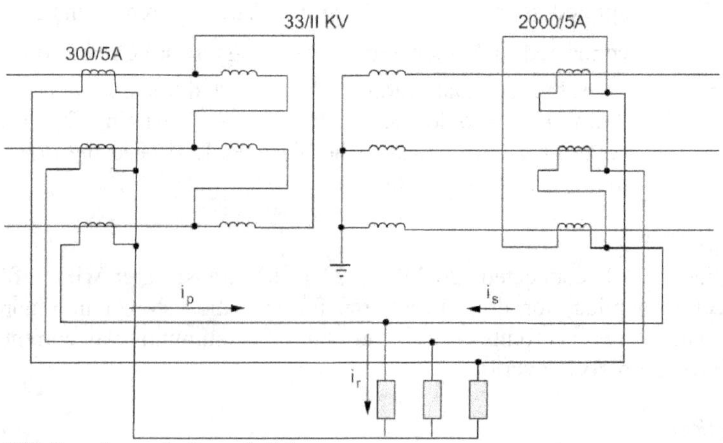


**Figure 7.37** Transformer for Example 7.9.

$$I_s = \frac{20 \times 10^6}{(\sqrt{3})(11 \times 10^3)} = 1049.73 \text{ A}$$

The C.T. current on the primary side is thus

$$i_p = 349.91\left(\frac{5}{300}\right) = 5.832 \text{ A}$$

The C.T. current in the secondary side is

$$i_s = 1049.73\left(\frac{5}{2000}\right)\sqrt{3} = 4.545 \text{ A}$$

Note that we multiply by $\sqrt{3}$ to obtain the values on the line side of the Δ-connected C.T.'s. The relay current on normal load is therefore

$$i_r = i_p - i_s = 5.832 - 4.545 = 1.287 \text{ A}$$

With 1.25 overload ratio, the relay setting should be

$$I_r = (1.25)(1.287) = 1.61 \text{ A}$$

## Buchholz (Gas Relay) Protection

In addition to the preceding protection schemes, it is common practice in transformer protection to employ gas-actuated relays for alarm and tripping. One such a relay is the Buchholz relay.

Faults within a transformer will result in heating and decomposing of the oil in the transformer tank. The decomposition produces gases such as hydrogen, carbon monoxide, and light hydrocarbons, which are released slowly for minor faults and rapidly for severe arcing faults. The relay is connected into the pipe leading to the conservator tank. As the gas accumulates, the oil level falls and a float *F* is lowered and operates a mercury switch to sound an alarm. Sampling the gas and performing a chemical analysis provide a means for classifying the type of fault. In the case of a winding fault, the arc generates gas at a high release rate that moves a vane *V* to cause tripping through contacts attached to the vane.

Buchholz protection provides an alarm for a number of fault conditions including:

1. Interturn faults or winding faults involving only lower power levels.
2. Core hot spots due to short circuits on the lamination insulation.
3. Faulty joints.
4. Core bolt insulation failure.

## 7.12 TRANSMISSION LINE PROTECTION

The excessive currents accompanying a fault, are the basis of overcurrent protection schemes. For transmission line protection in interconnected systems, it is necessary to provide the desired selectivity such that relay operation results in the least service interruption while isolating the fault. This is referred to as *relay coordination.* Many methods exist to achieve the desired selectivity. Time/current gradings are involved in three basic methods discussed below for radial or loop circuits where there are several line sections in series.

### Three Methods of Relay Grading

#### A) Time Grading

Time grading ensures that the breaker nearest to the fault opens first, by choosing an appropriate time setting for each of the relays. The time settings increase as the relay gets closer to the source. A simple radial system shown in Figure 7.38 demonstrates this principle.

A protection unit consisting of a definite time-delay overcurrent relay is placed at each of the points 2, 3, 4, and 5. The time-delay of the relay provides the means for selectivity. The relay at circuit breaker 2 is set at the shortest possible time necessary for the breaker to operate (typically 0.25 second). The relay setting at 3 is chosen here as 0.5 second, that of the relay at 4 at 1 second, and so on. In the event of a fault at $F$, the relay at 2 will operate and the fault will be isolated before the relays at 3, 4, and 5 have sufficient time to operate. The shortcoming of the method is that the longest fault-clearing time is associated with the sections closest to the source where the faults are most severe.

#### B) Current Grading

Fault currents are higher the closer the fault is to the source and this is utilized in the current-grading method. Relays are set to operate at a suitably graded current setting that decreases as the distance from the source is increased. Figure 7.39 shows an example of a radial system with current grading.

#### C) Inverse-Time Overcurrent Relaying

The inverse-time overcurrent relay method evolved because of the limitations imposed by the use of either current or time alone. With this method, the time of operation is inversely proportional to the fault current level, and the actual characteristics are a function of both time and current settings. Figure 7.40 shows some typical inverse-time relay characteristics. Relay type CO-7 is in common use. Figure 7.41 shows a radial system with time-graded inverse

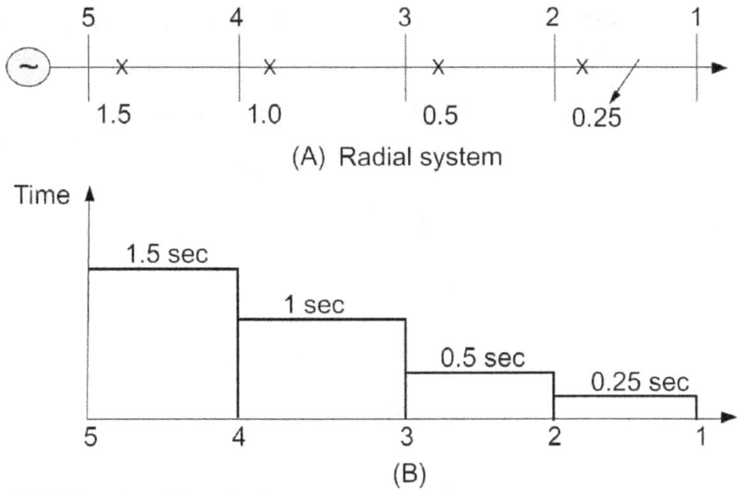


**Figure 7.38** Principles of Time Grading.

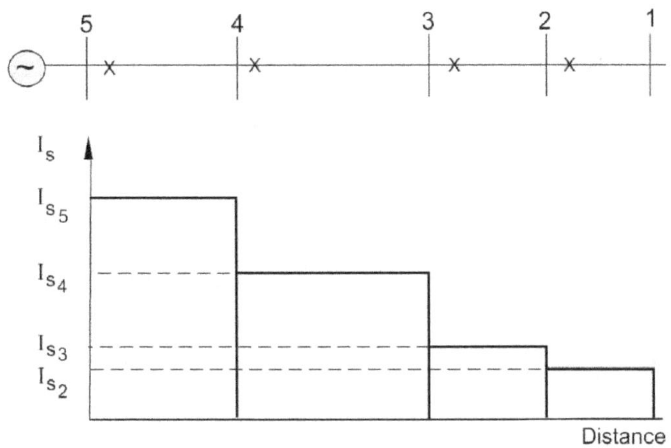


**Figure 7.39** Current Grading for a Radial System.

relays applied at breakers 1, 2, and 3.

For faults close to the relaying points, the inverse-time overcurrent method can achieve appreciable reductions in fault-clearing times.

The operating time of the time-overcurrent relay varies with the current magnitude. There are two settings for this type of relay:

1. *Pickup current* is determined by adjusted current coil taps or current tap settings (C.T.S.). The pickup current is the current that causes the relay to operate and close the contacts.
2. *Time dial* refers to the reset position of the moving contact, and it

varies the time of operation at a given tap setting and current magnitude.

The time characteristics are plotted in terms of time versus multiples of current tap (pickup) settings, for a given time dial position. There are five different curve shapes referred to by the manufacturer:

| | |
|---|---|
| CO-11 | Extreme inverse |
| CO-9 | Very inverse |
| CO-8 | Inverse |
| CO-7 | Moderately inverse |
| CO-6 | Definite minimum |

These shapes are given in Figure 7.40.

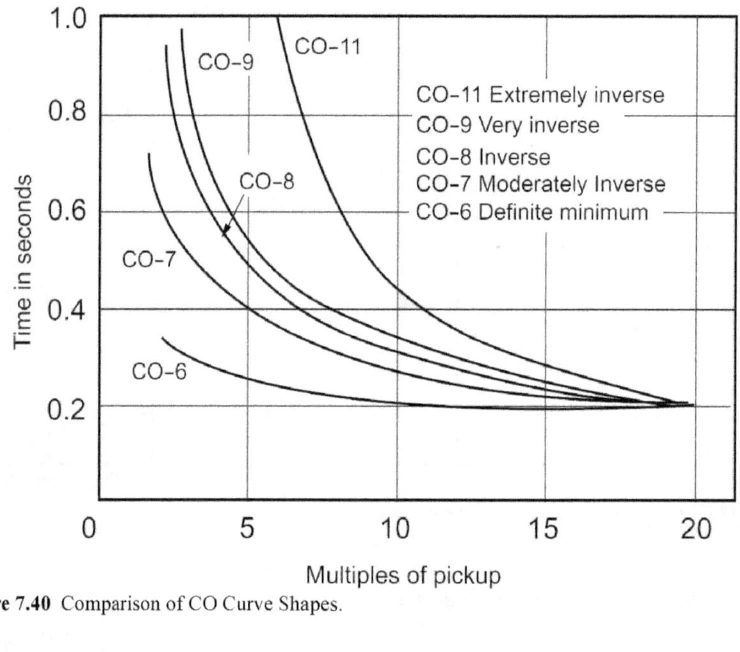


**Figure 7.40** Comparison of CO Curve Shapes.

**Example 7.10**

Consider the 11-kV radial system shown in Figure 7.42. Assume that all loads have the same power factor. Determine relay settings to protect the system assuming relay type CO-7 (with characteristics shown in Figure 7.43) is used.

*Solution*

The load currents are calculated as follows:

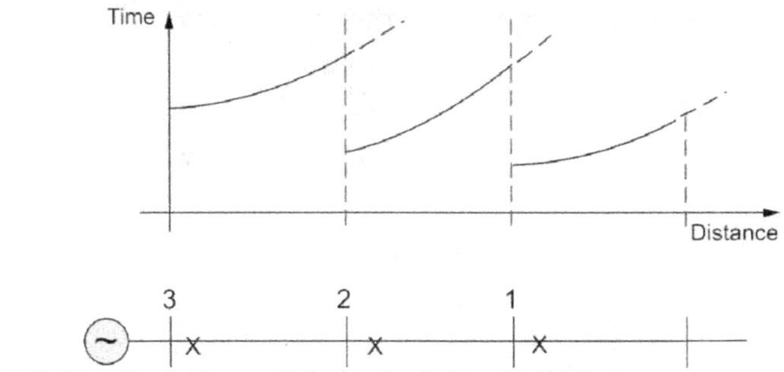


**Figure 7.41** Time-Graded Inverse Relaying Applied to a Radial System.

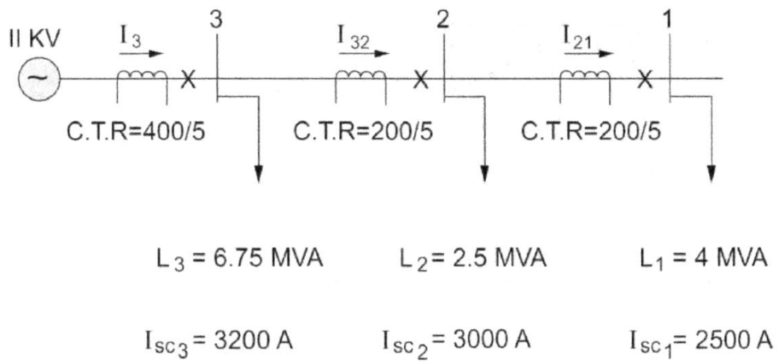


**Figure 7.42** An Example Radial System.

$$I_1 = \frac{4\times10^6}{\sqrt{3}\left(11\times10^3\right)} = 209.95 \text{ A}$$

$$I_2 = \frac{2.5\times10^6}{\sqrt{3}\left(11\times10^3\right)} = 131.22 \text{ A}$$

$$I_3 = \frac{6.75\times10^6}{\sqrt{3}\left(11\times10^3\right)} = 354.28 \text{ A}$$

The normal currents through the sections are calculated as

$$I_{21} = I_1 = 209.95 \text{ A}$$

$$I_{32} = I_{21} + I_2 = 341.16 \text{ A}$$

$$I_S = I_{32} + I_3 = 695.44 \text{ A}$$

With the current transformer ratios given, the normal relay currents are

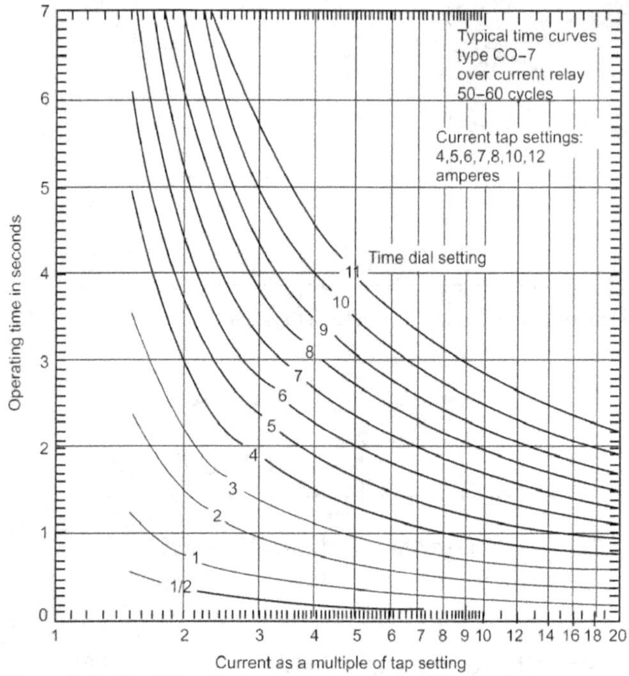


**Figure 7.43** CO-7 Time-Delay Overcurrent Relay Characteristics.

$$i_{21} = \frac{209.92}{\frac{200}{5}} = 5.25 \text{ A}$$

$$i_{32} = \frac{341.16}{\frac{200}{5}} = 8.53 \text{ A}$$

$$i_S = \frac{695.44}{\frac{400}{5}} = 8.69 \text{ A}$$

We can now obtain the current tap settings (C.T.S.) or pickup current in such a manner that the relay does not trip under normal currents. For this type of relay, the current tap settings available are 4, 5, 6, 7, 8, 10, and 12 amperes. For position 1, the normal current in the relay is 5.25 A; we thus choose

$$(\text{C.T.S.})_1 = 6 \text{ A}$$

For position 2, the normal relay current is 8.53 A, and we choose

$$(\text{C.T.S.})_2 = 10 \text{ A}$$

Similarly for position 3,

$$(\text{C.T.S.})_3 = 10 \text{ A}$$

Observe that we have chosen the nearest setting higher than the normal current.

The next task is to select the intentional delay indicated by the time dial setting (T.D.S.). We utilize the short-circuit currents calculated to coordinate the relays. The current in the relay at 1 on a short circuit at that location is

$$i_{SC_1} = \frac{2500}{\left(\frac{200}{5}\right)} = 62.5 \text{ A}$$

Expressed as a multiple of the pickup or C.T.S. value, we have

$$\frac{i_{SC_1}}{(\text{C.T.S.})_1} = \frac{62.5}{6} = 10.42$$

We choose the lowest T.D.S. for this relay for fastest action. Thus

$$(\text{T.D.S.})_1 = \frac{1}{2}$$

By reference to the relay characteristic, we get the operating time for relay 1 for a fault at 1 as

$$T_{1_1} = 0.15 \text{ s}$$

To set the relay at 2 responding to a fault at 1, we allow 0.1 second for breaker operation and an error margin of 0.3 second in addition to $T_{1_1}$. Thus,

$$T_{2_2} = T_{1_2} + 0.1 + 0.3 = 0.55 \text{ s}$$

The short circuit for a fault at 1 as a multiple of the C.T.S. at 2 is

$$\frac{i_{SC_1}}{(\text{C.T.S.})_2} = \frac{62.5}{10} = 6.25$$

From the characteristics for 0.55-second operating time and 6.25 ratio, we get

$$(\text{T.D.S.})_2 \cong 2$$

The final steps involve setting the relay at 3. For a fault at bus 2, the short-circuit current is 3000 A, for which relay 2 responds in a time $T_{22}$ obtained as follows:

$$\frac{i_{SC_2}}{(\text{C.T.S.})_2} = \frac{3000}{\left(\frac{200}{5}\right)10} = 7.5$$

For the $(\text{T.D.S.})_2 = 2$, we get from the relay's characteristic,

$$T_{22} = 0.50 \text{ s}$$

Thus allowing the same margin for relay 3 to respond to a fault at 2, as for relay 2 responding to a fault at 1, we have

$$\begin{aligned} T_{32} &= T_{22} + 0.1 + 0.3 \\ &= 0.90 \text{ s} \end{aligned}$$

The current in the relay expressed as a multiple of pickup is

$$\frac{i_{SC_2}}{(\text{C.T.S.})_3} = \frac{3000}{\left(\frac{400}{5}\right)10} = 3.75$$

Thus for $T_3 = 0.90$, and the above ratio, we get from the relay's characteristic,

$$(\text{T.D.S.})_3 \cong 2.5$$

We note here that our calculations did not account for load starting currents that can be as high as five to seven times rated values. In practice, this should be accounted for.

## Pilot-Wire Feeder Protection

Graded overcurrent feeder protection has two disadvantages. First, the grading settings may lead to tripping times that are too long to prevent damage and service interruption. Second, satisfactory grading for complex networks is quite difficult to attain. This led to the concept of "unit protection" involving the measurement of fault currents at each end of a limited zone of the feeder and the transmission of information between the equipment at zone boundaries. The principle utilized here is the differential (often referred to as Merz-Price) protection scheme. For short feeders, pilot-wire schemes are used to transmit the information. Pilot-wire differential systems of feeder protection are classified into three types: (1) the circulating-current systems, (2) the balanced-voltage systems, and (3) the phase-comparison (Casson-Last) system. All three systems depend on the fact that, capacitance current neglected, the instantaneous value of the current flowing into a healthy conductor at one end of the circuit is equal to the instantaneous current flowing out of the conductor at the other end, so that the net instantaneous current flowing into or out of the conductor is zero if the conductor is healthy. If, on the other hand, the conductor is short-circuited to earth or to another conductor at some point, then the net current flowing into or out of the conductor is equal to the instantaneous value of the current flowing out of or into the conductor at the point of fault.

## 7.13 IMPEDANCE-BASED PROTECTION PRINCIPLES

This section discusses the principles involved in protecting components such as transmission lines on the basis of measuring the input impedance of the component. We first discuss the idea of an *X-R* diagram which is an excellent graphical tool to demonstrate principles of impedance protection systems. The concept of relay comparators is then introduced. The specific parameter choices to allow for the creation of impedance relays based on either amplitude or phase comparisons are then discussed. The section concludes with a discussion of distance protection.

### A) The X-R Diagram

Consider a transmission line with series impedance $Z_L$ and negligible shunt admittance. At the receiving end, a load of impedance $Z_R$ is assumed. The phasor diagram shown in Figure 7.44 is constructed with $I$ taken as the reference. The phasor diagram represents the relation

$$V_S = IZ_L + V_r \tag{7.49}$$

giving rise to the heavy-lines diagram rather than the usual one shown by the dashed line. On the diagram, $d$ is the torque angle, which is the angle between $V_s$ and $V_r$.

If the phasor diagram, Eq. (7.49), is divided by the current $I$, we obtain the impedance equation

$$Z_s = Z_L + Z_r \tag{7.50}$$

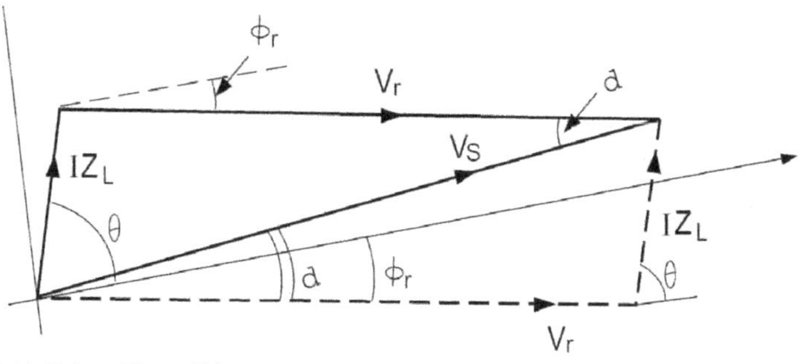


**Figure 7.44** Voltage Phasor Diagram.

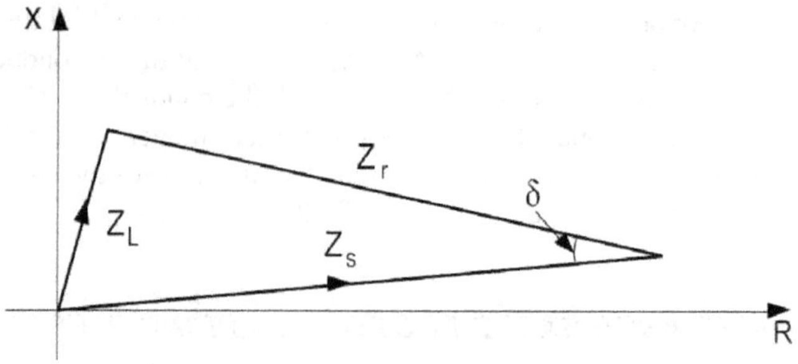


**Figure 7.45** Impedance Diagram.

where

$$Z_s = \frac{V_s}{I}$$

$$Z_r = \frac{V_r}{I}$$

An impedance diagram is shown in Figure 7.45. This is called the *X-R* diagram since the real axis represents a resistive component ($R$), and the imaginary axis corresponds to a reactive component ($X$). The angle $\delta$ appears on the impedance diagram as that between $Z_s$ and $Z_r$. The evaluation of $Z_r$ from complex power $S_R$ and voltage $V_r$ is straightforward.

## B) Relay Comparators

Relay comparators can have any number of input signals. However, we discuss the two-input comparator shown schematically in Figure 7.46. The input to the two transformer circuits 1 and 2 includes the line voltage $V_L$ and current $I_L$. The output of transformer 1 is $V_1$, and that of transformer 2 is $V_2$. Both $V_1$ and $V_2$ are input to the comparator, which produces a trip (operate) signal whenever $|V_2| > |V_1|$ in an amplitude comparison mode.

We assume that the line voltage $V_L$ is the reference phasor and that the line current lags $V_L$ by and angle $\phi_L$. Thus,

$$V_L = |V_L|\angle 0$$

$$I_L = |I_L|\angle -\phi_L$$

The impedance $Z_L$ is thus

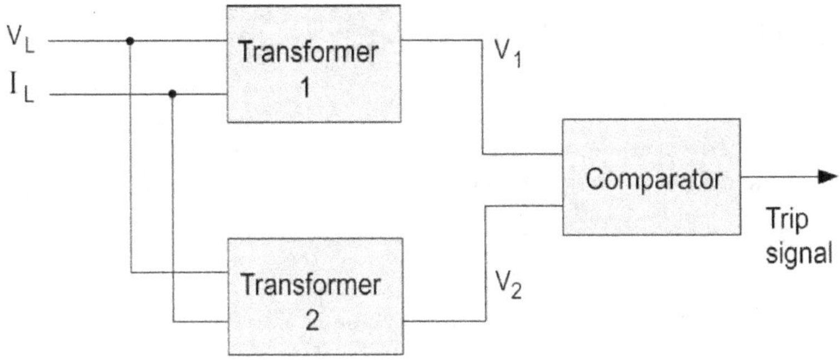


**Figure 7.46** Schematic of Relay Comparator Circuit.

$$Z_L = \frac{V_L}{I_L}$$
$$= |Z_L| \angle \phi_L$$

The transformers' output voltages $V_1$ and $V_2$ are assumed to be linear combinations of the input quantities

$$V_1 = k_1 V_L + Z_1 I_L \tag{7.51}$$

$$V_2 = k_2 V_L + Z_2 I_L \tag{7.52}$$

The impedances $Z_1$ and $Z_2$ are expressed in the polar form:

$$Z_1 = |Z_1| \angle \psi_1$$
$$Z_2 = |Z_2| \angle \psi_2$$

The comparator input voltages $V_1$ and $V_2$ are thus given by

$$V_1 = |I_L| \left( k_1 |Z_L| + |Z_1| \angle \psi_1 - \phi_L \right) \tag{7.53}$$

$$V_2 = |I_L| \left( k_2 |Z_L| + |Z_2| \angle \psi_2 - \phi_L \right) \tag{7.54}$$

### C) Amplitude Comparison

The trip signal is produced for an amplitude comparator when

$$|V_2| \geq |V_1| \tag{7.55}$$

The operation condition is obtained as

$$\left(k_1^2 - k_2^2\right)\left|Z_L\right|^2 + 2\left|Z_L\right|\left[k_1\left|Z_1\right|\cos(\psi_1 - \phi_L) - k_2\left|Z_2\right|\cos(\psi_2 - \phi_L)\right] + \left(\left|Z_1\right|^2 - \left|Z_2\right|^2\right) \leq 0 \tag{7.56}$$

This is the general equation for an amplitude comparison relay. The choices of $k_1$, $k_2$, $Z_1$, and $Z_2$ provide different relay characteristics.

### *Ohm Relay*

The following parameter choice is made:

$$k_1 = k \qquad k_2 = -k$$
$$Z_1 = 0 \qquad Z_2 = Z$$
$$\psi_1 = \psi_2 = \psi$$

The relay threshold equation becomes

$$R_L \cos\psi + X_L \sin\psi \leq \frac{|Z|}{2k} \tag{7.57}$$

This is a straight line in the $X_L$-$R_L$ plane as shown in Figure 7.47. The shaded area is the restrain area; an operate signal is produced in the nonshaded area.

### *Mho Relay*

The mho relay characteristic is obtained with the choice

$$k_1 = -k \qquad k_2 = 0$$
$$Z_1 = Z_2 = Z$$
$$\psi_1 = \psi_2 = \psi$$

$$\left(R_L - \frac{|Z|}{k}\cos\psi\right)^2 + \left(X_L - \frac{|Z|}{k}\sin\psi\right)^2 \leq \frac{|Z|^2}{k^2} \tag{7.58}$$

The threshold condition with equality sign is a circle as show in Figure 7.48.

### *Impedance Relay*

Here we set

$$k_1 = -k \qquad k_2 = 0$$
$$Z_1 \neq Z_2$$

The threshold equation is

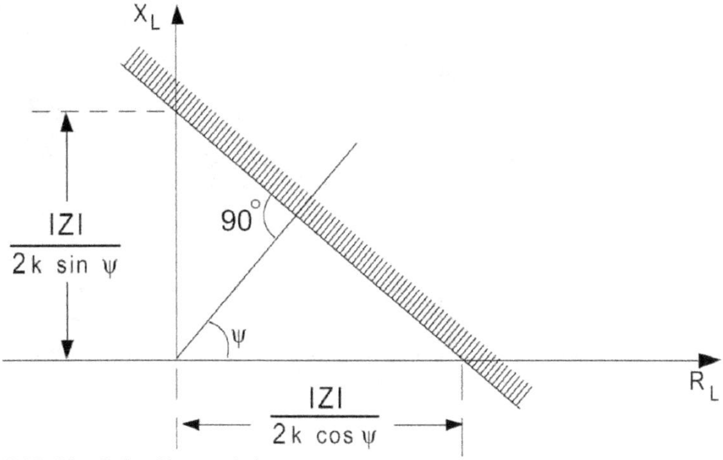


**Figure 7.47** Ohm Relay Characteristic.

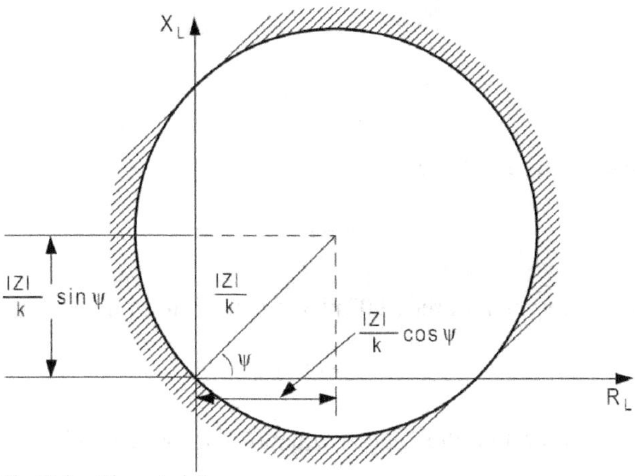


**Figure 7.48** Mho Relay Characteristic.

$$\left(R_L - \frac{|Z_1|\cos\psi}{k}\right)^2 + \left(X_L - \frac{|Z_1|\sin\psi}{k}\right)^2 \leq \frac{|Z_2|^2}{k^2} \tag{7.59}$$

The threshold condition is a circle with center at $|Z_1|/k\angle\psi$ and radius $|Z_2|/k$ as shown in Figure 7.49.

### *Phase Comparison*

Let us now consider the comparator operating in the phase comparison mode. Assume that

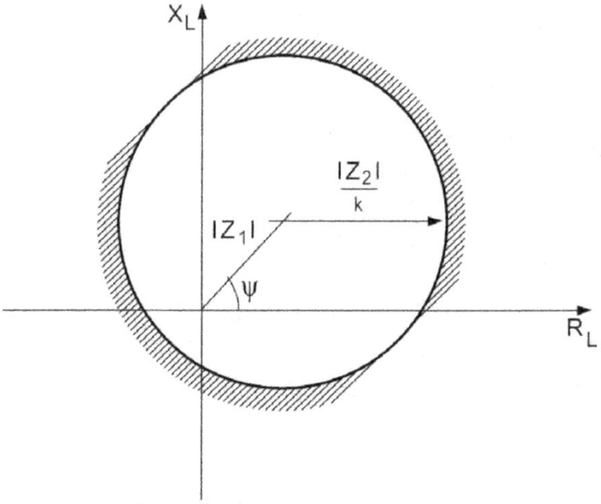


**Figure 7.49** Impedance Relay Characteristic.

$$V_1 = |V_1|\angle\theta_1$$
$$V_2 = |V_2|\angle\theta_2$$

Let the phase difference be defined as

$$\theta = \theta_1 - \theta_2$$

A criterion for operation of the ±90° phase comparator implies that

$$\cos\theta \geq 0$$

We can demonstrate that the general equation for the ±90° phase comparator is given by

$$k_2 k_2 |Z_L|^2 + |Z_L|[k_1|Z_2|\cos(\psi_1 - \phi_L) + k_2|Z_1|\cos(\psi_1 - \phi_L)] + |Z_1||Z_2|\cos(\psi_1 - \phi_L) \geq 0 \quad \textbf{(7.60)}$$

By assigning values to the parameters $k_2$, $k_3$, $Z_1$, and $Z_2$, different relay characteristics such as the ohm and mho relays are obtained.

## D) Distance Protection

Protection of lines and feeders based on comparison of the current values at both ends of the line can become uneconomical. Distance protection utilizes the current and voltage at the beginning of the line in a comparison scheme that essentially determines the fault position. Impedance measurement is performed using relay comparators. One input is proportional to the fault current and the other supplied by a current proportional to the fault loop voltage.

The characteristic of a plain impedance relay is shown in Figure 7.49. It will thus respond to faults behind it (third quadrant) in the *X-R* diagram as well as in front of it. One way to prevent this is to add a separate directional relay that will restrain tripping for faults behind the protected zone. The reactance or mho relay with characteristics as shown in Figure 7.48 combines the distance-measuring ability and the directional property. The term mho is given to the relay where the circumference of the circle passes through the origin, and the term was originally derived from the fact that the mho characteristic (ohm spelled backward) is a straight line in the admittance plane.

Early applications of distance protection utilized relay operating times that were a function of the impedance for the fault. The nearer the fault, the shorter the operating time. This is shown in Figure 7.50. This has the same disadvantages as overcurrent protection discussed earlier. Present practice is to set the relay to operate simultaneously for faults that occur in the first 80 percent of the feeder length (known as the first zone). Faults beyond this point and up to a point midway along the next feeder are cleared by arranging for the zone setting of the relay to be extended from the first zone value to the second zone value after a time delay of about 0.5 to 1 second. The second zone for the first relay should never be less than 20 percent of the first feeder length. The zone setting extension is done by increasing the impedance in series with the relay voltage coil current. A third zone is provided (using a starting relay) extending from the middle of the second feeder into the third feeder up to 25 percent of the length with a further delay of 1 or 2 seconds. This provides backup protection as well. The time-distance characteristics for a three-feeder system are shown in Figure 7.51.

Distance relaying schemes employ several relay units that are arranged to give response characteristics such as that shown in Figure 7.52. A typical system consists of:

1. Two offset mho units (with three elements each). The first operates as earth-fault starting and third zone measuring relay, and the second operates as phase-fault starting and third zone measuring relay.
2. Two polarized mho units (with three elements each). The first unit acts as first and second zone earth-fault measuring relay, and the second unit acts as first and second zone phase-fault measuring relay.
3. Two time-delay relays for second and third zone time measurement.

The main difference between earth-fault and phase-fault relays is in the potential transformer (P.T.) and C.T. connections, which are designed to cause the relay to respond to the type of fault concerned.

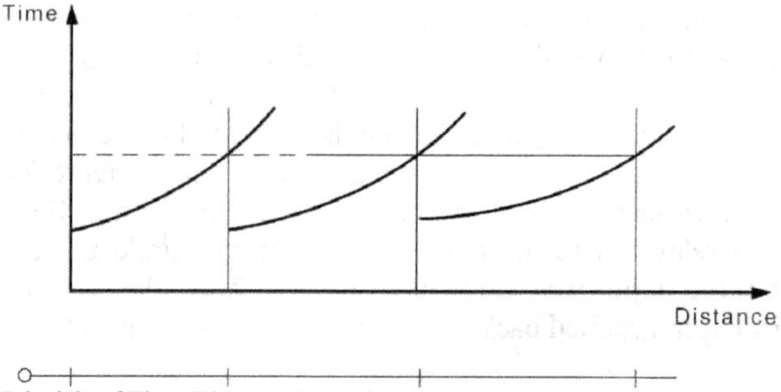


**Figure 7.50** Principle of Time-Distance Protection.

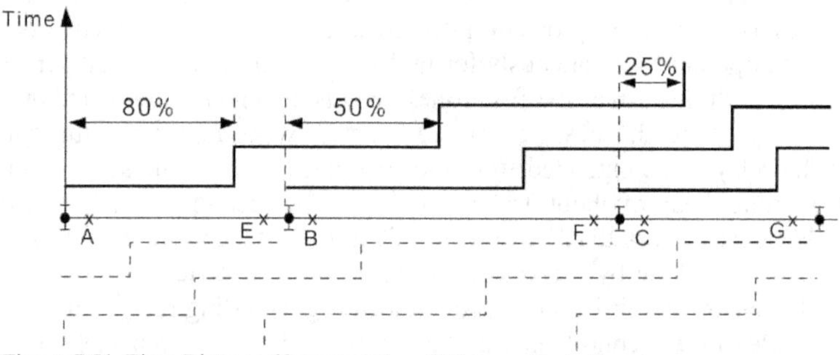


**Figure 7.51** Time-Distance Characteristics of Distance Protection.

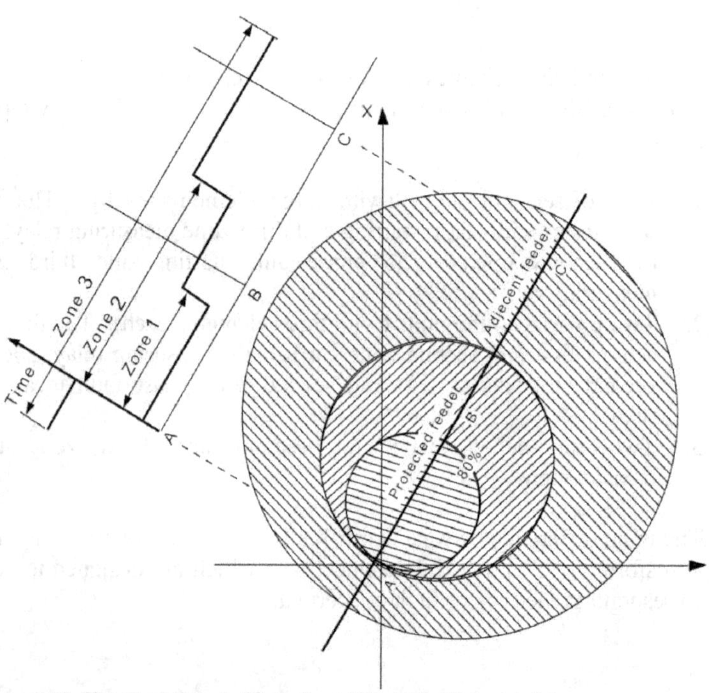


**Figure 7.52** Characteristics of a Three-Zone Offset Mho-Relaying Scheme.

## E) Power Line Carrier Protection

The overhead transmission lines are used as pilot circuits in carrier-current protection systems. A carrier-frequency signal (30-200 kHz) is carried by two of the line conductors to provide communication means between ends of the line. The carrier signal is applied to the conductors via carrier coupling into units comprising inductance/capacitor circuits tuned to the carrier signal frequency to perform a number of functions. The carrier signals thus travel mainly into the power line and not into undesired parts of the system such as the bus bars. The communication equipment that operates at impedance levels of the order of 50-150 Ω is to be matched to the power line that typically has a characteristic impedance is the range of 240-500 Ω.

Power line carrier systems are used for two purposes. The first involves measurements, and the second conveys signals from one end of the line to the other with the measurement being done at each end by relays. When the carrier channel is used for measurement, it is not practical to transmit amplitude measurements from one end to the other since signal attenuation beyond the control of the system takes place. As a result, the only feasible measurement carrier system compares the phase angle of a derived current at each end of the system in a manner similar to differential protection.

Radio and microwave links are applied in power systems to provide communication channels for teleprotection as well as for supervisory control and data acquisition.

## *7.14 COMPUTER RELAYING*

In the electric power industry computer-based systems have evolved to perform many complex tasks in energy control centers (treated in Chapter 8). Research efforts directed at the prospect of using digital computers to perform the tasks involved in power system protection date back to the mid-sixties and were motivated by the emergence of process-control computers. *Computer relaying* systems are now available. The availability of microprocessors used as a replacement for electromechanical and solid-state relays provides a number of advantages while meeting the basic protection philosophy requirement of decentralization.

There are many benefits of a digital relaying system:

1. *Economics*: With lower digital hardware cost, coupled with the increase in cost of conventional relaying, computer relaying is attractive. Software development cost is evened out by utilizing economies of scale in producing microprocessors dedicated to basic relaying tasks.
2. *Reliability*: A digital system is continuously active providing a high level of self-diagnosis to detect accidental failures within the

digital relaying system. The maintenance costs are low.

3. *Flexibility*: Revisions or modifications made necessary by changing operational conditions can be accommodated by utilizing the programmability features of a digital system. This would lead to reduced inventories of parts for repair and maintenance purposes.
4. *System interaction*: The availability of digital hardware that monitors continuously the system performance at remote substations can enhance the level of information available to the control center. Postfault analysis of transient data can be performed on the basis of system variables monitored by the digital relay and recorded by the peripherals.

The main elements of a digital computer-based relay include:

1. Analog input subsystem
2. Digital input subsystem
3. Digital output subsystem
4. Relay logic and settings
5. Digital filters

The input signals to the relay are analog (continuous) and digital power system variables. The digital inputs are of the order of five to ten and include status changes (on-off) of contacts and changes in voltage levels in a circuit. The analog signals are the 60-Hz currents and voltages. The number of analog signals needed depends on the relay function but is in the range of 3 to 30 in all cases. The analog signals are scaled down to acceptable computer input levels (±10 volts maximum) and then converted to digital (discrete) form through analog/digital converters (ADC). These functions are performed in the "Analog Input Subsystem" block.

The digital output of the relay is available through the computer's parallel output port. Five-to-ten digital outputs are sufficient for most applications. The analog signals are sampled at a rate between 240 Hz to about 2000 Hz. The sampled signals are entered into the scratch pad [random access memory (RAM)] and are stored in a secondary data file for historical recording. A digital filter removes noise effects from the sampled signals. The relay logic program determines the functional operations of the relay and uses the filtered sampled signals to arrive at a trip or no trip decision, which is then communicated to the system.

The heart of the relay logic program is a relaying algorithm that is designed to perform the intended relay function such as overcurrent detection, differential protection, or distance protection, etc.

## PROBLEMS

### Problem 7.1

Consider the case of an open-line fault on phase $B$ of a three-phase system, such that

$$I_A = I$$
$$I_B = 0$$
$$I_C = \alpha I$$

Find the sequence currents $I_+$, $I_-$, and $I_0$.

### Problem 7.2

Consider the case of a three-phase system supplied by a two-phase source such that

$$V_A = V$$
$$V_B = jV$$
$$V_C = 0$$

Find the sequence voltages $V_+$, $V_-$, and $V_0$.

### Problem 7.3

Calculate the phase currents and voltages for an unbalanced system with the following sequence values:

$$I_+ = I_- = I_0 = -j1.0$$
$$V_+ = 0.50$$
$$V_- = -0.30$$
$$V_0 = -0.20$$

### Problem 7.4

Calculate the apparent power consumed in the system of Problem 7.3 using sequence quantities and phase quantities.

### Problem 7.5

The zero and positive sequence components of an unbalanced set of voltages are

$$V_+ = 2$$
$$V_0 = 0.5 - j0.866$$

The phase $A$ voltage is

$$V_A = 3$$

Obtain the negative sequence component and the *B* and *C* phase voltages.

**Problem 7.6**

The following are the line currents at the terminals of a Y connected three phase source operating under unbalanced conditions.

$$I_A = 1$$
$$I_B = 1e^{-j30}$$
$$I_C = 1e^{j30}$$

The sequence values of the line to neutral voltages are given by:

$$V_+ = 1$$
$$V_- = 0.5e^{j60}$$
$$V_C = 0.5e^{-j60}$$

A. Determine the positive, negative and zero sequence currents. Is the neutral of the source grounded? Why?
B. Determine the phase coordinate values of the line to neutral voltages.
C. Find the total apparent power delivered by the source using both phase and sequence values obtained in parts A and B.

**Problem 7.7**

Obtain the sequence networks for the system shown in Figure 7.53 in the case of a fault at *F*. Assume the following data in pu on the same base are given:

| | |
|---|---|
| Generator $G_1$: | $X_+ = 0.2$ p.u. |
| | $X_- = 0.12$ p.u. |
| | $X_0 = 0.06$ p.u. |
| Generator $G_2$: | $X_+ = 0.33$ p.u. |
| | $X_- = 0.22$ p.u. |
| | $X_0 = 0.066$ p.u. |
| Transformer $T_1$: | $X_+ = X_- = X_0 = 0.2$ p.u. |
| Transformer $T_2$: | $X_+ = X_- = X_0 = 0.225$ p.u. |
| Transformer $T_3$: | $X_+ = X_- = X_0 = 0.27$ p.u. |
| Transformer $T_4$: | $X_+ = X_- = X_0 = 0.16$ p.u. |
| Line $L_1$: | $X_+ = X_- = 0.14$ p.u. |
| | $X_0 = 0.3$ p.u. |
| Line $L_2$: | $X_+ = X_- = 0.35$ p.u. |
| | $X_0 = 0.6$ p.u. |

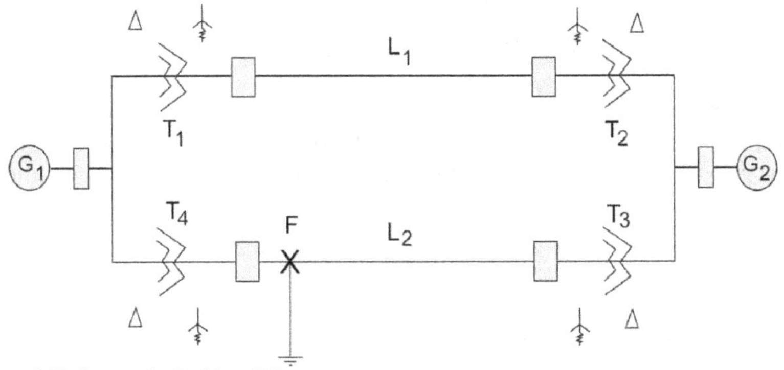


**Figure 7.53** System for Problem 7.7.

### Problem 7.8

Assume an unbalanced fault occurs on the line bus of transformer $T_3$ in the system of Problem 7.7. Find the equivalent sequence networks for this condition.

### Problem 7.9

Repeat Problem 7.8 for a fault on the generator bus of $G_2$.

### Problem 7.10

Repeat Problem 7.8 for the fault in the middle of the line $L_1$.

### Problem 7.11

Calculate the fault current for a single line-to-ground fault on phase $A$ for a fault location as in Problem 7.8.

### Problem 7.12

Repeat Problem 7.11 for a fault location in Problem 7.10.

### Problem 7.13

Calculate the fault current in phase $B$ for a double line-to-ground fault for a fault location as in Problem 7.8.

### Problem 7.14

Repeat Problem 7.13 for a fault location as in Problem 7.9.

### Problem 7.15

Repeat Problem 7.13 for a fault location in Problem 7.10.

### Problem 7.16

Calculate the fault current in phase $B$ for a line-to-line fault for a fault location as in Problem 7.8.

### Problem 7.17

Repeat Problem 7.16 for a fault location as in Problem 7.9.

## Problem 7.18

Repeat Problem 7.16 for a fault location as in Problem 7.10.

## Problem 7.19

Consider the system shown in the single-line diagram of Figure 7.54. All reactances are shown in per unit to the same base. It is required to find the fault current due to a bolted-three-phase short circuit at bus 4.

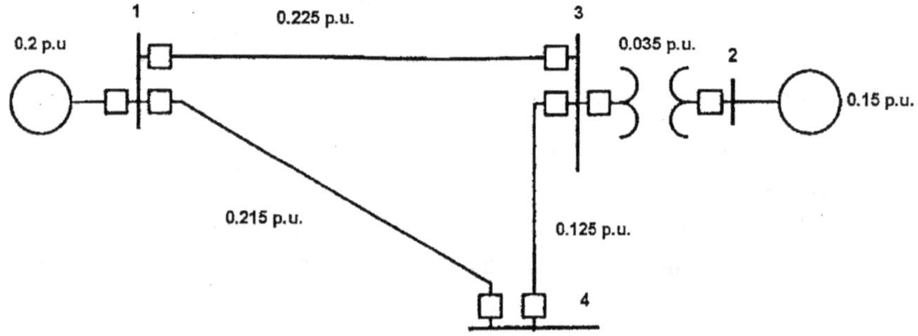


**Figure 7.54** Single-line diagram for Problem 7.19

## Problem 7.20

For the system of Figure 7.54

A. Find the fault current due to a bolted-three-phase short circuit in the middle of line 1-3.
B. Find the voltages at buses 1 and 2 under the fault conditions of part A above.

## Problem 7.21

The following sequence voltages were recorded on an unbalanced fault:

$$V_+ = 0.5 \text{ p.u.}$$
$$V_- = -0.4 \text{ p.u.}$$
$$V_0 = -0.1 \text{ p.u.}$$

Given that the positive sequence fault current is $-j1$, calculate the sequence impedances. Assume $E = 1$.

## Problem 7.22

The positive sequence current for a double line-to-ground fault in a system is $-j1$ p.u., and the corresponding negative sequence current is $j0.333$ p.u. Given that the positive sequence impedance is 0.8 p.u., find the negative and zero sequence impedances.

**Problem 7.23**

The positive sequence current on a single line-to-ground fault on phase *A* at the load end of a radial transmission system is $-j2$ p.u. For a double line-to-ground fault on phases *B* and *C*, the positive sequence current is $-j3.57$ p.u., and for a double-line fault between phases *B* and *C*, its value is $-j2.67$. Assuming the sending-end voltage $E = 1.2$, find the sequence impedances for this system.

**Problem 7.24**

A turbine generator has the following sequence reactances:

$$X_+ = 0.1$$
$$X_- = 0.13$$
$$X_0 = 0.04$$

Compare the fault currents for a three-phase fault and a single line-to-ground fault. Find the value of an inductive reactance to be inserted in the neutral connection to limit the current for a single line-to-ground fault to that for a three-phase fault.

**Problem 7.25**

A simultaneous fault occurs at the load end of a radial line. The fault consists of a line-to-ground fault on phase *A* and a line-to-line fault on phases *B* and *C*. The current in phase *A* is $-j5$ p.u., whereas that in phase *B* is $I_B = -3.46$ p.u. Given that $E = 1\angle 0$ and $Z_+ = j0.25$, find $Z_-$ and $Z_0$.

**Problem 7.26**

Repeat Example 7.9, for a transformer rating of 12-MVA.

**Problem 7.27**

Consider the system of Example 7.10. Assume now that the load at the far end of the system is increased to

$$L_1 = 6 \text{ MVA}$$

Determine the relay settings to protect the system using relay type CO-7.

**Problem 7.28**

Consider the radial system of Example 7.10. It is required to construct the relay response time-distance characteristics on the basis of the design obtained as follows:

A. Assuming the line's impedance is purely reactive, calculate the source reactance and the reactances between bus bars 3 and 2, and 2 and 1.
B. Find the current on a short circuit midway between buses 3 and 2 and between 2 and 1.
C. Calculate the relay response times for faults identified in Example

7.10 and part (B) above and sketch the relay response time-distance characteristics.

### Problem 7.29

Consider a system with $|V_r| = 1$ p.u.. Assume that the load is given by

$$S_r = 1 + j0.4 \text{ p.u.}$$

Find $Z_r$, $Z_s$, and the angle $\delta$ for this operating condition.

### Problem 7.30

Assume that a line has an impedance $Z_L = 0.1 + j0.3$ p.u. The load is $S_r = 2 + j0.8$ p.u., $|V_r| = 1$ p.u. This line is to be provided with 80 percent distance protection using an ohm relay with $\psi = 45°$. Find the relay's impedance $Z$ assuming $k = 1$ and that magnitude comparison is used.

### Problem 7.31

The line of Problem 7.30 is to be provided with 80 percent distance protection using either a resistance or a reactance ohm relay. Find the relay design parameters in each case, assuming that magnitude comparison is used.

# Chapter 8

# THE ENERGY CONTROL CENTER

## 8.1 INTRODUCTION

The following criteria govern the operation of any electric power system:

- Safety
- Quality
- Reliability
- Economy

The first criterion is the most important consideration and aims to ensure the safety of personnel, environment, and property in every aspect of system operations. Quality is defined in terms of variables, such as frequency and voltage, that must conform to certain standards to accommodate the requirements for proper operation of all loads connected to the system. Reliability of supply does not have to mean a constant supply of power, but it means that any break in the supply of power is one that is agreed to and tolerated by both supplier and consumer of electric power. Making the generation cost and losses at a minimum motivates the economy criterion while mitigating the adverse impact of power system operation on the environment.

Within an operating power system, the following tasks are performed in order to meet the preceding criteria:

- Maintain the balance between load and generation.
- Maintain the reactive power balance in order to control the voltage profile.
- Maintain an optimum generation schedule to control the cost and environmental impact of the power generation.
- Ensure the security of the network against credible contingencies. This requires protecting the network against reasonable failure of equipment or outages.

The fact that the state of the power network is ever changing because loads and networks configuration change, makes operating the system difficult. Moreover, the response of many power network apparatus is not instantaneous. For example, the startup of a thermal generating unit takes a few hours. This essentially makes it not possible to implement normal feed-forward control. Decisions will have to be made on the basis of predicted future states of the system.

Several trends have increased the need for computer-based operator support in interconnected power systems. Economy energy transactions, reliance on external sources of capacity, and competition for transmission resources have all resulted in higher loading of the transmission system. Transmission lines bring large quantities of bulk power. But increasingly, these same circuits are being used for other purposes as well: to permit sharing surplus generating capacity between adjacent utility systems, to ship large blocks of power from low-energy-cost areas to high-energy cost areas, and to provide emergency reserves in the event of weather-related outages. Although such transfers have helped to keep electricity rates lower, they have also added greatly to the burden on transmission facilities and increased the reliance on control.

Heavier loading of tie-lines which were originally built to improve reliability, and were not intended for normal use at heavy loading levels, has increased interdependence among neighboring utilities. With greater emphasis on economy, there has been an increased use of large economic generating units. This has also affected reliability.

These trends allow systems to be operated much closer to security limits (thermal, voltage and stability). On some systems, transmission links are being operated at or near limits 24 hours a day, with the following consequences

- The trends have adversely affected system dynamic performance. A power network stressed by heavy loading has a substantially different response to disturbances from that of a non-stressed system.
- The potential size and effect of contingencies has increased dramatically. When a power system is operated closer to the limit, a relatively small disturbance may cause a system upset. The situation is further complicated by the fact that the largest size contingency is increasing. Thus, to support operating functions many more scenarios must be anticipated and analyzed. In addition, bigger areas of the interconnected system may be affected by a disturbance.
- Where adequate bulk power system facilities are not available, special controls are employed to maintain system integrity. Overall, systems are more complex to analyze to ensure reliability and security.
- Some scenarios encountered cannot be anticipated ahead of time. Since they cannot be analyzed off-line, operating guidelines for these conditions may not be available, and the system operator may have to "improvise" to deal with them (and often does). As a result, there is an ever increasing need for mechanisms to support dispatchers in the decision making process. Indeed, there is a risk of human operators being unable to manage certain functions unless their awareness and understanding of the network state is enhanced.

To automate the operation of an electric power system electric utilities rely on a highly sophisticated integrated system for monitoring and control.

Such a system has a multi-tier structure with many levels of elements. The bottom tier (level 0) is the high-reliability switchgear, which includes facilities for remote monitoring and control. This level also includes automatic equipment such as protective relays and automatic transformer tap-changers. Tier 1 consists of telecontrol cabinets mounted locally to the switchgear, and provides facilities for actuator control, interlocking, and voltage and current measurement. At tier 2, is the data concentrators/master remote terminal unit which typically includes a man/machine interface giving the operator access to data produced by the lower tier equipment. The top tier (level 3) is the supervisory control and data acquisition (SCADA) system. The SCADA system accepts telemetered values and displays them in a meaningful way to operators, usually via a one-line mimic diagram. The other main component of a SCADA system is an alarm management subsystem that automatically monitors all the inputs and informs the operators of abnormal conditions.

Two control centers are normally implemented in an electric utility, one for the operation of the generation-transmission system, and the other for the operation of the distribution system. We refer to the former as the energy management system (EMS), while the latter is referred to as the distribution management system (DMS). The two systems are intended to help the dispatchers in better monitoring and control of the power system. The simplest of such systems perform data acquisition and supervisory control, but many also have sophisticated power application functions available to assist the operator. Since the early 1960s, electric utilities have been monitoring and controlling their power networks via SCADA, EMS, and DMS. These systems provide the "smarts" needed for optimization, security, and accounting. Today's EMS software captures and archives live data and records information especially during emergencies and system disturbances.

An energy control center represents a large investment by the power system ownership. Major benefits flowing from the introduction of this system include more reliable system operation and improved efficiency of usage of generation resources. In addition, power system operators are offered more in-depth information quickly. It has been suggested that at Houston Lighting & Power Co., system dispatchers' use of network application functions (such as Power Flow, Optimal Power Flow, and Security Analysis) has resulted in considerable economic and intangible benefits. A specific example of $ 70,000 in savings achieved through avoiding field crew overtime cost, and by leaving equipment out of service overnight is reported for 1993. This is part of a total of $ 340,000 savings in addition to increased system safety, security and reliability has been achieved through regular and extensive use of just some network analysis functions.

## *8.2 OVERVIEW OF EMS FUNCTIONS*

System dispatchers at the EMS are required to make short-term (next day) and long-term (prolonged) decisions on operational and outage scheduling on a daily basis. Moreover, they have to be always alert and prepared to deal with contingencies that may arise. Many software and hardware functions are required as operational support tools for the operator. Broadly speaking, we can classify these functions in the following manner:

- Base functions
- Generation functions
- Network functions

Each of these functions is discussed briefly in this section.

## Base Functions

The required base functions of the EMS include:

- The ability to acquire real time data from monitoring equipment throughout the power system.
- Process the raw data and distribute the processed data within the central control system.

The data acquisition function (DA) acquires data from remote terminal units (RTUs) installed throughout the system using special hardware connected to the real time data servers installed at the control center. Alarms that occur at the substations are processed and distributed by the DA function. In addition, protection and operation of main circuit breakers, some line isolators, transformer tap changers and other miscellaneous substation devices are provided with a sequence of events time resolution.

### *Data Acquisition*

The data acquisition function collects, manages, and processes information from the RTUs by periodically scanning the RTUs and presenting the raw analog data and digital status points to a data processing function. This function converts analog values into engineering units and checks the digital status points for change since the previous scan so that an alarm can be raised if status has changed. Computations can be carried out and operating limits can be applied against any analog value such that an alarm message is created if a limit is violated.

### *Supervisory Control*

Supervisory control allows the operator to remotely control all circuit breakers on the system together with some line isolators. Control of devices can be performed as single actions or a line circuit can be switched in or out of service.

### *Alarm Processor*

The alarm processor software is responsible to notify the operator of changes in the power system or the computer control system. Many classification and detection techniques are used to direct the alarms to the appropriate operator with the appropriate priorities assigned to each alarm.

### *Logical Alarming*

This predetermines a typical set of alarm operations, which would result from a single cause. For example, a faulted transmission line would be automatically taken out of service by the operation of protective and tripping relays in the substation at each end of the line and the automatic opening of circuit breakers. The coverage would identify the protection relays involved, the trip relays involved and the circuit breakers that open. If these were defined to the system in advance, the alarm processor would combine these logically to issue a priority 1 alarm that the particular power circuit had tripped correctly on protection. The individual alarms would then be given a lower priority for display. If no logical combination is viable for the particular circumstance, then all the alarms are individually presented to the dispatcher with high priority. It is also possible to use the output of a logical alarm as the indicator for a sequence-switching procedure. Thus, the EMS would read the particular protection relays which had operated and restore a line to service following a transient fault.

### *Sequence of Events Function*

The sequence of events function is extremely useful for post-mortem analysis of protection and circuit breaker operations. Every protection relay, trip relay, and circuit breaker is designated as a sequence of events digital point. This data is collected, and time stamped accurately so that a specified resolution between points is possible within any substation and across the system. Sequence of events data is buffered on each RTU until collected by data acquisition automatically or on demand.

### *Historical Database*

This function takes any data obtained by the system and stores it in a historical database. It then can be viewed by a tabular or graphical trend display. The data is immediately stored within the on-line system and transferred to a standard relational data base system periodically. Generally, this function allows all features of such database to be used to perform queries and provide reports.

### *Automatic Data Collection*

This function is specified to define the process taken when there is a major system disturbance. Any value or status monitored by the system can be defined as a trigger. This will then create a disturbance archive, which will contain a pre-disturbance and a post-disturbance snapshots to be produced.

### *Load Shedding Function*

This facility makes it possible to identify that particular load block and instruct the system to automatically open the correct circuit breakers involved. It is also possible to predetermine a list of load blocks available for load shedding. The amount of load involved within each block is monitored so that when a particular amount of load is required to shed in a system emergency, the operator can enter this value and instruct the system to shed the appropriate blocks.

### *Safety Management*

Safety management provided by an EMS is specific to each utility. A system may be specified to provide the equivalent of diagram labeling and paper based system on the operator's screen. The software allows the engineer, having opened isolators and closed ground switches on the transmission system, to designate this as safety secured. In addition, free-placed ground symbols can be applied to the screen-based diagram. A database is linked to the diagram system and records the request for plant outage and safety document details. The computer system automatically marks each isolator and ground switch being presently quoted on a safety document and records all safety documents using each isolator or ground switch. These details are immediately available at any operating position when the substation diagram is displayed.

## Generation Functions

The main functions that are related to operational scheduling of the generating subsystem involve the following:

- Load forecasting
- Unit commitment
- Economic dispatch and automatic generation control (AGC)
- Interchange transaction scheduling

Each of these functions is discussed briefly here.

### *Load Forecasting*

The total load demand, which is met by centrally dispatched generating units, can be decomposed into base load and controlled load. In some systems, there is significant demand from storage heaters supplied under an economy tariff. The times at which these supplies are made available can be altered using radio tele-switching. This enables the utility to shape the total demand curve by altering times of supply to these customers. This is done with the objective of making the overall generation cost as economic and environmentally compatible as possible. The other part of the demand consists of the uncontrolled use of electricity, which is referred to as the natural demand. It is necessary to be able to predict both of these separately. The base demand is predicted using historic load and weather data and a weather forecast.

### *Unit Commitment*

The unit commitment function determines schedules for generation operation, load management blocks and interchange transactions that can be dispatched. It is an optimization problem, whose goal is to determine unit startup and shutdown and when on-line, what is the most economic output for each unit during each time step. The function also determines transfer levels on interconnections and the schedule of load management blocks. The software takes into account startup and shutdown costs, minimum up and down times and constraints imposed by spinning reserve requirements.

The unit commitment software produces schedules in advance for the next time period (up to as many as seven days, at 15-minute intervals). The algorithm takes the predicted base demand from the load forecasting function and the predicted sizes of the load management blocks. It then places the load management blocks onto the base demand curve, essentially to smooth it optimally. The operator is able to use the software to evaluate proposed interchange transactions by comparing operating costs with and without the proposed energy exchange. The software also enables the operator to compute different plant schedules where there are options on plant availability

### *Economic Dispatch and AGC*

The economic dispatch (ED) function allocates generation outputs of the committed generating units to minimize fuel cost, while meeting system constraints such as spinning reserve. The ED functions to compute recommended economic base points for all manually controlled units as well as economic base points for units which may be controlled directly by the EMS.

The Automatic Generation Control (AGC) part of the software performs dispatching functions including the regulation of power output of generators and monitoring generation costs and system reserves. It creates control commands to change generation set points in response to changes in system frequency brought about by load fluctuations.

### *Interchange Transaction Scheduling Function*

This function allows the operator to define power transfer schedules on tie-lines with neighboring utilities. In many instances, the function evaluates the economics and loading implications of such transfers.

### *Current Operating Plan (COP)*

As part of the generation and fuel dispatch functions on the EMS at a typical utility is a set of information called the Current Operating Plan (COP) which contains the latest load forecast, unit commitment schedule, and hourly average generation for all generating units with their forecast operating status. The COP is typically updated every 4 to 8 hours, or as needed following major changes in load forecast and/or generating unit availability.

## Network Analysis Functions

Network applications can be subdivided into real-time applications and study functions. The real time functions are controlled by real time sequence control that allows for a particular function or functions to be executed periodically or by a defined event manually. The network study functions essentially duplicate the real time function and are used to study any number of "what if" situations. The functions that can be executed are:

- Topology Processing (Model Update) Function.
- State Estimation Function.
- Network Parameter Adaptation Function.
- Dispatcher Power Flow (DPF.)
- Network Sensitivity Function.
- Security Analysis Function.
- Security Dispatch Function.
- Voltage Control Function.
- Optimal Power Flow Function.

### *Topology Processing (Model Update) Function*

The topology processing (model-updating) module establishes the current configuration of the network, by processing the telemetered switch (breakers and isolators) status to determine existing connections and thus establish a node-branch representation of the system.

### *State Estimation Function*

The state estimator function takes all the power system measurements telemetered via SCADA, and provides an accurate power flow solution for the network. It then determines whether bad or missing measurements using redundant measurements are present in its calculation. The output from the state estimator is given on the one-line diagram and is used as input to other applications such as Optimal Power Flow.

### *Network Parameter Adaptation Function*

This module generates forecasts of busbar voltages and loads. The forecasts are updated periodically in real time. This allows the state estimator to schedule voltages and loads at busbars where no measurements are available.

### *Dispatcher Power Flow (DPF)*

A DPF is employed to examine the steady state conditions of the network. The solution provides information on network bus voltages (kV), and transmission line and transformer flows (MVA). The control center dispatchers use this information to detect system violations (over/under-voltages, branch overloads) following load, generation, and topology changes in the system.

### *Network Sensitivity Function*

In this function, the output of the state estimator is used to determine the sensitivity of network losses to changes in generation patterns or tie-line exchanges. The sensitivity parameters are then converted to penalty factors for economic dispatch purposes.

### *Security Analysis Function*

The SA is one of the main applications of the real time network analysis set. It is designed to assist system dispatchers in determining system security under specified single contingency and multiple contingency criteria. It helps the operator study system behavior under contingency conditions. The security analysis function performs a power flow solution for each contingency and advises of possible overloads or voltage limit violations. The function automatically reviews a list of potential problems, rank them as to their effect and advise on possible reallocation of generation. The objective of SA is to operate the network closer to its full capability and allow the proper assessment of risks during maintenance or unexpected outages.

### *Security Dispatch Function*

The security dispatch function gives the operator the capability of reducing or eliminating overloads by rearranging the generation pattern. The tool operates in real-time on the network in its current state, rather than for each contingency. The function uses optimal power flow and constrains economic dispatch to offer a viable security dispatch of the generating resources of the system.

### *Voltage Control Function*

The voltage control (VC) study is used to eliminate or reduce voltage violations, MVA overloads and/or minimize transmission line losses using transformer set point controls, generator MVAR, capacitor/reactor switching, load shedding, and transaction MW.

### *Optimal Power Flow Function*

The purpose of the Optimal Power Flow (OPF) is to calculate recommended set points for power system controls that are a trade-off between security and economy. The primary task is to find a set of system states within a region defined by the operating constraints such as voltage limits and branch flow limits. The secondary task is to optimize a cost function within this region. Typically, this cost function is defined to include economic dispatch of active power while recognizing network-operating constraints. An important limitation of OPF is that it does not optimize switching configurations.

Optimal power flow can be integrated with other EMS functions in either a preventive or corrective mode. In the preventive mode, the OPF is used to provide suggested improvements for selected contingency cases. These may be the worst cases found by contingency analysis or planned outages.

In the corrective mode, an OPF is run after significant changes in the topology of the system. This is the situation when the state estimation output indicates serious violations requiring the OPF to reschedule the active and reactive controls.

It is important to recognize that optimization is only possible if the network is controllable, i.e., the control center must have control of equipment such as generating units or tap-changer set points. This may present a challenge to an EMS that does not have direct control of all generators. To obtain the full benefit of optimization of the reactive power flows and the voltage profile, it is important to be able to control all voltage regulating devices as well as generators.

The EMS network analysis functions (e.g., Dispatcher Power Flow and Security Analysis) are the typical tools for making many decisions such as outage scheduling. These tools can precisely predict whether the outage of a specific apparatus (i.e., transformer, generator, or transmission line) would cause any system violations in terms of abnormal voltages or branch overloads.

In a typical utility system, outage requests are screened based on the system violation indications from DPF and SA studies. The final approval for crew scheduling is granted after the results from DPF and SA are reviewed.

## **Operator Training Simulator**

An energy management system includes a training simulator that allows system operators to be trained under normal operating conditions and simulated power system emergencies. System restoration may also be exercised. It is important to realize that major power system events are relatively rare, and usually involve only one shift team out of six, real experience with emergencies builds rather slowly.

The interface to the operator appears identical to the normal control interface. The simulator relies on two models: one of the power system and the other represents the control center. Other software is identical to that used in real time. A scenario builder is available such that various contingencies can be simulated through a training session. The instructor controls the scenarios and plays the role of an operator within the system.

## 8.3 POWER FLOW CONTROL

The power system operator has the following means to control system power flows:

1. Prime mover and excitation control of generators.
2. Switching of shunt capacitor banks, shunt reactors, and static var systems.
3. Control of tap-changing and regulating transformers.
4. FACTS based technology.

A simple model of a generator operating under balanced steady-state conditions is given by the Thévenin equivalent of a round rotor synchronous machine connected to an infinite bus as discussed in Chapter 3. $V$ is the generator terminal voltage, $E$ is the excitation voltage, $\delta$ is the power angle, and $X$ is the positive-sequence synchronous reactance. We have shown that:

$$P = \frac{EV}{X}\sin\delta$$

$$Q = \frac{V}{X}[E\cos\delta - V]$$

The active power equation shows that the active power $P$ increases when the power angle $\delta$ increases. From an operational point of view, when the operator increases the output of the prime mover to the generator while holding the excitation voltage constant, the rotor speed increases. As the rotor speed increases, the power angle $\delta$ also increases, causing an increase in generator active power output $P$. There is also a decrease in reactive power output $Q$, given by the reactive power equation. However, when $\delta$ is less than 15°, the increase in $P$ is much larger than the decrease in $Q$. From the power-flow point of view, an increase in prime-mover power corresponds to an increase in $P$ at the constant-voltage bus to which the generator is connected. A power-flow program will compute the increase in $\delta$ along with the small change in $Q$.

The reactive power equation demonstrates that reactive power output $Q$ increases when the excitation voltage $E$ increases. From the operational point of view, when the generator exciter output increases while holding the prime-mover power constant, the rotor current increases. As the rotor current increases, the excitation voltage $E$ also increases, causing an increase in generator reactive power output $Q$. There is also a small decrease in $\delta$ required to hold $P$ constant in the active power equation. From the power-flow point of view, an increase in generator excitation corresponds to an increase in voltage magnitude at the infinite bus (constant voltage) to which the generator is connected. The power-flow program will compute the increase in reactive power $Q$ supplied by the generator along with the small change in $\delta$.

The effect of adding a shunt capacitor bank to a power-system bus can be explained by considering the Thévenin equivalent of the system at that bus. This is simply a voltage source $V_{Th}$ in series with the impedance $Z_{sys}$. The bus voltage V before connecting the capacitor is equal to $V_{Th}$. After the bank is connected, the capacitor current $I_C$ leads the bus voltage $V$ by 90°. Constructing a phasor diagram of the network with the capacitor connected to the bus reveals that $V$ is larger than $V_{Th}$. From the power-flow standpoint, the addition of a shunt capacitor bank to a load bus corresponds to the addition of a reactive generating source (negative reactive load), since a capacitor produces positive reactive power (absorbs negative reactive power). The power-flow program computes the increase in bus voltage magnitude along with a small change in $\delta$. Similarly, the addition of a shunt reactor corresponds to the addition of a positive reactive load, wherein the power flow program computes the decrease in voltage magnitude.

Tap-changing and voltage-magnitude-regulating transformers are used to control bus voltages as well as reactive power flows on lines to which they are connected. In a similar manner, phase-angle-regulating transformers are used to control bus angles as well as real power flows on lines to which they are connected. Both tap changing and regulating transformers are modeled by a transformer with an off-nominal turns ratio. From the power flow point of view, a change in tap setting or voltage regulation corresponds to a change in tap ratio. The power-flow program computes the changes in $Y_{bu}$, bus voltage magnitudes and angles, and branch flows.

FACTS is an acronym for flexible AC transmission systems. They use power electronic controlled devices to control power flows in a transmission network so as to increase power transfer capability and enhance controllability. The concept of flexibility of electric power transmission involves the ability to accommodate changes in the electric transmission system or operating conditions while maintaining sufficient steady state and transient margins.

A FACTS controller is a power electronic-based system and other static equipment that provide control of one or more ac transmission system parameters. FACTS controllers can be classified according to the mode of their connection to the transmission system as:

1. Series-Connected Controllers.
2. Shunt-Connected Controllers.
3. Combined Shunt and Series-Connected Controllers.

The family of series-connected controllers includes the following devices:

1. The Static Synchronous Series Compensator ($S^3C$) is a static, synchronous generator operated without an external electric energy source as a series compensator whose output voltage is in quadrature with, and controllable independently of, the line current for the purpose of increasing or decreasing the overall reactive voltage drop across the line and thereby controlling the transmitted electric power. The $S^3C$ may include transiently rated energy storage or energy absorbing devices to enhance the dynamic behavior of the power system by additional temporary real power compensation, to increase or decrease momentarily, the overall real (resistive) voltage drop across the line.
2. Thyristor Controlled Series Compensation is offered by an impedance compensator, which is applied in series on an ac transmission system to provide smooth control of series reactance.
3. Thyristor Switched Series Compensation is offered by an impedance compensator, which is applied in series on an ac transmission system to provide step-wise control of series reactance.
4. The Thyristor Controlled Series Capacitor (TCSC) is a capacitive reactance compensator which consists of a series capacitor bank shunted by thyristor controlled reactor in order to provide a smoothly variable series capacitive reactance.
5. The Thyristor Switched Series Capacitor (TSSC) is a capacitive reactance compensator which consists of a series capacitor bank shunted by thyristor controlled reactor in order to provide a stepwise control of series capacitive reactance.
6. The Thyristor Controlled Series Reactor (TCSR) is an inductive reactance compensator which consists of a series reactor shunted by thyristor controlled reactor in order to provide a smoothly variable series inductive reactance.
7. The Thyristor Switched Series Reactor (TSSR) is an inductive reactance compensator which consists of a series reactor shunted by thyristor controlled reactor in order to provide a stepwise control of series inductive reactance.

Shunt-connected Controllers include the following categories:

1. A Static Var Compensator (SVC) is a shunt connected static var generator or absorber whose output is adjusted to exchange capacitive or inductive current so as to maintain or control specific parameters of the electric power system (typically bus voltage). SVCs have been in use since the early 1960s. The SVC application for transmission voltage control began in the late 1970s.
2. A Static Synchronous Generator (SSG) is a static, self-commutated switching power converter supplied from an appropriate electric energy source and operated to produce a set of adjustable multi-phase output voltages, which may be coupled to an ac power

system for the purpose of exchanging independently controllable real and reactive power.

3. A Static Synchronous Compensator (SSC or STATCOM) is a static synchronous generator operated as a shunt connected static var compensator whose capacitive or inductive output current can be controlled independent of the ac system voltage.
4. The Thyristor Controlled Braking Resistor (TCBR) is a shunt-connected, thyristor-switched resistor, which is controlled to aid stabilization of a power system or to minimize power acceleration of a generating unit during a disturbance.
5. The Thyristor Controlled Reactor (TCR) is a shunt-connected, thyristor-switched inductor whose effective reactance is varied in a continuous manner by partial conduction control of the thyristor valve.
6. The Thyristor Switched Capacitor (TSC) is a shunt-connected, thyristor-switched capacitor whose effective reactance is varied in a stepwise manner by full or zero-conduction operation of the thyristor valve.

The term Combined Shunt and Series-Connected Controllers is used to describe controllers such as:

1. The Unified Power Flow Controller (UPFC) can be used to control active and reactive line flows. It is a combination of a static synchronous compensator (STATCOM) and a static synchronous series compensator ($S^3C$) which are coupled via a common dc link. This allows bi-directional flow of real power between the series output terminals of the $S^3C$ and the shunt output terminals of the STATCOM, and are controlled to provide concurrent real and reactive series line compensation without an external electric energy source. The UPFC, by means of angularly unconstrained series voltage injection, is capable of controlling, concurrently or selectively, the transmission line voltage, impedance, and angle or, alternatively, the real and reactive power flow in the line. The UPFC may also provide independently controllable shunt reactive compensation.
2. The Thyristor Controlled Phase Shifting Transformer (TCPST) is a phase shifting transformer, adjusted by thyristor switches to provide a rapidly variable phase angle.
1. The Interphase Power Controller (IPC) is a series-connected controller of active and reactive power consisting, in each phase, of inductive and capacitive branches subjected to separately phase-shifted voltages. The active and reactive power can be set independently by adjusting the phase shifts and/or the branch impedances, using mechanical or electronic switches. In the particular case where the inductive and capacitive impedances form a conjugate pair, each terminal of the IPC is a passive current source dependent on the voltage at the other terminal.

The significant impact that FACTS devices will make on transmission systems arises because of their ability to effect high-speed control. Present control actions in a power system, such as changing transformer taps, switching current or governing turbine steam pressure, are achieved through the use of mechanical devices, that impose a limit on the speed at which control action can be made. FACTS devices are capable of control actions at far higher speeds. The three parameters that control transmission line power flow are line impedance and the magnitude and phase of line end voltages. Conventional control of these parameters is not fast enough for dealing with dynamic system conditions. FACTS technology enhances the capability to control the system.

The accelerated use of FACTS is motivated by the deregulation/ competitive environment in contemporary utility business. FACTS allow operators to control the path of the flow of electric power, and the ability to effectively join electric power networks that are not well interconnected.

## 8.4 POWER FLOW

Earlier chapters of this book treated modeling major components of an electric power system for analysis and design purposes. In this section we consider the system as a whole. An ubiquitous EMS application software is the power flow program, which solves for network state given specified conditions throughout the system. While there are many possible ways for formulating the power flow equations, the most popular formulation of the network equations is based on the nodal admittance form. The nature of the system specifications dictates that the network equations are nonlinear and hence no direct solution is possible. Instead, iterative techniques have to be employed to obtain a solution. As will become evident, good initial estimates of the solution are important, and a technique for getting started is discussed. There are many excellent numerical solution methods for solving the power flow problem. We choose here to introduce the Newton-Raphson method.

### Network Nodal Admittance Formulation

Consider a power system network shown in Figure 8.1 with generating capabilities as well as loads indicated. Buses 1, 2, and 3 are buses having generation capabilities as well as loads. Bus 3 is a load bus with no real generation. Bus 4 is a net generation bus.

Using the $\pi$ equivalent representation for each of the lines, we obtain

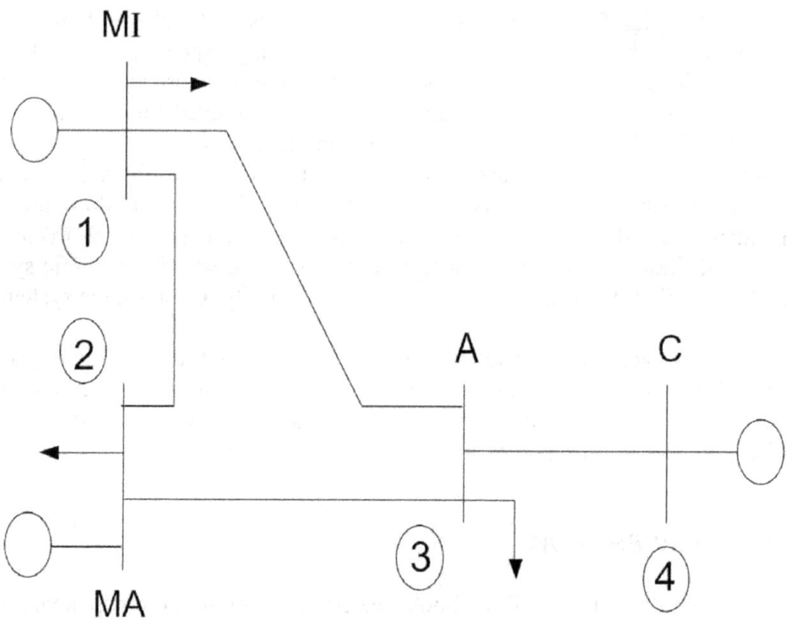


**Figure 8.1** Single-Line Diagram to Illustrate Nodal Matrix Formulation.

the network shown in Figure 8.2. Let us examine this network in which we exclude the generator and load branches. We can write the current equations as

$$I_1 = V_1Y_{10} + (V_1 - V_2)Y_{L_{12}} + (V_1 - V_3)Y_{L_{13}}$$
$$I_2 = V_2Y_{20} + (V_2 - V_1)Y_{L_{12}} + (V_2 - V_3)Y_{L_{23}}$$
$$I_3 = V_3Y_{30} + (V_3 - V_1)Y_{L_{13}} + (V_3 - V_4)Y_{L_{34}} + (V_3 - V_2)Y_{L_{23}}$$
$$I_4 = V_4Y_{40} + (V_4 - V_3)Y_{L_{34}}$$

We introduce the following admittances:

$$Y_{11} = Y_{10} + Y_{L_{12}} + Y_{L_{13}}$$
$$Y_{22} = Y_{20} + Y_{L_{12}} + Y_{L_{23}}$$
$$Y_{33} = Y_{30} + Y_{L_{13}} + Y_{L_{23}} + Y_{L_{34}}$$
$$Y_{44} = Y_{40} + Y_{L_{34}}$$
$$Y_{12} = Y_{21} = -Y_{L_{12}}$$
$$Y_{13} = Y_{31} = -Y_{L_{13}}$$
$$Y_{23} = Y_{32} = -Y_{L_{23}}$$
$$Y_{34} = Y_{43} = -Y_{L_{34}}$$

Thus the current equations reduce to

$$I_1 = Y_{11}V_1 + Y_{12}V_2 + Y_{13}V_3 + 0V_4$$
$$I_2 = Y_{21}V_1 + Y_{22}V_2 + Y_{23}V_3 + 0V_4$$
$$I_3 = Y_{13}V_1 + Y_{23}V_2 + Y_{33}V_3 + Y_{34}V_4$$
$$I_4 = 0V_1 + 0V_2 + Y_{43}V_3 + Y_{44}V_4$$

Note that $Y_{14} = Y_{41} = 0$, since buses 1 and 4 are not connected; also $Y_{24} = Y_{42} = 0$ since buses 2 and 4 are not connected.

The preceding set of equations can be written in the nodal-matrix current equation form:

$$\mathbf{I}_{\text{bus}} = \mathbf{Y}_{\text{bus}}\mathbf{V}_{\text{bus}} \tag{8.1}$$

where the current vector is defined as

$$\mathbf{I}_{\text{bus}} = \begin{bmatrix} I_1 \\ I_2 \\ I_3 \\ I_4 \end{bmatrix}$$

The voltage vector is defined as

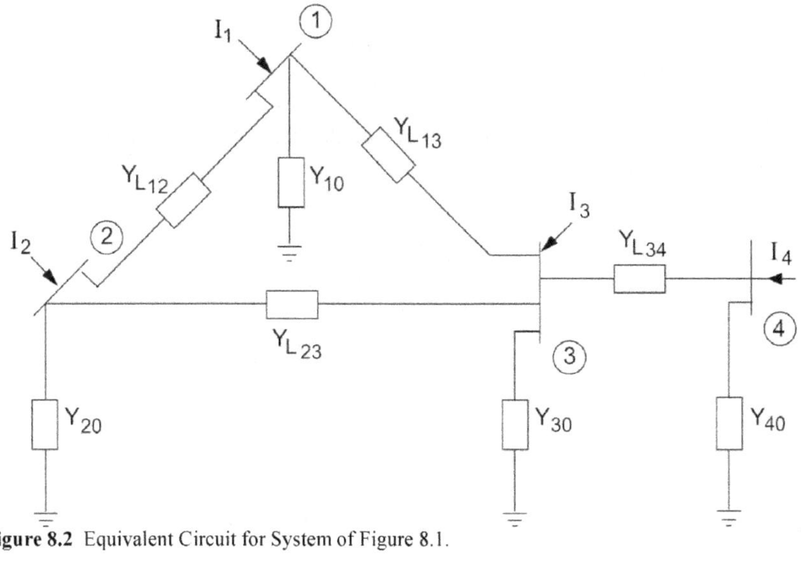


**Figure 8.2** Equivalent Circuit for System of Figure 8.1.

$$\mathbf{V}_{\text{bus}} = \begin{bmatrix} V_1 \\ V_2 \\ V_3 \\ V_4 \end{bmatrix}$$

The admittance matrix is defined as

$$\mathbf{Y}_{\text{bus}} = \begin{bmatrix} Y_{11} & Y_{12} & Y_{13} & Y_{14} \\ Y_{12} & Y_{22} & Y_{23} & Y_{24} \\ Y_{13} & Y_{23} & Y_{33} & Y_{34} \\ Y_{14} & Y_{24} & Y_{34} & Y_{44} \end{bmatrix}$$

We note that the bus admittance matrix $\mathbf{Y}_{\text{bus}}$ is symmetric.

## The General Form of the Load-Flow Equations

The result obtained for the 4 bus network can be generalized to the case of $n$ buses. Here, each of vectors $\mathbf{I}_{\text{bus}}$ and $\mathbf{V}_{\text{bus}}$ are $n \times 1$ vectors. The bus admittance matrix becomes an $n \times n$ matrix with elements

$$Y_{ij} = Y_{ji} = -Y_{L_{ij}} \qquad \textbf{(8.2)}$$

$$Y_{ii} = \sum_{j=0}^{n} Y_{L_{ij}} \qquad \textbf{(8.3)}$$

The summation is over the set of all buses connected to bus $i$ including the ground (node 0).

We recall that bus powers $S_i$ rather than the bus currents $I_i$ are, in practice, specified. We thus use

$$I_i^* = \frac{S_i}{V_i}$$

As a result, we have

$$\frac{P_i - jQ_i}{V_i^*} = \sum_{j=1}^{n} \left(Y_{ij} V_j\right) \qquad (i = 1, \ldots, n) \qquad \textbf{(8.4)}$$

These are the static power flow equations. Each equation is complex, and therefore we have $2n$ real equations.

The nodal admittance matrix current equation can be written in the power form:

$$P_i - jQ_i = \left(V_i^*\right)\sum_{j=1}^{n}\left(Y_{ij}V_j\right) \tag{8.5}$$

The bus voltages on the right-hand side can be substituted for using either the rectangular form,

$$V_i = e_i + jf_i$$

or the polar form:

$$\begin{aligned} V_i &= |V_i|e^{j\theta_i} \\ &= |V_i|\angle\theta_i \end{aligned}$$

### *Rectangular Form*

If we choose the rectangular form, then we have by substitution,

$$P_i = e_i\left(\sum_{j=1}^{n}\left(G_{ij}e_j - B_{ij}f_i\right)\right) + f_i\left(\sum_{j=1}^{n}\left(G_{ij}f_j + B_{ij}e_i\right)\right) \tag{8.6}$$

$$Q_i = f_i\left(\sum_{j=1}^{n}\left(G_{ij}e_j - B_{ij}f_i\right)\right) - e_i\left(\sum_{j=1}^{n}\left(G_{ij}f_j + B_{ij}e_i\right)\right) \tag{8.7}$$

where the admittance is expressed in the rectangular form:

$$Y_{ij} = G_{ij} + jB_{ij} \tag{8.8}$$

### *Polar Form*

On the other hand, if we choose the polar form, then we have

$$P_i = |V_i|\sum_{j=1}^{n}|Y_{ij}||V_j|\cos\left(\theta_i - \theta_j - \psi_{ij}\right) \tag{8.9}$$

$$Q_i = |V_i|\sum_{j=1}^{n}|Y_{ij}||V_j|\sin\left(\theta_i - \theta_j - \psi_{ij}\right) \tag{8.10}$$

where the admittance is expressed in the polar form:

$$Y_{ij} = |Y_{ij}| \angle \psi_{ij} \tag{8.11}$$

### *Hybrid Form*

The hybrid form is essentially the polar form with the admittances expressed in rectangular form. Expanding the trigonometric functions, we have

$$P_i = |V_i| \sum_{j=1}^{n} |Y_{ij}||V_j| \left[ \cos(\theta_i - \theta_j) \cos \psi_{ij} + \sin(\theta_i - \theta_j) \sin \psi_{ij} \right] \tag{8.12}$$

$$Q_i = |V_i| \sum_{j=1}^{n} |Y_{ij}||V_j| \left[ \sin(\theta_i - \theta_j) \cos \psi_{ij} - \cos(\theta_i - \theta_j) \sin \psi_{ij} \right] \tag{8.13}$$

Now we use

$$\begin{aligned} Y_{ij} &= |Y_{ij}| (\cos \psi_{ij} + j \sin \psi_{ij}) \\ &= G_{ij} + jB_{ij} \end{aligned} \tag{8.14}$$

Separating the real and imaginary parts of (8.14), we obtain

$$G_{ij} = |Y_{ij}| \cos \psi_{ij} \tag{8.15}$$

$$B_{ij} = |Y_{ij}| \sin \psi_{ij} \tag{8.16}$$

so that the power-flow equations become

$$P_i = |V_i| \sum_{j=1}^{n} |V_j| \left[ G_{ij} \cos(\theta_i - \theta_j) + B_{ij} \sin(\theta_i - \theta_j) \right] \tag{8.17}$$

$$Q_i = |V_i| \sum_{j=1}^{n} |V_j| \left[ G_{ij} \sin(\theta_i - \theta_j) - B_{ij} \cos(\theta_i - \theta_j) \right] \tag{8.18}$$

## The Power Flow Problem

The power-flow (or load-flow) problem is concerned with finding the static operating conditions of an electric power transmission system while satisfying constraints specified for power and/or voltage at the network buses.

Generally, buses are classified as follows:

6. A *load bus* (*P-Q* bus) is one at which $S_i = P_i + jQ_i$ is specified.
7. A *generator bus* (*P-V* bus) is a bus with specified injected active power and a fixed voltage magnitude.
8. A *system reference or slack* (swing) bus is one at which both the magnitude and phase angle of the voltage are specified. It is customary to choose one of the available *P-V* buses as slack and to regard its active power as the unknown.

As we have seen before, each bus is modeled by two equations. In all, we have $2n$ equations in $2n$ unknowns. These are $|V|$ and $\theta$ at the load buses, $Q$ and $\theta$ at the generator buses, and the $P$ and $Q$ at the slack bus.

Let us emphasize here that due to the bus classifications, it is not necessary for us to solve the $2n$ equations simultaneously. We can reduce the required number of equations by designating the unknown voltage magnitudes $|V_i|$ and angles $\theta_i$ at load buses and $\theta_i$ at generator buses as primary unknowns. Once these values are obtained, then we can evaluate the secondary unknowns $P_i$ and $Q_i$ at the slack bus and the reactive powers for the generator buses. This leads us to specifying the necessary equations for a full solution:

1. At load buses, two equations for active and reactive powers are needed.
2. At generator buses, with $|V_j|$ specified, only the active power equation is needed.

### *Nonlinearity of the Power Flow Problem*

Consider the power flow problem for a two-bus system with bus 1 being the reference bus and bus 2 is the load bus. The unknown is $|V_2|$, and is replaced by $x$.

$$x = |V_2|$$

$$\alpha = |Y_{22}|^2 \tag{8.19}$$

$$\beta = 2\left(B_{22}Q_2^{\text{sp}} - G_{22}P_2^{\text{sp}}\right) - |Y_{12}|^2 \tag{8.20}$$

$$\gamma = \left(S_2^{\text{sp}}\right)^2 \tag{8.21}$$

$$\left(S_2^{\text{sp}}\right)^2 = \left(P_2^{\text{sp}}\right)^2 + \left(Q_2^{\text{sp}}\right)^2 \tag{8.22}$$

We can demonstrate that the power flow equations reduce to the following equation:

$$\alpha x^4 + \beta x^2 + \gamma = 0 \tag{8.23}$$

The solution to the fourth order equation is straightforward since we can solve first for $x^2$ as

$$x^2 = \frac{-\beta \pm \sqrt{\beta^2 - 4\alpha\gamma}}{2\alpha} \tag{8.24}$$

Since $x^2$ cannot be imaginary, we have a first condition for the existence of a solution requiring that

$$\beta^2 - 4\alpha\gamma \geq 0$$

From the definitions of $\alpha$, $\beta$, and $\gamma$, we can show that for a meaningful solution to exist, we need to satisfy the condition:

$$|Y_{12}|^4 \geq 4\left[\left(B_{22}P_2^{\text{sp}} + G_{22}Q_2^{\text{sp}}\right)^2 + |Y_{12}|^2\left(B_{22}P_2^{\text{sp}} - G_{22}Q_2^{\text{sp}}\right)\right] \tag{8.25}$$

A second condition can be obtained if we observe that $x$ cannot be imaginary, requiring that $x^2$ be positive. Observing that $\alpha$ and $\gamma$ are positive by their definition leads us to conclude that

$$\left|\sqrt{\beta^2 - 4\alpha\gamma}\right| \leq |\beta|$$

For $x^2$ to be positive, we need

$$\beta \leq 0$$

or

$$2\left(B_{22}Q_2^{\text{sp}} - G_{22}P_2^{\text{sp}}\right) - |Y_{12}|^2 \leq 0 \tag{8.26}$$

We can therefore conclude the following:

- There may be some specified operating conditions for which no solution exists.
- More than one solution can exist. The choice can be narrowed down to a practical answer using further considerations.

Except for very simple networks, the load-flow problem results in a set of simultaneous algebraic equations that cannot be solved in closed form. It is necessary to employ numerical iterative techniques that start by assuming a set of values of the unknowns and then repeatedly improve on their values in an organized fashion until (hopefully) a solution satisfying the power flow equations is reached. The next section considers the question of getting estimates (initial guess) for the unknowns.

## Generating Initial Guess Solution

It is important to have a good approximation to the load-flow solution, which is then used as a starting estimate (or initial guess) in the iterative procedure. A fairly simple process can be used to evaluate a good approximation to the unknown voltages and phase angles. The process is implemented in two stages: the first calculates the approximate angles, and the second calculates the approximate voltage magnitudes.

### *Busbar Voltage Angles Approximation*

In this stage we make the following assumptions:

1. All angles are small, so that sin $\theta \cong \theta$, cos $\theta \cong 1$.
2. All voltage magnitudes are 1 p.u.

Applying these assumptions to the active power equations for the generator buses and load buses in hybrid form, we obtain

$$P_i = \sum_{j=1}^{N} \left(G_{ij}\right) + B_{ij}\left(\theta_i - \theta_j\right)$$

This is a system of $N$-1 simultaneous linear equations in $\theta_i$, which is then solved to obtain the busbar voltage angle approximations.

### *Busbar Voltage Magnitude Approximation*

The calculation of voltage magnitudes employs the angles provided by the above procedure. The calculation is needed only for load buses. We represent each unknown voltage magnitude as

$$\left|V_i\right| = 1 + \Delta V_i$$

We also assume that

$$\frac{1}{1 + \Delta V_i} \cong 1 - \Delta V_i$$

By considering all load buses we obtain a linear system of simultaneous equations in the unknowns $\Delta V_i$. The results are much more reliable than the commonly used flat-start process where all voltages are assumed to be $1\angle 0$.

## Newton-Raphson Method

The Newton-Raphson (NR) method is widely used for solving nonlinear equations. It transforms the original nonlinear problem into a sequence of linear problems whose solutions approach the solution of the original problem. The method can be applied to one equation in one unknown or to a system of simultaneous equations with as many unknowns as equations.

### *One-Dimensional Case*

Let $F(x)$ be a nonlinear equation. Any value of x that satisfies $F(x) = 0$ is a root of $F(x)$. To find a particular root, an initial guess for $x$ in the vicinity of the root is needed. Let this initial guess be $x_0$. Thus

$$F(x_0) = \Delta F_0$$

where $\Delta F_0$ is the error since $x_0$ is not a root. A tangent is drawn at the point on the curve corresponding to $x_0$, and is projected until it intercepts the $x$-axis to determine a second estimate of the root. Again the derivative is evaluated, and a tangent line is formed to proceed to the third estimate of $x$. The line generated in this process is given by

$$y(x) = F(x_n) + F'(x_n)(x - x_n) \tag{8.27}$$

which, when $y(x) = 0$, gives the recursion formula for iterative estimates of the root:

$$x_{n+1} = x_n - \frac{F(x_n)}{F'(x_n)} \tag{8.28}$$

### *N-Dimensional Case*

The single dimensional concept of the Newton-Raphson method can be extended to $N$ dimensions. All that is needed is an $N$-dimensional analog of the first derivative. The Jacobian matrix provides this. Each of the $n$ rows of the Jacobian matrix consists of the partial derivatives of one of the equations of the system with respect to each of the $N$ variables.

An understanding of the general case can be gained from the specific example $N = 2$. Assume that we are given the two nonlinear equations $F_1$, $F_2$. Thus,

$$F_1(x_1, x_2) = 0 \qquad F_2(x_1, x_2) = 0 \tag{8.29}$$

The Jacobian matrix for this $2 \times 2$ system is

$$\begin{bmatrix} \dfrac{\partial F_1}{\partial x_1} & \dfrac{\partial F_1}{\partial x_2} \\ \dfrac{\partial F_2}{\partial x_1} & \dfrac{\partial F_2}{\partial x_2} \end{bmatrix} \tag{8.30}$$

If the Jacobian matrix is numerically evaluated at some point ( $x_1^{(k)}$ , $x_2^{(k)}$ ), the following linear relationship is established for small displacements ( $\Delta x_1$ , $\Delta x_2$ ):

$$\begin{bmatrix} \dfrac{\partial F_1^{(k)}}{\partial x_1} & \dfrac{\partial F_1^{(k)}}{\partial x_2} \\ \dfrac{\partial F_2^{(k)}}{\partial x_1} & \dfrac{\partial F_2^{(k)}}{\partial x_2} \end{bmatrix} \begin{bmatrix} \Delta x_1^{(k+1)} \\ \\ \Delta x_2^{(k+1)} \end{bmatrix} = \begin{bmatrix} \Delta F_1^{(k)} \\ \\ \Delta F_2^{(k)} \end{bmatrix} \tag{8.31}$$

A recursive algorithm can be developed for computing the vector displacements ( $\Delta x_1$ , $\Delta x_2$ ). Each displacement is a solution to the related linear problem. With a good initial guess and other favorable conditions, the algorithm will converge to a solution of the nonlinear problem. We let ( $x_1^{(0)}$ , $x_2^{(0)}$ ) be the initial guess. Then the errors are

$$\Delta F_1^{(0)} = -F_1\left[x_1^{(0)}, x_2^{(0)}\right],\ \Delta F_2^{(0)} = -F_2\left[x_1^{(0)}, x_2^{(0)}\right] \tag{8.32}$$

The Jacobian matrix is then evaluated at the trial solution point [ $x_1^{(0)}$ , $x_2^{(0)}$ ]. Each element of the Jacobian matrix is computed from an algebraic formula for the appropriate partial derivative using $x_1^{(0)}$ , $x_2^{(0)}$ . Thus,

$$\begin{bmatrix} \dfrac{\partial F_1^{(0)}}{\partial x_1} & \dfrac{\partial F_1^{(0)}}{\partial x_2} \\ \dfrac{\partial F_2^{(0)}}{\partial x_1} & \dfrac{\partial F_2^{(0)}}{\partial x_2} \end{bmatrix} \begin{bmatrix} \Delta x_1^{(1)} \\ \\ \Delta x_2^{(1)} \end{bmatrix} = \begin{bmatrix} \Delta F_1^{(0)} \\ \\ \Delta F_2^{(0)} \end{bmatrix} \tag{8.33}$$

This system of linear equations is then solved directly for the first correction. The correction is then added to the initial guess to complete the first iteration:

$$\begin{bmatrix} x_1^{(1)} \\ x_2^{(1)} \end{bmatrix} = \begin{bmatrix} x_1^{(0)} \\ x_2^{(0)} \end{bmatrix} + \begin{bmatrix} \Delta x_1^{(1)} \\ \Delta x_2^{(1)} \end{bmatrix} \tag{8.34}$$

Equations (8.33) and (8.34) are rewritten using matrix symbols and a general superscript $h$ for the iteration count;

$$\left[J^{h-1}\right]\left[\Delta x^{h}\right]=\left[\Delta F^{h-1}\right] \tag{8.35}$$

$$x^{h}=x^{h-1}+\Delta x^{h} \tag{8.36}$$

The algorithm is repeated until $\Delta F^h$ satisfies some tolerance. In most solvable problems it can be made practically zero.

## The Newton-Raphson Method for Load-Flow Solution

There are different ways to apply the Newton-Raphson method to solving the load-flow equations. We illustrate a popular version employing the polar form. For each generator bus (except for the slack bus), we have the active power equation and the corresponding unknown phase $\theta_i$. We write this equation in the form

$$\Delta P_i = P_i^{\text{sch}} - P_i = 0$$

For each load bus we have the active and reactive equations and the unknowns $|V_i|$ and $\theta_i$. We write the two equations in the form

$$\Delta P_i = P_i^{\text{sch}} - P_i = 0$$
$$\Delta Q_i = Q_i^{\text{sch}} - Q_i = 0$$

In the above equations, the superscript "sch" denotes the scheduled or specified bus active or reactive powers. We use the polar form to illustrate the process.

$$P_i = |V_i| \sum_{j=1}^{n} |Y_{ij}||V_j| \cos(\theta_i - \theta_j - \psi_{ij})$$
$$Q_i = |V_i| \sum_{j=1}^{n} |Y_{ij}||V_j| \sin(\theta_i - \theta_j - \psi_{ij})$$

We show the application of the Newton-Raphson method to solve the power flow problem. The incremental corrections to estimates of the unknowns are obtained as the solution to the linear system of equations. Thus, for the example network we have:

$$\frac{\partial P_2}{\partial \theta_2}(\Delta\theta_2)+\frac{\partial P_2}{\partial \theta_3}(\Delta\theta_3)+\frac{\partial P_2}{\partial |V_3|}\Delta|V_3| = \Delta P_2$$
$$\frac{\partial P_3}{\partial \theta_2}(\Delta\theta_2)+\frac{\partial P_3}{\partial \theta_3}(\Delta\theta_3)+\frac{\partial P_3}{\partial |V_3|}\Delta|V_3| = \Delta P_3$$
$$\frac{\partial Q_3}{\partial \theta_2}(\Delta\theta_2)+\frac{\partial Q_3}{\partial \theta_3}(\Delta\theta_3)+\frac{\partial Q_3}{\partial |V_3|}\Delta|V_3| = \Delta Q_3$$

To simplify the calculation, the third term in each of the equations is modified so that we solve for $\left(\Delta|V_3|/|V_3|\right)$. We therefore have in matrix notation:

$$\begin{bmatrix} \frac{\partial P_2}{\partial \theta_2} \frac{\partial P_2}{\partial \theta_3} \left(|V_3|\frac{\partial P_2}{\partial |V_3|}\right) \\ \frac{\partial P_3}{\partial \theta_2} \frac{\partial P_3}{\partial \theta_3} \left(|V_3|\frac{\partial P_3}{\partial |V_3|}\right) \\ \frac{\partial Q_3}{\partial \theta_2} \frac{\partial Q_3}{\partial \theta_3} \left(|V_3|\frac{\partial Q_3}{\partial |V_3|}\right) \end{bmatrix} \begin{bmatrix} \Delta\theta_2 \\ \Delta\theta_3 \\ \frac{\Delta|V_3|}{|V_3|} \end{bmatrix} = \begin{bmatrix} \Delta P_2 \\ \Delta P_3 \\ \Delta Q_3 \end{bmatrix}$$

Solving for $\Delta\theta_2$, $\Delta\theta_3$ and $\left(\Delta|V_3|/|V_3|\right)$, we thus obtain the new estimates at the $(h+1)^{\text{th}}$ iteration:

$$\theta_2^{(h+1)} = \theta_2^{(h)} + \Delta\theta_2$$
$$\theta_3^{(h+1)} = \theta_3^{(h)} + \Delta\theta_3$$
$$|V_3|^{(h+1)} = |V_3|^{(h)} + \Delta|V_3|$$

The application in the general case assumes that bus 1 is the slack bus, that buses 2, . . ., $m$ are generator buses, and that buses $m + 1$, $m + 2$, . . ., $n$ are load buses. We introduce the Van Ness variables:

$$H_{ij} = \frac{\partial P_i}{\partial \theta_j} \qquad N_{ij} = \frac{\partial P_i}{\partial |V_j|}|V_j|$$
$$J_{ij} = \frac{\partial Q_i}{\partial \theta_j} \qquad L_{ij} = \frac{\partial Q_i}{\partial |V_j|}|V_j|$$

In condensed form, we have

$$\left[\begin{array}{c|c} \mathbf{H} & \mathbf{N} \\ \hline \mathbf{J} & \mathbf{L} \end{array}\right]\left[\begin{array}{c} \Delta\theta \\ \hline \left(\frac{\Delta \mathbf{V}}{\mathbf{V}}\right) \end{array}\right]=\left[\begin{array}{c} \Delta \mathbf{P} \\ \hline \Delta \mathbf{Q} \end{array}\right]$$

For the standpoint of computation, we use the rectangular form of the power equations. We introduce

$$a_{ij} = G_{ij}e_j - B_{ij}f_j$$
$$b_{ij} = G_{ij}f_j + B_{ij}e_j$$

In terms of the $a_{ij}$ and $b_{ij}$ variables, we have

$$P_i = (e_i)\sum_{j=1}^{n} a_{ij} + (f_i)\sum_{j=1}^{n} b_{ij}$$
$$Q_i = (f_i)\sum_{j=1}^{n} a_{ij} - (e_i)\sum_{j=1}^{n} b_{ij}$$

To summarize the expressions for the Van Ness variables are given by:

$$\begin{aligned} H_{ij} &= L_{ij} = a_{ij}f_i - b_{ij}e_i \\ N_{ij} &= -J_{ij} = a_{ij}e_i + b_{ij}f_i \\ H_{ii} &= -Q_i - B_{ii}V_i^2 \\ L_{ii} &= Q_i - B_{ii}V_i^2 \\ N_{ii} &= P_i + G_{ii}V_i^2 \\ J_{ii} &= P_i - G_{ii}V_i^2 \end{aligned} \qquad (i \neq j \text{ off diagonals})$$

A tremendous number of iterative techniques have been proposed to solve the power flow problem. It is beyond the scope of this text to outline many of the proposed variations. The Newton-Raphson method has gained a wide acceptability in industry circles, and as a result there are a number of available computer packages that are based on this powerful method and sparsity-directed programming.

## 8.5 STABILITY CONSIDERATIONS

We are interested in the behavior of the system immediately following a disturbance such as a short circuit on a transmission line, the opening of a line, or the switching on of a large block of loads. Studies of this nature are called *transient stability analysis*. The term *stability* is used in the sense of the ability of the system machines to recover from small random perturbing forces and still maintain synchronism. In this section we give an introduction to transient stability in electric power systems. We treat the case of a single machine operating to supply an infinite bus. We do not deal with the analysis of the more complex problem of large electric power networks with the interconnections taken into consideration.

## The Swing Equation

The dynamic equation relating the inertial torque to the net accelerating torque of the synchronous machine rotor is called the *swing equation.* This simply states

$$J\left(\frac{d^2\theta}{dt^2}\right) = T_a \mathrm{N}\cdot\mathrm{m} \tag{8.37}$$

The left-hand side is the inertial torque, which is the product of the inertia (in kg. m$^2$) of all rotating masses attached to the rotor shaft and the angular acceleration. The accelerating torque $T_a$ is in Newton meters and can be expressed as

$$T_a = T_m - T_e \tag{8.38}$$

In the above, $T_m$ is the driving mechanical torque, and $T_e$ is the retarding or load electrical torque.

The angular position of the rotor $\theta$ may be expressed as:

$$\theta = \alpha + \omega_R t + \delta \tag{8.39}$$

The angle $\alpha$ is a constant that is needed if the angle $\delta$ is measured from an axis different from the angular reference. The angle $\omega_R t$ is the result of the rotor angular motion at rated speed. The angle $\delta$ is time varying and represents deviations from the rated angular displacements. This is the basis for the new relation

$$J\left(\frac{d^2\delta}{dt^2}\right) = T_m - T_e \tag{8.40}$$

It is more convenient to make the following substitution of the dot notation:

$$\ddot{\delta} = \frac{d^2\delta}{dt^2}$$

Therefore we have

$$J\ddot{\delta} = T_m - T_e \quad \textbf{(8.41)}$$

An alternative form of Eq. (8.41) is the power form obtained by multiplying both sides of Eq. (8.41) by $\omega$ and recalling that the product of the torque $T$ and angular velocity is the shaft power. This results in

$$J\omega\ddot{\delta} = P_m - P_e$$

The quantity $J\omega$ is called the *inertia constant* and is truly an angular momentum denoted by $M$ (Js/rad). As a result,

$$M = J\omega \quad \textbf{(8.42)}$$

Thus, the power form is

$$M\ddot{\delta} = P_m - P_e \quad \textbf{(8.43)}$$

## Concepts in Transient Stability

In order to gain an understanding of the concepts involved in transient stability prediction, we will concentrate on the simplified network consisting of a series reactance $X$ connecting the machine and the infinite bus. Under these conditions the active power expression is given by:

$$P_e = \frac{EV}{X}\sin\delta \quad \textbf{(8.44)}$$

This yields the power angle curve shown in Figure 8.3.

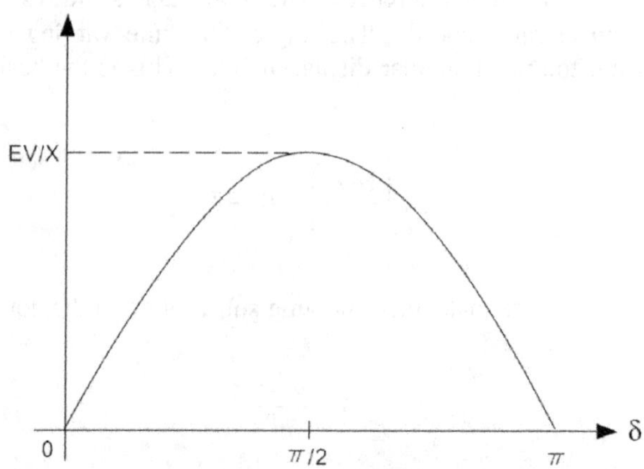


**Figure 8.3** Power Angle Curve Corresponding to Eq. (8.44).

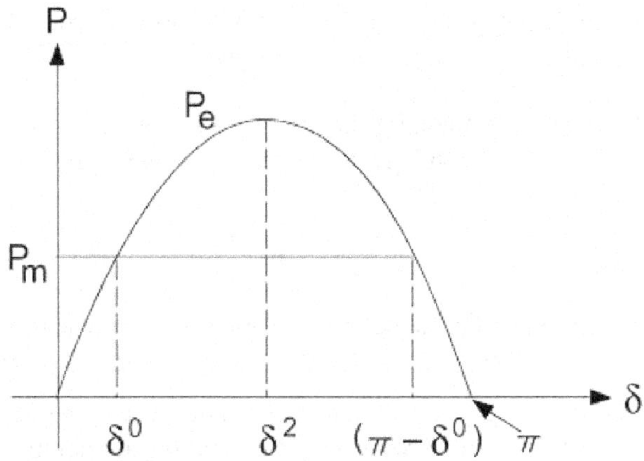


**Figure 8.4** Stable and Unstable Equilibrium Points.

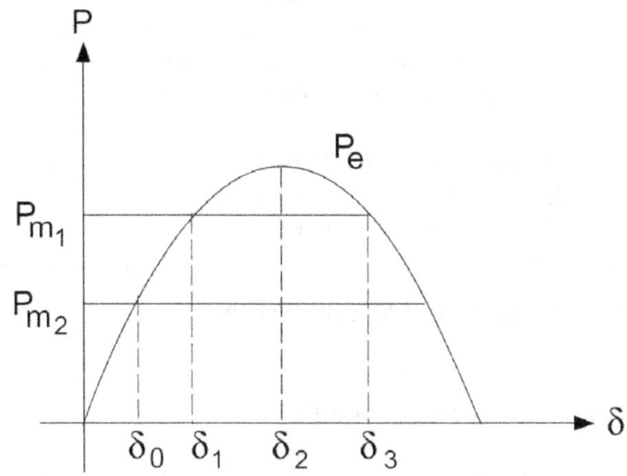


**Figure 8.5** System Reaction to Sudden Change.

We assume that the electric changes involved are much faster than the resulting mechanical changes produced by the generator/turbine speed control. Thus we assume that the mechanical power is a constant for the purpose of transient stability calculations. The functions $P_m$ and $P_e$ are plotted in Figure 8.4. The intersection of these two functions defines two values for $\delta$. The lower value is denoted $\delta^0$; consequently, the higher is $\pi - \delta^0$ according to the symmetry of the curve. At both points $P_m = P_e$; that is, $d^2\delta/dt^2 = 0$, and we say that the system is *in equilibrium.*

Assume that a change in operation of the system occurs such that $\delta$ is increased by a small amount $\Delta\delta$. Now for operation near $\delta^0$, $P_e > P_m$ and $d^2\delta/dt^2$ becomes negative according to the swing equation, Eq. (8.43). Thus $\delta$ is decreased, and the system responds by returning to $\delta^0$. We refer to this as a *stable equilibrium point.* On the other hand, operating at $\pi - \delta^0$ results in a system response that will increase $\delta$ and moving further from $\pi - \delta^0$. For this reason, we call $\pi - \delta^0$ an *unstable equilibrium point.*

If the system is operating in an equilibrium state supplying an electric power $P_{e_0}$ with the corresponding mechanical power input $P_{m_0}$, then

$$P_{m_0} = P_{e_0}$$

and the corresponding rotor angle is $\delta_0$. Suppose the mechanical power $P_m$ is changed to $P_{m_1}$ at a fast rate, which the angle $\delta$ cannot follow as shown in Figure 8.5. In this case, $P_m > P_e$ and acceleration occurs so that $\delta$ increases. This goes on until the point $\delta_1$ where $P_m = P_e$, and the acceleration is zero. The speed, however, is not zero at that point, and $\delta$ continues to increase beyond $\delta_1$. In this region, $P_m < P_e$ and rotor retardation takes place. The rotor will stop at $\delta_2$, where the speed is zero and retardation will bring $\delta$ down. This process continues on as oscillations around the new equilibrium point $\delta_1$. This serves to illustrate what happens when the system is subjected to a sudden change in the power balance of the right-hand side of the swing equation.

Changes in the network configuration between the two sides (sending and receiving) will alter the value of $X_{\text{eq}}$ and hence the expression for the electric power transfer. For example, opening one circuit of a double circuit line increases the equivalent reactance between the sending and receiving ends and therefore reduces the maximum transfer capacity $\dfrac{EV}{X_{eq}}$.

## A Method for Stability Assessment

In order to predict whether a particular system is stable after a disturbance it is necessary to solve the dynamic equation describing the behavior of the angle $\delta$ immediately following an imbalance or disturbance to the system. The system is said to be unstable if the angle between any two machines tends to increase without limit. On the other hand, if under disturbance effects, the angles between every possible pair reach maximum value and decrease thereafter, the system is deemed stable.

Assuming as we have already done that the input is constant, with negligible damping and constant source voltage behind the transient reactance, the angle between two machines either increases indefinitely or oscillates after all disturbances have occurred. Therefore, in the case of two machines, the two machines either fall out of step on the first swing or never. Here the observation that the machines' angular differences stay constant can be taken as an indication of system stability. A simple method for determining stability known as the *equal-area method* is available, and is discussed in the following.

## *The Equal-Area Method*

The swing equation for a machine connected to an infinite bus can be written as

$$\frac{d\omega}{dt} = \frac{P_a}{M} \tag{8.45}$$

where $\omega = d\delta/dt$ and $P_a$ is the accelerating power. We would like to obtain an expression for the variation of the angular speed $\omega$ with $P_a$. We observe that Eq. (8.45) can be written in the alternate form

$$d\omega = \frac{p_a}{M}\left(\frac{d\delta}{d\delta}\right)dt$$

or

$$\omega d\omega = \frac{P_a}{M}(d\delta)$$

Integrating, we obtain

$$\int_{\omega_0}^{\omega} \omega d\omega = \frac{1}{M}\int_{\delta_0}^{\delta} P_a(d\delta)$$

Note that we may assume $\omega_0 = 0$; consequently,

$$\omega^2 = \frac{2}{M}\int_{\delta_0}^{\delta} P_a(d\delta)$$

or

$$\frac{d\delta}{dt} = \left[\frac{2}{M}\int_{\delta_0}^{\delta} P_a(d\delta)\right]^{1/2} \tag{8.46}$$

The above equation gives the relative speed of the machine with respect to a reference from moving at a constant speed (by the definition of the angle $\delta$).

If the system is stable, then the speed must be zero when the acceleration is either zero or is opposing the rotor motion. Thus for a rotor that is accelerating, the condition for stability is that a value of $\delta_s$ exists such that

$$P_a(\delta_s) \leq 0$$

and

$$\int_{\delta_0}^{\delta_s} P_a(d\delta) = 0$$

This condition is applied graphically in Figure 8.6 where the net area under the $P_a$ - $\delta$ curve reaches zero at the angle $\delta_s$ as shown. Observe that at $\delta_s$, $P_a$ is negative, and consequently the system is stable. Also observe that area $A_1$ equals $A_2$ as indicated.

The accelerating power need not be plotted to assess stability. Instead, the same information can be obtained from a plot of electrical and mechanical powers. The former is the power angle curve, and the latter is assumed constant. In this case, the integral may be interpreted as the area between the $P_e$ curve and the curve of $P_m$, both plotted versus $\delta$. The area to be equal to zero must consist of a positive portion $A_1$, for which $P_m > P_e$, and an equal and opposite negative portion $A_2$, for which $P_m < P_e$. This explains the term *equal-area criterion for transient stability*. This situation is shown in Figure 8.7.

If the accelerating power reverses sign before the two areas $A_1$ and $A_2$ are equal, synchronism is lost. This situation is illustrated in Figure 8.8. The area $A_2$ is smaller that $A_1$, and as $\delta$ increases beyond the value where $P_a$ reverses sign again, the area $A_3$ is added to $A_1$.

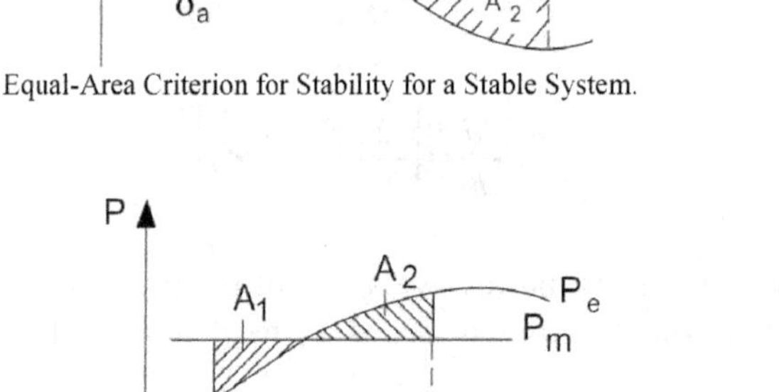


**Figure 8.6** The Equal-Area Criterion for Stability for a Stable System.

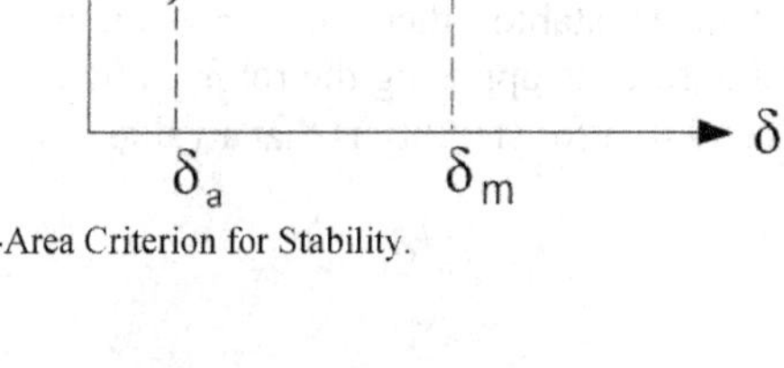


**Figure 8.7** The Equal-Area Criterion for Stability.

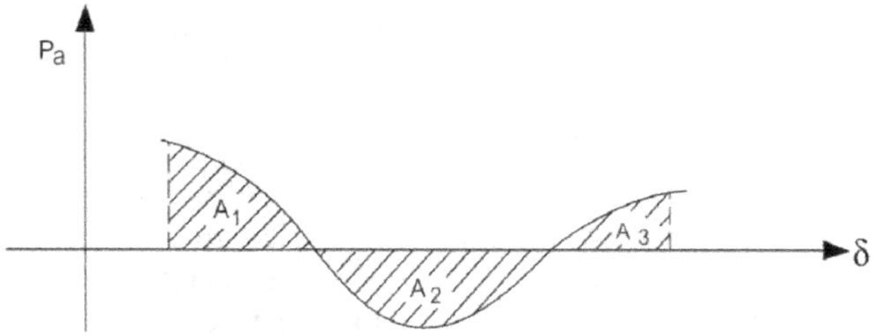


**Figure 8.8** The Equal-Area Criterion for an Unstable System.

## Improving System Stability

The stability of the electric power system can be affected by changes in the network or changes in the mechanical (steam or hydraulic) system. Network changes that adversely affect system stability can either decrease the amplitude of the power curve or raise the load line. Examples of events that decrease the amplitude of the power curve are: short circuits on tie lines, connecting a shunt reactor, disconnecting a shunt capacitor, or opening a tie line. Events that raise the load line include: disconnecting a resistive load in a sending area, connecting a resistive load in a receiving area, the sudden loss of a large load in a sending area, or the sudden loss of a generator in a receiving area. Changes in a steam or hydraulic system that can affect stability include raising the load line by either closing valves or gates in receiving areas or opening valves or gates in sending areas.

There are several corrective actions that can be taken in order to enhance the stability of the system following a disturbance. These measures can be classified according to the type of disturbance – depending on whether it is a loss of generation or a loss of load.

In the case of a loss of load, the system will have an excess power supply. Among the measures that can be taken are:

1. Resistor braking.
2. Generator dropping.
3. Initiation along with braking, fast steam valve closures, bypassing of steam, or reduction of water acceptance for hydro units.

In the case of loss of generation, countermeasures are:

1. Load shedding.
2. Fast control of valve opening in steam electric plants; and in the case of hydro, increasing the water acceptance.

The measures mentioned above are taken at either the generation or the load sides in the system. Measures that involve the interties (the lines) can be taken to enhance the stability of the system. Among these we have the switching of series capacitors into the lines, the switching of shunt capacitors or reactors, or the boosting of power on HVDC lines.

Resistor braking relies on the connection of a bank of resistors in shunt with the three-phase bus in a generation plant in a sending area through a suitable switch. This switch is normally open and will be closed only upon the activation of a control device that detects the increase in kinetic energy exceeding a certain threshold. Resistive brakes have short time ratings to make the cost much less than that of a continuous-duty resistor of the same rating. If the clearing of the short circuit is delayed for more than the normal time (about three cycles), the brakes should be disconnected and some generation should be dropped.

Generator dropping is used to counteract the loss of a large load in a sending area. This is sometimes used as a cheap substitute for resistor braking to counteract short circuits in sending systems. It should be noted that better control is achieved with resistor braking than with generator dropping.

To counteract the loss of generation, load shedding is employed. In this instance, a rapid opening of selected feeder circuit breakers in selected load areas is arranged. This disconnects the customer's premises with interruptible loads such as heating, air conditioning, air compressors, pumps where storage is provided in tanks, or reservoirs. Aluminum reduction plants are among loads that can be interrupted with only minor inconvenience. Load shedding by temporary depression of voltage can also be employed. This reduction of voltage can be achieved either by an intentional short circuit or by the connection of a shunt reactor.

The insertion of switched series capacitors can counteract faults on ac interties or permanent faults on dc interties in parallel with ac lines. In either case, the insertion of the switched series capacitor decreases the transfer reactance between the sending and receiving ends of the interconnection and consequently increases the amplitude of the sine curve and therefore enhances the stability of the system. It should be noted that the effect of a shunt capacitor inserted in the middle of the intertie or the switching off of a shunt reactor in the middle of the intertie is equivalent to the insertion of a series capacitor (this can be verified by means of a Y-Δ transformation).

To relieve ac lines of some of the overload and therefore provide a larger margin of stability, the power transfer on a dc line may be boosted. This is one of the major advantages of HVDC transmission.

## 8.6 POWER SYSTEM STATE ESTIMATION

Within the framework of an energy control center, there are three types of real-time measurements:

- Analog measurements that include real and reactive power flows through transmission lines, real and reactive power injections (generation or demand at buses), and bus voltage magnitudes.
- Logic measurements that consist of the status of switches/breakers, and transformer LTC positions.
- Pseudo-measurements that may include predicted bus loads and generation.

Analog and logic measurements are telemetered to the control center. Errors and noise may be contained in the data. Data errors are due to failures in measuring and telemetry equipment, noise in the communication system, and delays in the transmission of data.

The state of a system is described by a set of variables that at time $t_0$ contains all information about the system, which allows us to determine completely the system behavior at a future time $t_1$. A convenient choice is the selection of a minimum set of variables, thus defining a minimum, but sufficient set of state variables. Note that the state variables are not necessarily directly accessible, measurable, or observable. Since the system model used is based on a nodal representation, the choice of the state variables is rather obvious. Assuming that line impedances are known, the state variables are the voltage magnitudes and angles. This follows because all other values can be uniquely defined once the state values are known.

State estimation is a mathematical procedure to yield a description of the power system by computing the best estimate of the state variables (bus voltages and angles) of the power system based on the received noisy data. Once state variables are estimated, secondary quantities (e.g., line flows) can readily be derived. The network topology module processes the logic measurements to determine the network configuration. The state estimator processes the set of analog measurements to determine the system state; it also uses data such as the network parameters (e.g., line impedance), network configuration supplied by the network topology, and sometimes, pseudo-measurements. Since it not practical to make extensive measurements of network parameters in the field, manufacturers data and one line drawings are used to determine parameter values. This may then introduce another source of error.

The mathematical formulation of the basic power system state estimator assumes that the power system is static. Consider a system, which is characterized by $n$ state variables, denoted by $x_i$, with $i = 1, \ldots, n$. Let $m$ measurements be available. The measurement vector is denoted $z$ and the state vector is $x$. If the noise is denoted by $v$, then the relation between measurements and states denoted by $h$ is given by:

$$z_i = h_i(x) + v_i \tag{8.47}$$

or in compact form:

$$z = h(x) + v$$

Let us linearize $h(x)$, and we thus deal with:

$$z = Hx + v \tag{8.48}$$

$H$ is called the measurement matrix and is independent of the state variables.

There are many techniques for finding the best estimate of $x$, denoted by $\hat{x}$. We discuss the most popular approach based on the weighted least squares WLS concept. The method aims to minimize the deviations between the measurements and the corresponding equations. This requires minimizing the following objective function:

$$J(x) = \sum_{1}^{m} k_i [z_i - h_i(x)]^2 \tag{8.49}$$

We can demonstrate that the optimal estimates are obtained using the following recursive equation:

$$x_{k+1} = x_k + \left[H_k^T W H_k\right]^{-1} H_k^T W [z - h(x_k)] \tag{8.50}$$

This means that the state variables are successively approximated closer and closer to some value and a convergence criterion determines when the iteration is stopped. The matrix $W$ is called the weighting matrix, and relates the measurements individually to each other. The results are influenced by the choice of the elements of $W$. If one chooses $W = I$, all measurements are of equal quality.

## Observability

If the number of measurements is sufficient and well-distributed geographically, the state estimator will give an estimate of system state (i.e., the state estimation equations are solvable). In this case, the network is said to be observable. Observability depends on the number of measurements available and their geographic distribution. Usually a measurement system is designed to be observable for most operating conditions. Temporary unobservability may still occur due to unexpected changes in network topology or failures in the telecommunication systems.

Before applying state estimation in power system operation, we need to conduct an observability analysis study. The aim here is to ensure that there are enough real-time measurements to make state estimation possible. If not, we need to determine where additional meters should be placed so that state estimation is possible. Moreover, we need to determine how are the states of these observable islands are estimated, and how additional pseudo-measurements are included in the measurement set to make state estimation possible. Finally, we need to be able to guarantee that the inclusion of the additional pseudo-measurements will not contaminate the result of the state estimation. Observability analysis includes observability testing, identification of observable islands, and measurement placement.

## Bad Data Detection and Identification

State estimation is formulated as a weighted least square error problem, and implicitly assumes that the errors are small. Large errors or bad data occasionally occur. The residual (the Weighted Least Square error) will be large if bad data or structural error is present. Action is needed to detect the bad data; identify which measurements are bad; and to remove all bad data so that they do not corrupt the state estimates. Detecting bad data is based on techniques of hypothesis testing to determine when the residual or the error is too large. Note, however, that a switch indicating other than its true position can cause larger error and hence we may end up discarding a valid analog reading. In practice, a major benefit of state estimation is identifying bad data in the system.

## Benefits of Implementing a State Estimator

Implementation of a state estimator establishes the following data:

- The correct impedance data for all modeled facilities. This might seem to be information which should be readily available from the system plans of any given power system. Note, however, that between the time a facility is planned and placed in service, distances for transmission lines change due to right-of-way realignment, or the assumed conductor configuration is changed, or the conductor selected is not as assumed, etc. The net result is that the impedance according to the system plan may be up to 10 percent off from the present actual values.
- The correct fixed tap position for all transformers in the modeled network.
- The correct load tap changing information for all modeled Load Tap Changing (LTC) transformers.
- The correct polarity of all MW and MVAR flow meters.
- Detect bad meters as they go bad. As a result, more confidence is established in the entire active meter set if meters are corrected as they are detected to be bad.
- When an unusual event occurs, the active meter set can be believed before the power system security process has been rerun. This saves time for the system operators.

## External Network Modeling

In an interconnected system, the responsibility of each energy control center is to operate its own part of the system. A control center receives telemetered real-time measurements of its own system; referred to as the internal system. Neighboring systems are called the external system. Any unmonitored portions of the internal system such as distribution/sub-transmission networks or unmonitored substations must also be incorporated in the "external" mode. Data exchange between utilities is often a difficult and sensitive issue, and this impacts the state estimation function. It is not always clear how much of the neighboring systems need to be measured for satisfactory performance of the state estimator.

To determine the current status of the internal system using a state estimator, it is not necessary to know more about the external system. It is important, however, to include the response of the external system in evaluating the consequence of various contingencies for security assessment. An external model is constructed either on-line or off-line, or using a combination of both. This model can be a full or a reduced power flow model, or a combination of both. The external model is then attached to the internal system as the power flow model to evaluate the response of the internal system to various contingencies. A reduced power flow model of the external system is called an external equivalent.

There may be portions of the transmission or sub-transmission system for which there is no direct telemetry. The choices are whether to neglect this portion, or put it into an equivalent form. It is often practical to eliminate a portion of the network if its most direct through-path directly paralleling a modeled transmission path is ten times or more the impedance of the modeled path. If, on the other hand, the step-down transformers to that portion are to be monitored, then the underlying system must be at least modeled as an equivalent path.

## 8.7 POWER SYSTEM SECURITY

By power system security, we understand a qualified absence of risk of disruption of continued system operation. Security may be defined from a control point of view as the probability of the system's operating point remaining in a viable state space, given the probabilities of changes in the system (contingencies) and its environment (weather, customer demands, etc.). Security can be defined in terms of how it is monitored or measured, as the ability of a system to withstand without serious consequences any one of a pre-selected list of "credible" disturbances ("contingencies"). Conversely, insecurity at any point in time can be defined as the level of risk of disruption of a system's continued operation.

Power systems are interconnected for improved economy and availability of supplies across extensive areas. Small individual systems would be individually more at risk, but widespread disruptions would not be possible. On the other hand, interconnections make widespread disruptions possible.

Operation of interconnected power systems demands nearly precise synchronism in the rotational speed of many thousands of large interconnected generating units, even as they are controlled to continuously follow significant changes in customer demand. There is considerable rotational energy involved, and the result of any cascading loss of synchronism among major system elements or subsystems can be disastrous. Regardless of changes in system load or sudden disconnection of equipment from the system, synchronized operation requires proper functioning of machine governors, and that operating conditions of all equipment remain within physical capabilities.

The risk of cascading outages still exists, despite improvements made because many factors increase the risks involved in interconnected system operation:

- Wide swings in the costs of fuels result in significant changes in the geographic patterns of generation relative to load. This leads to transmission of electric energy over longer distances in patterns other than those for which the transmission networks had been originally designed.
- Rising costs due to inflation and increasing environmental concerns constrain any relief through further transmission construction. Thus, transmission, as well as generation, must be operated closer to design limits, with smaller safety (security) margins.
- Relaxation of energy regulation to permit sales of electric energy by independent power producers, together with increasing pressure for essentially uncontrolled access to the bulk power transmission network.

## Development of the Concept of Security

Prior to the 1965 Northeast blackout, system security was part of reliability assured at the system planning stage by providing a strong system that could ride out any "credible" disturbances without serious disruption. It is no longer economically feasible to design systems to this standard. At that time, power system operators made sure that sufficient spinning reserve was on line to cover unexpected load increases or potential loss of generation and to examine the impact of removing a line or other apparatus for maintenance. Whenever possible, the operator attempted to maintain a desirable voltage profile by balancing VARs in the system.

Security monitoring is perceived as that of monitoring, through contingency analysis, the conditional transition of the system into an emergency state.

## Two Perspectives of Security Assessment

There is a need to clarify the roles of security assessment in the planning and real-time operation environments. The possible ambiguity is the result of the shift of focus from that of system robustness designed at the planning stage as part of reliability, to that of risk avoidance that is a matter operators must deal with in real time. The planner is removed from the time-varying real world environment within which the system will ultimately function. The term "security" within a planning context refers to those aspects of reliability analysis that deal with the ability of the system, as it is expected to be constituted at some future time, to withstand unexpected losses of certain system components. Reliability has frequently been considered to consist of adequacy and security. Adequacy is the ability to supply energy to satisfy load demand. Security is the ability to withstand sudden disturbances. This perspective overlooks the fact that the most reliable system will ultimately experience periods of severe insecurity from the operator's perspective. System operations is concerned with security as it is constituted at the moment, with a miscellaneous variety of elements out for maintenance, repair, etc., and exposed to environmental conditions that may be very different from the normal conditions considered in system planning. In operations, systems nearly always have less than their full complement of equipment in service. As a result, an operator must often improvise to improve security in ways that are outside the horizon of planners.

## Security Assessment Defined

Security assessment involves using available data to estimate the relative security level of the system currently or at some near-term future state. Approaches to security assessment are classified as either direct or indirect.

- **The direct approach:** This approach evaluates the likelihood of the system operating point entering the emergency state. It calculates the probability that the power System State will move from normal state to emergency state, conditioned on its current state, projected load variations, and ambient conditions. It is common practice to assess security by analyzing a fixed set of contingencies. The system is declared as insecure if any member of the set would result in transition to the emergency state. This is a limiting form of direct assessment, since it implies a probability of the system being in the emergency state conditioned on the occurrence of any of the defined contingencies.
- **The indirect approach:** Here a number of reserve margins are tracked relative to predetermined levels deemed adequate to maintain system robustness vis-a-vis pre-selected potential disturbances. An indirect method of security assessment defines a set of system "security" variables that should be maintained with predefined limits to provide adequate reserve margins. Appropriate variables might include, MW reserves, equipment emergency ratings (line, transformer, etc.), or VAR reserves within defined regions. The reserve margins to be maintained for each of the security variables could be determined by offline studies for an

appropriate number of conditions with due consideration to the degree to which random events can change the security level of a system in real time. Security assessment then would consist of tracking all such reserve margins relative to system conditions.

For a number of years, security concerns dealt with potential post-contingency line overloads and because line MW loading can be studied effectively by means of a linear system network model, it was possible to study the effects of contingencies using linear participation or distribution factors. Once derived for a given system configuration, they could be applied without further power flow analysis to determine post-contingency line loading even, by superposition, for multiple contingencies. Such a computationally simple method of analysis made on-line contingency assessment practicable for "thermal security," where reactive flows were not of concern.

More recently, post-contingency voltage behavior has become a prominent element in security assessment. Assessment of "voltage security" is a complex process because the behavior of a system undergoing voltage collapse cannot be completely explained on the basis of static analysis alone.

## Implications of Security

The trend towards reducing the costs associated with robust systems has lead to heightened requirements of active security control. This necessitates an increase in the responsibilities of the system operator. Accordingly, it requires operator training and the development and provision of tools that will enable the operator to function effectively in the new environment.

## Security Analysis

On-line security analysis and control involve the following three ingredients:

- Monitoring
- Assessment
- Control

The following framework relates the three modules:

Step 1. **Security Monitoring**: Identify whether the system is in the normal state or not using real-time system measurements. If the system is in an emergency state, go to step 4. If load has been lost, go to step 5.

Step 2. **Security Assessment**: If the system is in the normal state, determine whether the system is secure or insecure with respect to a set of next contingencies.

Step 3. **Security Enhancement**: If insecure, i.e., there is at least one contingency, which can cause an emergency, determine what action to take to make the system secure through preventive actions.

Step 4. **Emergency Control (remedial action):** Perform proper corrective action to bring the system back to the normal state following a contingency, which causes the system to enter an emergency state.

Step 5. **Restorative Control:** Restore service to system loads.

Security analysis and control have been implemented in modem energy control centers. The major components of on-line security analysis are shown in Fig. 8.9.

The monitoring module starts with real-time measurements of physical quantities such as line power and current flows, power injections, bus voltage magnitudes, and the status of breakers and switches. Measured data are telemetered from various locations to the control center computer. A simple check of the transmitted data for reasonability and consistency allows filtering of bad measurement data, which are then rejected. Healthy data are systematically processed to determine network topology, i.e., the system configuration. The available data are further processed to obtain an estimate of the system state variables (bus voltage magnitudes and phase angles for normal steady state). State estimation is a mathematical procedure for computing the "best" estimate of the state variables of the system based on the available data, which are in general corrupted with errors. Prior to state estimation one would

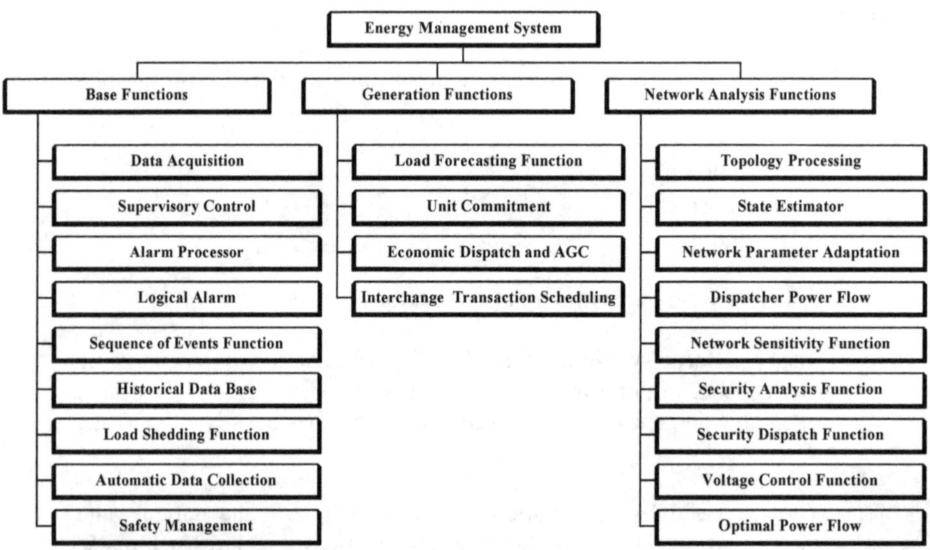


**Figure 8.9** Functional Structure of Energy Management Centers.

like to know:

1) Whether state estimation of the system is possible (enough of the network is observable), and if not, for which part of the system state estimation is still possible.
2) Whether there is any bad data present, and if so, which data is bad and should be discarded.

Observability analysis and bad data detection and identification are parts of the state estimation function.

We need a set of contingencies to assess whether a normal operating state is secure or not. The contingency selection process employs a scheme to select a set of important and plausible disturbances. Security assessment involves primarily steady-state power flow analysis. Stability constraints are expressed in terms of the limits on line flows and bus voltages. As a result, to assess system response to contingencies, a contingency evaluation is carried out using on-line power flows. The on-line power flow uses the actual power flow model of the system under study (from the state estimation solution) together with a system representation of the unmonitored network and neighboring systems, i.e., an external network model.

Since the contingencies are future events, a bus-load forecast is needed. Certain implementations of the state estimator render the external model observable by strategic placement of pseudo-measurements. Then the state estimate is performed on the entire model in one step.

## *8.8 CONTINGENCY ANALYSIS*

Contingency analysis indicates to the operator what might happen to the system in the event of unplanned equipment outage. It essentially offers answers to questions such as "What will be the state of the system if an outage on part of the major transmission system takes place?" The answer might be that power flows and voltages will readjust and remain within acceptable limits, or that severe overloads and under-voltages will occur with potentially severe consequences should the outage take place.

A severe overload, persisting long enough, can damage equipment of the system, but usually relays are activated to isolate the affected equipment once it fails. The outage of a second component due to relay action is more serious and often results in yet more readjustment of power flows and bus voltages. This can in turn cause more overloads and further removal of equipment. An uncontrollable cascading series of overloads and equipment removals may then take place, resulting in the shutting down of a significant portion of the system.

The motivation to use contingency analysis tools in an EMS is that when forewarned the operator can initiate preventive action before the event to avoid problems should an outage take place. From an economic point of view, the operator strives to avoid overloads that might directly damage equipment, or worse, might cause the system to lose a number of components due to relay action and then cause system-wide outages.

Insulation breakdown, over-temperature relay action, or simply incorrect operation of relay devices are internal causes of contingencies. External contingencies are caused by environmental effects such as lightning, high winds and ice conditions or else are related to some non-weather related events such as vehicle or aircraft coming into contact with equipment, or even human or animal direct contact. These causes are treated as unscheduled, random events, which operators can not anticipate, but for which they must be prepared.

The operator must play an active role in maintaining system security. The first step is to perform contingency analysis studies frequently enough to assure that system conditions have not changed significantly from the last execution. The outcome of contingency analysis is a series of warnings or alarms to the operators alerting them that loss of component *A* will result in an overload of *X*% on line T1. To achieve an accurate picture of the system's exposure to outage events several points need to be considered:

### *A) System Model*

Contingency analysis is carried out using a power flow model of the system. Additional information about system dynamics is needed to assess stability as well. Voltage levels and the geographic extent to include in the model are issues to be considered. In practice, all voltage levels that have any possibility of connecting circuits in parallel with the high voltage system are included. This leaves out those that are radial to it such as distribution networks. While the geographical extent is difficult to evaluate, it is common to model the system to the extent real-time measurement data is available to support the model.

### *B) Contingency Definition*

Each modeled contingency has to be specified on its own. The simplest definition is to name a single component. This implies that when the model of the system is set up, this contingency will be modeled by removing the single component specified. Another important consideration is the means of specifying the component outage. The component can be specified by name, such as a transmission line name, or more accurately, a list of circuit breakers can be specified as needing to be operated to correctly model the outage of the component. Contingencies that require more than one component to be taken out together must be defined as well. There is an advantage here to using a "list of breakers" in that the list is simply expanded to include all breakers necessary to remove all relevant equipment.

### *C) Double Contingencies*

A double contingency is the overlapping occurrence of two independent contingent events. To be specific, one outside event causes an outage and while this outage is still in effect, a second totally independent event causes another component to be taken out. The overlap of the two outages often causes overloads and under-voltages that would not occur if either happened separately. As a result, many operators require that a contingency analysis program be able to take two independent contingencies and model them as if they had happened in an overlapping manner.

### *D) Contingency List*

Generally, contingency analysis programs are executed based on a list of valid contingencies. The list might consist of all single component outages including all transmission lines, transformers, substation buses, and all generator units. For a large interconnected power system just this list alone could result in thousands of contingency events being tested. If the operator wished to model double contingencies, the number becomes millions of possible events. Methods of selecting a limited set of priority contingencies are then needed.

### *E) Speed*

Generally, operators need to have results from a contingency analysis program in the order of a few minutes up to fifteen minutes. Anything longer means that the analysis is running on a system model that does not reflect current system status and the results may not be meaningful.

### *F) Modeling Detail*

The detail required for a contingency case is usually the same as that used in a study power flow. That is, each contingency case requires a fully converged power flow that correctly models each generator's VAR limits and each tap adjusting transformer's control of voltage.

## Historical Methods of Contingency Analysis

There is a conflict between the accuracy with which the power system is modeled and the speed required for modeling all the contingencies specified by the operator. If the contingencies can be evaluated fast enough, then all cases specified on the contingency list are run periodically and alarms reported to the operators. This is possible if the computation for each outage case can be performed very fast or else the number of contingencies to be run is very small. The number of contingency cases to be solved in common energy management systems is usually a few hundred to a few thousand cases. This coupled with the fact that the results are to be as accurate as if run with a full power flow program make the execution of a contingency analysis program within an acceptable time frame extremely difficult.

## Selection of Contingencies to be Studied

A full power flow must be used to solve for the resulting flows and voltages in a power system with serious reactive flow or voltage problems when an outage occurs. In this case, the operators of large systems looking at a large number of contingency cases may not be able to get results soon enough. A significant speed increase could be obtained by simply studying only the important cases, since most outages do not cause overloads or under-voltages.

### *1) Fixed List*

Many operators can identify important outage cases and they can get acceptable performance. The operator chooses the cases based on experience and then builds a list for the contingency analysis program to use. It is possible that one of the cases that were assumed to be safe may present a problem because some assumptions used in making the list are no longer true.

### *2) Indirect Methods (Sensitivity-Based Ranking Methods)*

An alternative way to produce a reduced contingency list is to perform a computation to indicate the possible bad cases and perform it as often as the contingency analysis itself is run. This builds the list of cases dynamically and the cases that are included in the list may change as conditions on the power system change. This requires a fast approximate evaluation to discover those outage cases that might present a real problem and require further detailed evaluation by a full power flow. Normally, a sensitivity method based on the concept of a network performance index is employed. The idea is to calculate a scalar index that reflects the loading on the entire system.

### *3) Comparison of Direct and Indirect Methods*

Direct methods are more accurate and selective than the indirect ones at the expense of increased CPU requirements. The challenge is to improve the efficiency of the direct methods without sacrificing their strengths. Direct methods assemble severity indices using monitored quantities (bus voltages, branch flows, and reactive generation), that have to be calculated first. In contrast, the indirect methods calculate severity indices explicitly without evaluating the individual quantities. Therefore, indirect methods are usually less computationally demanding. Knowing the individual monitored quantities enables one to calculate severity indices of any desired complexity without significantly affecting the numerical performance of direct methods. Therefore, more attention has been paid recently to direct methods for their superior accuracy (selectivity). This has lead to drastic improvements in their efficiency and reliability.

### *4) Fast Contingency Screening Methods*

To build a reduced list of contingencies one uses a fast solution (normally an approximate one) and ranks the contingencies according to its results. Direct contingency screening methods can be classified by the imbedded modeling assumptions. Two distinct classes of methods can be identified:

a) Linear methods specifically intended to screen contingencies for possible real power (branch MW overload) problems.
b) Nonlinear methods intended to detect both real and reactive power problems (including voltage problems).

Bounding methods offer the best combination of numerical efficiency and adaptability to system topology changes. These methods determine the parts of the network in which branch MW flow limit violations may occur. A linear incremental solution is performed only for the selected system areas rather than for the entire network. The accuracy of the bounding methods is limited only by the accuracy of the incremental linear power flow.

Nonlinear methods are designed to screen the contingencies for reactive power and voltage problems. They can also screen for branch flow problems (both MW and MVA/AMP). Recent proposed enhancements include attempts to localize the outage effects, and speeding the nonlinear solution of the entire system.

An early localization method is the "concentric relaxation" which solves a small portion of the system in the vicinity of the contingency while treating the remainder of the network as an "infinite expanse." The area to be solved is concentrically expanded until the incremental voltage changes along the last solved tier of buses are not significantly affected by the inclusion of an additional tier of buses. The method suffered from unreliable convergence, lack of consistent criteria for the selection of buses to be included in the small network; and the need to solve a number of different systems of increasing size resulting from concentric expansion of the small network (relaxation).

Different attempts have been made at improving the efficiency of the large system solution. They can be classified as speeding up the solution by means of:

1) Approximations and/or partial (incomplete) solutions.
2) Using network equivalents (reduced network representation).

The first approach involves the "single iteration" concept to take advantage of the speed and reasonably fast convergence of the Fast Decoupled Power Flow to limit the number of iterations to one. The approximate, first iteration solution can be used to check for major limit violations and the calculation of different contingency severity measures. The single iteration approach can be combined with other techniques like the use of the reduced network representations to improve numerical efficiency.

An alternative approach is based upon bounding of outage effects.

Similar to the bounding in linear contingency screening, an attempt is made to perform a solution only in the stressed areas of the system. A set of bounding quantities is created to identify buses that can potentially have large reactive mismatches. The actual mismatches are then calculated and the forward solution is performed only for those with significant mismatches. All bus voltages are known following the backward substitution step and a number of different severity indices can be calculated.

The zero mismatch (ZM) method extends the application of localization ideas from contingency screening to full iterative simulation. Advantage is taken of the fact that most contingencies significantly affect only small portions (areas) of the system. Significant mismatches occur in only very few areas of the system being modeled. There is a definite pattern of very small mismatches throughout the rest of the system model. This is particularly true for localizable contingencies, e.g., branch outages, bus section faults. Consequently, it should be possible to utilize this knowledge and significantly speed up the solution of such contingencies. The following is a framework for the approach:

1) Bound the outage effects for the first iteration using for example a version of the complete boundary.
2) Determine the set of buses with significant mismatches resulting from angle and magnitude increments.
3) Calculate mismatches and solve for new increments.
4) Repeat the last two steps until convergence occurs.

The main difference between the zero mismatch and the concentric relaxation methods is in the network representation. The zero mismatch method uses the complete network model while a small cutoff representation is used in the latter one. The zero mismatch approach is highly reliable and produces results of acceptable accuracy because of the accuracy of the network representation and the ability to expand the solution to any desired bus.

## 8.9 OPTIMAL PREVENTIVE AND CORRECTIVE ACTIONS

For contingencies found to cause overloads, voltage limit violations, or stability problems, preventive actions are required. If a feasible solution exists to a given security control problem, then it is highly likely that other feasible solutions exist as well. In this instance, one solution must be chosen from among the feasible candidates. If a feasible solution does not exist (which is also common), a solution must be chosen from the infeasible candidates. Security optimization is a broad term to describe the process of selecting a preferred solution from a set of (feasible or infeasible) candidate solutions. The term Optimal Power Flow (OPF) is used to describe the computer application that performs security optimization within an Energy Management System.

## Optimization in Security Control

To address a given security problem, an operator will have more than one control scheme. Not all schemes will be equally preferred and the operator will thus have to choose the best or "optimal" control scheme. It is desirable to find the control actions that represent the optimal balance between security, economy, and other operational considerations. The need is for an optimal solution that takes all operational aspects into consideration. Security optimization programs may not have the capability to incorporate all operational considerations into the solution, but this limitation does not prevent security optimization programs from being useful.

The solution of the security optimization program is called an "optimal solution" if the control actions achieve the balance between security, economy, and other operational considerations. The main problem of security optimization seeks to distinguish the preferred of two possible solutions. A method that chooses correctly between any given pair of candidate solutions is capable of finding the optimal solution out of the set of all possible solutions. There are two categories of methods for distinguishing between candidate solutions: one class relies on an objective function, the other class relies on rules.

### *1) The Objective Function*

The objective function method assumes that it is possible to assign a single numerical value to each possible solution, and that the solution with the lowest value is the optimal solution. The objective function is this numerical assignment. In general, the objective function value is an explicit function of the controls and state variables, for all the networks in the problem. Optimization methods that use an objective function typically exploit its analytical properties, solving for control actions that represent the minimum. The conventional optimal power flow (OPF) is an example of an optimization method that uses an objective function.

The advantages of using an objective function method are:

- Analytical expressions can be found to represent MW production costs and transmission losses, which are, at least from an economic view point, desirable quantities to minimize.
- The objective function imparts a value to every possible solution. Thus all candidate solutions can, in principle, be compared on the basis of their objective function value.
- The objective function method assures that the present optimal solution can be recognized by virtue of its having the minimum value.

Typical objective functions used in OPF include MW production costs or expressions for active (or reactive) power transmission losses. However, when the OPF is used to generate control strategies that are intended to keep the power system secure, it is typical for the objective function to be an expression of the MW production costs, augmented with fictitious control costs that represent other operational considerations. This is especially the case when security against contingencies is part of the problem definition. Thus when security constrained OPF is implemented to support real-time operations, the objective function tends to be a device whose purpose is to guide the OPF to find the solution that is optimal from an operational perspective, rather than one which represents a quantity to be minimized.

Some examples of non-economic operational considerations that a utility might put into its objective function are:

- a preference for a small number of control actions;
- a preference to keep a control away from its limit;
- the relative preference or reluctance for preventive versus post-contingent action when treating contingencies; and
- a preference for tolerating small constraint violations rather than taking control action.

The most significant shortcoming of the objective function method is that it is difficult (sometimes impossible) to establish an objective function that consistently reflects true production costs and other non-economic operational considerations.

### *2) Rules*

Rules are used in methods relying on expert systems techniques. A rule-based method is appropriate when it is possible to specify rules for choosing between candidate solutions easier than by modeling these choices via an objective function. Optimization methods that use rules typically search for a rule that matches the problem addressed. The rule indicates the appropriate decision (e.g., control action) for the situation. The main weakness of a rule-based approach is that the rule base does not provide a continuum in the solution space. Therefore, it may be difficult to offer guidance for the OPF from the rule base when the predefined situations do not exist in the present power system state.

Rules can play another important role when the OPF is used in the real-time environment. The real-time OPF problem definition itself can be ill defined and rules may be used to adapt the OPF problem definition to the current state of the power system.

## Optimization Subject to Security Constraints

The conventional OPF formulation seeks to minimize an objective function subject to security constraints, often presented as "hard constraints," for which even small violations are not acceptable. A purely analytical formulation might not always lead to solutions that are optimal from an operational perspective. Therefore, the OPF formulation should be regarded as a framework in which to understand and discuss security optimization problems, rather than as a fundamental representation of the problem itself.

### 1) *Security Optimization for the Base Case State*

Consider the security optimization problem for the base case state ignoring contingencies. The power system is considered secure if there are no constraint violations in the base case state. Thus any control action required will be corrective action. The aim of the OPF is to find the optimal corrective action.

When the objective function is defined to be the MW production costs, the problem becomes the classical active and reactive power constrained dispatch. When the objective function is defined to be the active power transmission losses, the problem becomes one of active power loss minimization.

### 2) *Security Optimization for Base Case and Contingency States*

Now consider the security optimization problem for the base case and contingency states. The power system is considered secure if there are no constraint violations in the base case state, and all contingencies are manageable with post-contingent control action. In general, this means that base case control action will be a combination of corrective and preventive actions and that post-contingent control action will be provided in a set of contingency plans. The aim of the OPF is then to find the set of base case control actions plus contingency plans that is optimal.

Dealing with contingencies requires solving OPF involving multiple networks, consisting of the base case network and each contingency network. To obtain an optimal solution, these individual network problems must be formulated as a multiple network problem and solved in an integrated fashion. The integrated solution is needed because any base case control action will affect all contingency states, and the more a given contingency can be addressed with post-contingency control action, the less preventive action is needed for that contingency.

When an operator is not willing to take preventive action, then all contingencies must be addressed with post-contingent control action. The absence of base case control action decouples the multiple network problems into a single network problem for each contingency. When an operator is not willing to rely on post-contingency control action, then all contingencies must be addressed with preventive action. In this instance, the cost of the preventive action is preferred over the risk of having to take control action in the post-contingency state. The absence of post-contingency control action means that the multiple network problem may be represented as the single network problem for the base case, augmented with post-contingent constraints.

Security optimization for base case and contingency states will involve base case corrective and preventive action, as well as contingency plans for post-contingency action. To facilitate finding the optimal solution, the objective function and rules that reflect operating policy are required. For example, if it is preferred to address contingencies with post-contingency action rather than preventive action, then post-contingent controls may be modeled as having a lower cost in the objective function. Similarly, a preference for preventive action over contingency plans could be modeled by assigning the post-contingent controls a higher cost than the base case controls. Some contingencies are best addressed with post-contingent network switching. This can be modeled as a rule that for a given contingency, switching is to be considered before other post-contingency controls.

### *3) Soft Constraints*

Another form of security optimization involves "soft" security constraints that may be violated but at the cost of incurring a penalty. This is a more sophisticated method that allows a true security/economy trade-off. Its disadvantage is requiring a modeling of the penalty function consistent with the objective function. When a feasible solution is not possible, this is perhaps the best way to guide the algorithm toward finding an "optimal infeasible" solution.

### *4) Security versus Economy*

As a general rule, economy must be compromised for security. However, in some cases security can be traded off for economy. If the constraint violations are small enough, it may be preferable to tolerate them in return for not having to make the control moves. Many constraint limits are not truly rigid and can be relaxed. Thus, in general, the security optimization problem seeks to determine the proper balance of security and economy. When security and economy are treated on the same basis, it is necessary to have a measure of the relative value of a secure, expensive state relative to a less secure, but also less expensive state.

### *5) Infeasibility*

If a secure state cannot be achieved, there is still a need for the least insecure operating point. For OPF, this means that when a feasible solution cannot be found, it is still important that OPF reach a solution, and that this solution be "optimal" in some sense, even though it is infeasible. This is especially appropriate for OPF problems that include contingencies in their definition. The OPF program needs to be capable of obtaining the "optimal infeasible" solution. There are several approaches to this problem. Perhaps the best approach is one that allows the user to model the relative importance of specific violations, with this modeling then reflected in the OPF solution. This modeling may involve the objective function (i.e., penalty function) or rules, or both.

## The Time Variable

The preceding discussion assumes that all network states are based on the same (constant) frequency, and all transient effects due to switching and outages are assumed to have died out. While bus voltages and branch flows are, in general, sinusoidal functions of time, only the amplitudes and phase relationships are used to describe network state. Load, generation, and interchange schedules change slowly with time, but are treated as constant in the steady state approximation. There are still some aspects of the time variable that need to be accounted for in the security optimization problem.

### 1) *Time Restrictions on Violations and Controls*

The limited amount of time to correct constraint violations is a security concern. This is because branch flow thermal limits typically have several levels of rating (normal, emergency, etc.), each with its maximum time of violation. (The higher the rating, the shorter the maximum time of violation.) Voltage limits have a similar rating structure and there is very little time to recover from a violation of an emergency voltage rating.

Constraint violations need to be corrected within a specific amount of time. This applies to violations in contingency states as well as actual violations in the base case state. Base case violations, however, have the added seriousness of the elapsed time of violation: a constraint that has been violated for a period of time has less time to be corrected than a constraint that has just gone into violation.

The situation is further complicated by the fact that controls cannot move instantaneously. For some controls, the time required for movement is significant. Generator ramp rates can restrict the speed with which active power is rerouted in the network. Delay times for switching capacitors and reactors and transformer tap changing mechanisms can preclude the immediate correction of serious voltage violations. If the violation is severe enough, slow controls that would otherwise be preferred may be rejected in favor of fast, less preferred controls. When the violation is in the contingency state, the time criticality may require the solution to select preventive action even though a contingency plan for post-contingent corrective action might have been possible for a less severe violation.

### 2) *Time in the Objective Function*

It is common for the MW production costs to dominate the character of the objective function for OPF users. The objective function involves the time variable to the extent that the OPF is minimizing a time rate of change. This is also the case when the OPF is used to minimize the cost of imported power or active power transmission losses. Not all controls in the OPF can be "costs" in terms of dollars per hour. The start-up cost for a combustion turbine, for example, is expressed in dollars, not dollars per hour. The costing of reactive controls is even more difficult, since the unwillingness to move these controls is not easily expressed in either dollars or dollars per hour. OPF technology requires a single objective function, which means that all control costs must be expressed in the same units. There are two approaches to this problem:

- Convert dollar per hour costs into dollar costs by specifying a time interval for which the optimization is to be valid. Thus control costs in dollars per hour multiplied by the time interval, yield control costs in dollars. This is now in the same units as controls whose costs are "naturally" in dollars. This approach thus "integrates" the time variable out of the objective function completely. This may be appropriate when the OPF solution is intended for a well-defined (finite) period of time.
- Regard all fixed control costs (expressed in dollars) as occurring repeatedly in time and thus having a justified conversion into dollars per hour. For example, the expected number of times per year that a combustion turbine is started defines a cost per unit time for the start-up of the unit. Similarly, the unwillingness to move reactive controls can be thought of as reluctance over and above an acceptable amount of movement per year. This approach may be appropriate when the OPF is used to optimize over a relatively long period of time.
- Simply adjust the objective function empirically so that the OPF provides acceptable solutions. This method can be regarded as an example of either of the first two approaches.

## Using an Optimal Power Flow Program

OPF programs are used both in on-line and in off-line (study mode) studies. The two modes are not the same.

### *1) On-line Optimal Power Flow*

The solution speed of an on-line OPF should be high enough so that the program converges to a solution before the power system configuration has changed appreciably. Thus the on-line OPF should be fast enough to run several times per hour. The values of the algorithm's input parameters should be valid over a wide range of operating states, such that the program continues to function as the state of the system changes. Moreover, the application needs to address the correct security optimization problem and that the solutions conform to current operating policy.

### *2) Advisory Mode versus Closed Loop Control*

On-line OPF programs are implemented in either advisory or closed loop mode. In advisory mode, the control actions that constitute the OPF solution are presented as recommendations to the operator. For closed loop OPF, the control actions are actually implemented in the power system, typically via the SCADA subsystem of the Energy Management System. The advisory mode is appropriate when the control actions need review by the dispatcher before their implementation. Closed loop control for security optimization is appropriate for problems that are so well defined that dispatcher review of the control actions is not necessary. An example of closed loop on-line OPF is the Constrained Economic Dispatch (CED) function. Here, the constraints are the active power flows on transmission lines, and the controls are the MW output of generators on automatic generation control (AGC). When the conventional Economic Dispatch would otherwise tend to overload the transmission lines in its effort to minimize production costs, the CED function supplies a correction to the controls to avoid the overloads. Security optimization programs that include active and reactive power constraints and controls, in contingency states as well as in the base case, are implemented in an advisory mode. Thus the results of the on-line OPF are communicated to the dispatchers via EMS displays. Considering the typical demands on the dispatchers' time and attention in the control center, the user interface for on-line OPF needs to be designed such that the relevant information is communicated to the dispatchers "at-a-glance."

### *3) Defining the Real-time Security Optimization Problem*

As the power system state changes through time, the various aspects of the security optimization problem definition can change their relative importance. For example, concern for security against contingencies may be a function of how secure the base case is. If the base case state has serious constraint violations, one may prefer to concentrate on corrective action alone, ignoring the risk of contingencies. In addition, the optimal balance of security and economy may depend on the current security state of the power system. During times of emergency, cost may play little or no role in determining the optimal control action. Thus the security optimization problem definition itself can be dynamic and sometimes not well defined.

## *8.10 DYNAMIC SECURITY ANALYSIS*

The North American Electric Reliability Council (NERC) defines security as "the prevention of cascading outages when the bulk power supply is subjected to severe disturbances." To assure that cascading outages will not take place, the power system is planned and operated such that the following conditions are met at all times in the bulk power supply:

- No equipment or transmission circuits are overloaded;
- No buses are outside the permissible voltage limits (usually within +5 percent of nominal); and
- When any of a specified set of disturbances occurs, the transient will be followed by acceptable steady-state conditions (i.e., instability will not occur).

Security analysis is carried out to ensure that these conditions are met. The first two require only steady-state analysis; but the third requires transient analysis (e.g., using a transient stability application). It has also been recognized that some of the voltage instability phenomena are dynamic in nature, and require new analysis tools.

Generally, security analysis is concerned with the system's response to disturbances. In steady-state analysis the transition to a new operating condition is assumed to have taken place, and the analysis ascertains that operating constraints are met in this condition (thermal, voltage, etc.). In dynamic security analysis the transition itself is of interest, i.e., the analysis checks that the transition will lead to an acceptable operating condition. Examples of possible concern include loss of synchronism by some generators, transient voltage at a key bus (e.g., a sensitive load) falling below a certain level, and operation of an out-of-step relay resulting in the opening of a heavily loaded tie-line.

The computational capability of some control centers may limit security analysis to steady state calculations. The post-contingency steady-state conditions are computed and limit checked for flow or voltage violations. The dynamics of the system may then be ignored and whether the post-contingency state was reached without losing synchronism in any part of the system remains unknown. As a result, instead of considering actual disturbances, the contingencies are defined in terms of outages of equipment and steady-state analysis is done for these outages. This assumes that the disturbance did not cause any instability and that simple protective relaying caused the outage. Normally, any loss of synchronism will cause additional outages thus making the present steady-state analysis of the post-contingency condition inadequate for unstable cases. It is clear that dynamic analysis is needed.

In practice, we define a list of equipment losses for static analysis. Such a list usually consists of all single outages and a careful choice of multiple outages. Ideally, the outages should be chosen according to their probability of occurrence but these probabilities are usually not known. In some instance the available probabilities are so small that comparisons are usually meaningless. The choice of single outages is reasonable because they are more likely to occur than multiple ones. Including some multiple outages is needed because certain outages are likely to occur together because of proximity (e.g., double lines on the same tower) or because of protection schemes (e.g., a generator may be relayed out when a line is on outage). The size of this list is usually several hundred and can be a couple of thousand.

For dynamic security analysis, contingencies are considered in terms of the total disturbance. All faults can be represented as three phase faults, with or without impedances, and the list of contingencies is a list of locations where this can take place. This is a different way of looking at contingencies where the post-contingency outages are determined by the dynamics of the system including the protection system. Obviously, if all possible locations are considered, this list can be very large.

In steady-state security analysis, it is not necessary to treat all of the hundreds of outage cases using power flow calculations, because the operator is interested in worst possibilities rather than all possibilities. It is practical to use some approximate but faster calculations to filter out these worst outages, which can then be analyzed by a power flow. This screening of several hundred outages to find the few tens of the worst ones has been the major breakthrough that made steady-state security analysis feasible. Generally, this contingency screening is done for the very large list of single outages while the multiple outages are generally included in the short list for full power flow analysis. The trend is to use several different filters (voltage filter versus line overload filter) for contingency screening. It is also necessary to develop fast filtering schemes for dynamic security analysis to find the few tens of worst disturbances for which detailed dynamic analysis will have to be done. The filters are substantially different from those used for static security.

From a dispatcher's point of view, static and dynamic security analyses are closely related. The worst disturbances and their effects on the system are to be examined. The effects considered include the resulting outages and the limit violations in the post-contingency condition. In addition, it would be useful to know the mechanism that caused the outages, whether they were due to distance relay settings or loss of synchronism or other reasons. This latter information is particularly useful for preventive action.

The stability mechanism that causes the outages is referred to as the "mode of disturbance." A number of modes exist. A single generating unit may go out of synchronism on the first swing (cycle). A single unit may lose synchronism after several cycles, up to a few seconds. Relays may operate to cause transmission line outages. Finally, periodic oscillations may occur between large areas of load and/or generation. These oscillations may continue undamped to a point of loss of synchronism. All of these types of events are called modes of disturbances.

## Motivation for Dynamic Security Analysis

Ascertaining power system security involves considering all possible (and credible) conditions and scenarios; analysis is then performed on all of them to determine the security limits for these conditions. The results are given to the operating personnel in the form of "operating guides," establishing the "safe" regions of operation. The key power system parameter or quantity is monitored (in real time) and compared with the available (usually pre-computed) limit. If the monitored quantity is outside the limit, the situation is alerted or flagged for some corrective action.

Recent trends in operating power systems close to their security limits (thermal, voltage and stability) have added greatly to the burden on transmission facilities and increased the reliance on control. Simultaneously, they have increased the need for on-line dynamic security analysis.

For on-line dynamic security analysis, what is given is a base case steady-state solution (the real time conditions are obtained from the state estimator and external model computation, or a study case as set up by the operator) and a list of fault locations. The effects of these faults have to be determined and, specifically, the expected outages have to be identified. Examining the dynamic behavior of the system can do this. Some form of fast approximate screening is required such that the few tens of worst disturbances can be determined quickly.

Traditionally, for off-line studies, a transient stability program is used to examine the dynamic behavior. This program, in the very least, models the dynamic behavior of the machines together with their interconnection through the electrical network. Most production grade programs have elaborate models for the machines and their controls together with dynamic models of other components like loads, dc lines, static VAR compensators, etc. These models are simulated in time using some integration algorithm and the dynamic behavior of the system can be studied. If instability (loss of synchronism) is detected, the exact mode of instability (the separation boundary) can be identified. Many programs have relay models that can also pinpoint the outages caused by relay operation due to the dynamic behavior.

To perform the analysis in on-line mode the time required for the computation is a crucial consideration. That is, the analysis itself by a pure time domain simulation is known to be feasible but whether this analysis can be completed within the time frame needed in the control center environment is the real challenge. The time taken for time domain analysis of power system dynamics depends on many factors. The most obvious one is the length of simulation or the time period for which the simulation needs to be done so that all the significant effects of the disturbance can be captured. Other factors include the size of the power system, and the size and type of the models used. Additional factors like the severity of the disturbance and the solution algorithm used also influence the computation time.

Determining the vulnerability of the present system conditions to disturbances does not complete the picture because the solution to any existing problems must also be found. Quite often the post-contingency overloads and out-of limit voltage conditions are such that they can be corrected after the occurrence of the fault. Sometimes, and especially for unstable faults, the post-contingency condition is not at all desirable and preventive remedial action is needed. This usually means finding new limits for operating conditions or arming of special protective devices. Although remedial action is considered, as a separate function from security analysis, operators of stability limited systems need both.

## Approaches to DSA

A number of approaches to the on-line dynamic stability analysis problem have been studied. To date, engineers perform a large number of studies off-line to establish operating guidelines, modified by judgement and experience. Conventional wisdom has it that computer capability will continue to make it more economically feasible to do on-line dynamic security assessment (DSA) as long as the appropriate methods are developed.

The most obvious method for on-line DSA is to implement the off-line time domain techniques on faster, more powerful and less expensive computers. Equivalencing and localization techniques are ways to speed up the time domain solutions. Also parallel and array processors show promise in accelerating portions of the time domain solution.

Direct methods of transient stability, e.g., the transient energy function (TEF) method, have emerged with the potential of meeting some of the needs for DSA. They offer the possibility of conducting stability studies in near real-time, provide a qualitative judgement on stability, and they are suitable for use in sensitivity assessments. The TEF methods are limited to first swing analysis. An advantage, however, is that the TEF methods provide energy margins to indicate the margin to instability.

Eigenvalue and related methods, and frequency response methods are used as part of off-line studies, for example, using frequency response methods to design power system stabilizers, but are not currently thought of as part of an on-line DSA. Probabilistic methods have the advantage of providing a measure of the likelihood of a stability problem. Their application in dynamic security assessment appears to be in the areas of contingency screening and in quantifying the probability of the next state of the system.

Artificial intelligence techniques including computational neural networks, fuzzy logic, and expert systems have proven to be appropriate solutions to other power system operations problems, and there is speculation that these technologies will play a major role in DSA.

# Chapter 9

# THE PRESENT AND FUTURE OF ELECTRIC ENERGY SYSTEMS

## 9.1 INTRODUCTION

An electric power grid is a complex interconnected arrangement of systems that have to work together to deliver electricity, while maintaining synchronized operation of all its generating units. A power grid is a sprawling network involving many separate utilities; each independent in its own control area and regulated by local bodies, to meet the demand for electricity reliably and economically. These organizations have to coordinate to operate the interconnected system, and this coordination can take many forms, from a loose agreement of operational principles to a strong pooling arrangement of operating together. In a loosely coordinated grid, each balancing utility is responsible for operating its own portion while exchanging all relevant information with its partners. Alternatively, groups of utilities may set up a pool where operational decisions may be made centrally and then implemented by each utility.

The electrical power infrastructure in the United States represents an enormous investment, including over 15,000 generators in 10,000 power plants and hundreds of thousands of miles of transmission lines and distribution networks.

## 9.2 CHALLENGES FACING THE SYSTEM

Historically, the electric power business was structured such that generation and transmission, and in many cases distribution facilities, were owned and operated by the same company. This structure is known as a vertically integrated business, whose objective is to deliver bulk power from generation to load areas while centrally optimizing the generation schedules in a secure and reliable manner.

Many publicly and privately owned individual power companies were interconnected, often across national boundaries, to form electric power grids with the intent of improving network reliability by enabling neighboring utilities to share generation reserves to support each other during emergency (stressed) conditions, and making economic transactions when load diversities and generation scheduling plans create opportunities. The first interconnection in the United States was constructed in the late 1920s, in the Pennsylvania-New Jersey-Maryland area (now referred to as the PJM system). Electric power grids are hailed as the 20th century's engineering innovation most beneficial to human civilization.

## Deregulation

There has been a worldwide trend since the latter part of the twentieth century toward deregulating the electric power industry, which has had a profound impact on how the electric grid is operated. Under deregulation, many different organizations are responsible for different portions of the operation and control of the generation-transmission-distribution networks. Because deregulation involves separating ownership of generating, transmission and distribution facilities, new operational procedures are needed to ensure the reliability and security of electricity supply. Presently, many jurisdictions have deregulated ownership of generation sources. In such cases, generation companies have to compete in the electricity markets to sell their supply. As a result, market transactions determine generation production while the transmission operator is mainly concerned with the reliability of the grid.

In contrast with the case of a vertically integrated utility where the economic goal of minimizing the cost of supply is achieved by optimizing the dispatch and scheduling of its generators, in deregulated regions the electricity markets optimize the cost of the electricity supply. The market determines the levels of power sales and purchases for the long and short terms and after the markets clear, the generating companies can schedule hydro resources, fossil fuels, and equipment maintenance over many weeks, and the commitment (start-up and shutdown) of generating units over many hours. The computerized tools needed to run the market and those used by the generation companies to determine their bid strategies are getting more sophisticated all the time. The transmission operator has to run contingency (what-if) studies using these schedules to ensure that all reliability criteria are met or else the purchase-sale transactions have to be changed to meet reliability criteria requirements.

## The Nature of the Electric Power System

Power-system operations can be divided into three stages: operations planning, real-time control, and after-the-fact accounting. A basic requirement is the ability to forecast load demand in the next few hours, days, weeks, or months that allows the market to determine the schedules and the transmission operator to approve these schedules if they meet the reliability requirements. The aim of real-time control is to respond to the actual demand, which always deviates from the forecast, and any unforeseen contingencies (equipment outages). Maintaining reliability of the system so that a possible contingency cannot disrupt power supply is an integral part of real-time control function. After-the-fact accounting tracks the actual delivery of energy between organizations to allow billing. All of this requires significant data communication as well as engineering computation within an individual company and between companies.

Superior communication and computation technology play a vital role and the heart of monitoring and control of system operations in a utility is the energy control or management system, which became the operations headquarters for each utility. Although these centers have greatly advanced technology, their basic control objective of maintaining reliability and security remain the same.

Implementing deregulation would not be feasible without these advanced technologies. The trends in separating the economic functions from the reliability functions are happening at different rates in different parts of the world and best practices for operating these deregulated systems are still emerging.

## The Challenges of the Electric Business

The nature of all electric power grids is that transmission lines connect all the electric generation sources and distribution points within their respective territories. Although electric power grids are highly reliable, the same systems are subject to many challenges that have the potential to adversely impact their reliability. While some might argue that the challenges may be broadly classified as being caused either by natural evolution or by the advent of electric power deregulation, there are sufficient overlaps between the two categories that discourage such classification.

The power grid evolved over the past 100 years without a conscious awareness of how its evolution would affect its operation under deregulation, the digital economy, and interaction with other infrastructures. Competition and deregulation created multiple energy entities that must share the same regulated energy-delivery network, while imposing additional requirements of high transfers from new generation sources to the load areas. Grid planning faces a difficult task given the requirement to achieve resource adequacy in the electric power industry, where market economics and local concerns often drive short-term decisions.

### *Load Growth*

Demand is already out-pacing available resources in several regions because the economy, population, and technological achievements continue to grow. During the last decade of the twentieth century, actual US demand for electricity increased some 35%, while capacity increased only 18%. Meeting the growing load in the more densely populated heavy load areas requires transmission and generation expansions be added to handle load increases.

Public pressure based on environmental and other considerations such as the "not in my backyard" sentiment make it difficult to site transmission lines or major local generation sources. This makes system expansion very expensive and difficult. It is becoming increasingly harder for power generators and delivery entities to get permits and ensure that their return on investment is acceptable.

Thus fewer total circuit miles are being added and the complex systems that relieve bottlenecks and clear disturbances during peak demand are now closer to the edge and at greater risk of serious disruption.

### *Aging Infrastructure*

In many regions of the world the amortization/depreciation rate has exceeded utility construction expenditures. Consequently utility construction expenditures have lagged behind asset depreciation. As a result of these diminished "shock absorbers," the electric grid is becoming increasingly stressed, and whether the carrying capacity or safety margin will exist to support anticipated demand is questionable.

### *Knowledge Attrition*

Knowledgeable and skilled human resources require training and experience that take time to develop. With an aging power engineering force, a major challenge that faces the electric power industry is to replace the retirees. This situation is not helped by trends in electrical engineering education that have not favored the electric power engineering stream.

### *Requirement for Quality*

Digital computer technology users require a much higher quality of electricity. Some experts indicate that reliability will need to go from 99.9% (roughly 8 hours of power loss per year) to 99.99999999% reliability (32 seconds of power loss per year). The industry will also need new equipment to protect against sags and disruptions.

### *Grid Complexity*

The power grid involves many interaction points (operators, power consumers and producers, and several layers including power plants, control centers, and transmission, distribution, and corporate networks). Additional complexity developed because the interaction of these elements further increased the number of possible outcomes.

### *Regulatory or Legislative Issues*

Given the complexities of an interconnected power system, these challenges have arisen, as the bulk grids were originally not designed to transfer large amounts of power between neighboring systems. Enacting regulatory requirements for planning and operational standards have not kept pace with the changes in the nature of the electric business.

### *Outages and Blackouts*

Widespread outages and huge price spikes during the recent past have raised public concern about grid reliability at the national level. The potential for larger scale and more frequent power disruptions is considered higher now than at any time since the great Northeast blackout in 1965. The ramifications of network failure have never been greater, as all the other infrastructures depend more and more on the power grid to energize and control their operations.

### *Times and Scales*

The time and operational scales at which the electric grid operates are an important part of its complexity. The time scale for various power grid control and operation tasks can be anywhere from microseconds to a decade, which greatly complicates modeling, analysis, simulation, control, and operations tasks. The key elements and principles of operation for interconnected power systems were established prior to the emergence of extensive computer and communication networks.

Coordination across the network happens on slower timescales. Some coordination occurs under computer control, but much of it is still based on telephone calls between system operators at the utility control centers (even, or especially, during an emergency).

## *9.3 BLACKOUTS AND THEIR IMPACT*

The term blackout is used to describe a widespread cutoff or loss of supply of electrical power, especially as a result of a shortage, a mechanical failure, or overuse by consumers. It is well known that blackouts happen only rarely and they are usually caused by a sequence of low probability outages. Common electric power system design practice strives to have the system remain functional for ($N$-1) contingency (one major power system element taken out of service: a transmission line, major generation source, or transmission transformer) and not for the sequence of low probability disturbances. Most disturbances associated with major blackouts have occurred following a series of successive unscheduled equipment outages beyond ($N$-2) subsequent to the occurrence of extremely low-probability events. Blackouts are caused by a variety of phenomena. The causes include weaknesses in the infrastructure, storms, the general climate and human error. The consequences can be far ranging: disruption to industry, transport problems and a variety of social and political effects.

"2003 was the year of the blackout."[i] Major wide area blackouts included:

- 10th June - Athens

- 26th June - Rome
- 14th August – Northeast US and Central Canada
- 23rd August - Helsinki
- 27th August - Shanghai
- 28th August - London
- 23rd September - Sweden and Denmark
- 23rd September - Georgia
- 28th September - Italy

## Selected Blackout Events

### *The August 14, 2003 North American Blackout*

The U.S.-Canadian blackout of August 14, 2003 involved a general area covering eight U.S. states and two Canadian provinces and affected approximately 50 million people while interrupting 63 GW of load and tripping over 400 transmission lines and 531 generating units at 261 power plants. The earlier indicators of potential difficulties included:

- Many generators in FirstEnergy's (FE) Northern Ohio service area were required to generate high levels of reactive power which continued to increase throughout the day.
- The Midwest Independent System Operator (MISO) state estimator (SE) and real time contingency analysis software were not functioning properly from 12:15 to 16:04. This prevented the MISO from performing proper "early warning" assessments of the system.
- At the FirstEnergy (FE) control center, a number of computer software failures occurred on their Energy Management System (EMS) software starting at 14:14. This prevented FE from having adequate knowledge of the events taking place on its own system until approximately 15:45.

This major blackout evolved in the following manner:

- The first major event was that the voltage regulator of the FE's Eastlake unit 5 generator tripped to manual because of over-excitation. The generator tripped at 13:31:34 as the operator attempted to place the generator back under automatic voltage control.
- The Chamberlin-Harding 345-kV line tripped at 15:05 in FE's system due to tree contact; the line was loaded to only 44% of summer normal/emergency rating.
- The Hanna-Juniper 345-kV line loaded to 88% of its summer emergency rating and tripped due to tree contact at 15:32.
- The Star-Canton 345-kV line loaded to 93% of its rating and tripped due to tree contact at 15:41.
- During this period due to EMS failures at FE and MISO control centers, no proper operational actions were being taken due to lack of

situation awareness. This cascading loss of lines continued while little load was being shed or lost.

- At 16:05:57 the Sammis-Star 345-kV line tripped by Sammis end zone 3 relay operating on real and reactive current overload, and depressed voltage. This is deemed to be the critical event leading to widespread cascading in Ohio and beyond.
- Tripping of many additional lines in Ohio and Michigan by zone 3 relays or zone 2 relays set similar to zone 3 relays followed. Prior to the Sammis-Star tripping, the blackout could have been prevented by load shedding in northeast Ohio.
- The power transfer between the U.S. and Canada on the Michigan border shifted at approximately 16:10, due to the cascading loss of major tie lines in Ohio and Michigan. This means that power started flowing counterclockwise from Pennsylvania, through New York and then Ontario and finally into Michigan and Ohio. This huge (3700 MW) reverse power flow was to serve load in the Michigan and Ohio system, which was at this stage severed from all other systems except Ontario.

Voltage collapsed due to extremely heavily loaded transmission and a cascading outage of several hundred lines and generators ensued culminating in a blackout of the entire region.

### *The September 23, 2003 Blackout in Southern Sweden and Eastern Denmark*

During this blackout, a total of 4700 MW of load was lost in Sweden affecting 1.6 million people and 1850 MW in Denmark affecting 2.4 million people. It is customary during this time of the year to conduct a significant amount of annual maintenance work before the winter peak load period. Prior to the blackout two 400-kV lines and three HVDC links connecting the Nordel system with Poland and Germany were out of service for scheduled annual maintenance. The system was not stressed because the load levels were moderate.

The first major event was the loss of a 1200-MW nuclear unit at Oskarshamn in southern Sweden at 12:30 due to internal valve problems in the feedwater circuits. This increased power transfer from the north. System security was still acceptable after this contingency. Loss of a single 1200 MW unit on this system is a routine event and is regarded as a standard contingency. Real and reactive reserves and available transmission capacity are available to cope with this level of disturbance severity. These conditions were well under control and the system is able to handle this outage without any immediate serious consequences. Following a normal transient in frequency and automatic activation of the instantaneous spinning reserves the system returned to stable operating conditions within less than a minute. Voltages in the southern part had dropped around 5 kV, but remained within 405-409 kV, which is not critical. Frequency was automatically stabilized slightly below the normal operating limit of 49.90 Hz. Actions were therefore initiated to raise the frequency. The power flow was redistributed in the grid due to the loss of generation on the southeastern side. More power was flowing on the western side to supply the demand in the south.

Around 12:35 an unusual thermal overload overheated one of the mechanical joints of an isolator (disconnecting device) located in a 400 kV substation on the western coast of Sweden. Two 900 MW units in the nuclear power station of Ringhals normally supply this substation over two radial lines, connected to separate busbars. The busbars are connected to each other through interconnector bays, equipped with circuit breakers that are designed to sectionalize and split the substation in order to contain a fault on one busbar and leave the other intact with its connected lines in service. With the joint broken, the vertical structure of the isolator collapsed and it fell to the side in the direction of the other parallel busbar. Falling down, the contact with its own busbar opened with the load current still flowing. This created an arc initially between the busbar contact and the isolator parts. Eventually the arc flashed over to the nearest phase of the adjacent busbar. Two different phases from the two busbars were short-circuited through the arc. This fault was directly detected by the separate busbar protection devices, immediately tripping the circuit-breakers for all incoming lines to both busbars.

The disconnection of the busbars caused the two nuclear units to trip and the loss of a number of lines. The grid then lost its transmission capacity along the west coast. Initially this triggered heavy power oscillations in the system, very low voltages and a further drop in frequency down to a level slightly over 49.00 where under-frequency load-shedding schemes start to operate.

The grid was then heavily overloaded on the remaining southeast and south-central parts in terms of capability to sustain the voltages. This part of the grid had no major generation connected and thus the reactive power support was weak.

Within 90 seconds after the busbar fault the oscillations subsided and the system appeared to stabilize. Meanwhile the demand in the area recovered gradually from the initial reduction following the voltage drop by action of the numerous feeder transformer tap-changers. This lowered the voltage further on the 400 kV grid down to critical levels. Finally the situation developed into a voltage collapse in a section of the grid southwest of the area around Stockholm the capital of Sweden.

When the voltage collapsed to very low levels, circuit breakers in the critical grid section were tripped by distance protection action. Thus the grid split up into two parts. The southern part, consisting of Sweden and eastern Denmark, initially remained interconnected but sustained a considerable generation shortage. Within seconds, the frequency and voltage had dropped to levels where generator and other grid protection systems reacted and this entire subsystem collapsed.

### *The September 28, 2003 Italian Blackout*

On Sunday, September 28, 2003, the Italian Power System experienced a blackout that affected an area with close to 60 million people and load variation in the continental grid from about 24 GW in the early morning hours increasing to 50 GW during the day. The event that caused the blackout began at 3:01.00 a.m. and proceeded up to 3:28.00 a.m.

The security and reliability standards of the operation of the European interconnected electricity system are set by the Union for the Coordination of Electricity Transmission (UCTE), which is the association of Transmission System Operators (TSOs) in continental Europe. The main function of the European interconnected systems is to assure the security of supply at a reasonable cost. The system has been developed to assure mutual assistance between national subsystems including common use of reserve capacities and to optimize the use of energy resources by allowing exchanges between these systems. Recent market development with its high level of cross-border exchanges was out of the scope of the original system design. It has led the TSOs to operate the system close to its limits as allowed by the security criteria. The cross border exchange was improved recently by using several measures such as: computerized control and data acquisition, phase shifting transformers, coordination mechanisms and electronic data exchange among operators. The appropriate countermeasure for the loss of interconnection lines is the shutting down of the pumps, in the Italian pumped storage hydro plants. The pumping load in Italy is about 3500 MW. Another countermeasure is actual automatic load shedding.

The total power imported into Italy at 03:00 was 6.65 GW mainly generated in France, Germany and Poland with a highly loaded transmission grid in the proximity of Italy's northern borders. Before the blackout, because of the nightly energy trading requirements, the Swiss grid was also in a highly stressed condition, operated very close to the $N$-1 security limit. The Italian system had a large amount of active power reserve, greater than the power imported. Moreover, 1.2 GW of "without notice" interruptible load was available within 1 to 2 minute delay.

The most significant dynamic phenomena leading to the separation of Italy from the UCTE happened from 03:25:21 up to 03:26:39 and their description requires the subdivision of that time interval (lasting 78 s) into three stages (A, B, and C).

## Cascading Line Tripping Begins

The Swiss 380 kV line (Mettlen-Lavorgo) was loaded at approximately 86% of its maximum capacity. Due to tree contact caused by increased conductor sag a single phase to ground fault occurred at 03:01:21. Because the power imported into Italy was high, the phase angle difference across the line was too large which caused the automatic breaker controls to prevent reclosing. At 03:08 the operators attempt to manually restore this line into service failed again because of a too high phase angle (42°), caused by the high power flow to Italy.

Since the connection was not restored, this caused an overload (at around 110% of its nominal capacity) on a parallel path on the other Swiss 380 kV line (Sils-Soazza.)

To reduce the load on this line the following measures were taken:

- At 03:21, the Swiss and Italian operators decided to reduce the Italian imported power by about 300 MW.
- One 380/220 kV transformer in Soazza was switched off at 03:18.
- At 03:22:02 the transformer tap of 220 kV in Lavorgo was changed which slightly reduced the transformer load in Soazza.
- After 24 minutes this line also tripped at 03:25:21 due to tree contact which interrupted 1783 MW following a single phase to ground fault.
- Immediately after the second line tripping an internal 220kV line in Switzerland took over a significant part of the power flow and was highly overloaded. After 4 seconds at 03:25:25 the protection triggered at Airolo, interrupting a 740 MW power flow to Italy (Ponte).
- Immediately after the third line tripped, the nearest (West side) interconnection line 220kV between Riddes in Switzerland and Valpelline in Italy which was already overloaded, assumed an additional significant portion of the power flow toward Italy and subsequently tripped at 03:25:27.
- Simultaneously the other line from Riddes to Avise and Valpelline overloaded and tripped, thus interrupting about 600 MW of power flow to Italy. At the same time the 220kV line Cislago-Sondrio overloaded and tripped interrupting half of the total power flow from Sils to Italy of about 260 MW.
- The tripping of this fifth line concluded Stage A which involved a total power flow interruption from Switzerland to Italy of 2523 MW. While the frequency recovered to nominal value, the voltages in some areas near Turin became very low (350kV).

## Further Cascading Line Tripping due to Voltage Instability

The electrical link between the Italian and UCTE grids became weaker and relied mainly on the interconnection with France. Most of the remaining lines came very close to their safe limits. It is noted that during Stage A the voltage decreases stop at each frequency recovery, whereas during Stage B the voltage continued to decrease despite the frequency recovery to 50 Hz. This voltage drop increased the reactive power import from France and lowered the voltages at the border with Italy. Meanwhile, at the border with Slovenia, to the east of Italy, the voltages dropped but the imported reactive power from Croatia and Hungary partially moderated the progressive voltage reduction. The northwest Italian grid was in a very critical situation due to wide area very low voltages which caused the following events:

- At 03:25:30 a distance relay at Rondissone in Italy operated in 0.26 s opening the 380 kV France-Italy line number 1 from Albertville interrupting the flow of 841MW from France. The probable cause is low voltage. This event is deemed to have initiated Stage B involving voltage instability.
- At 03:25:32 a local distance relay opened the 400 kV La Coche-Pranz line in France probably due to lower voltage values, which continued to decrease under the voltage instability in progress. This was the seventh line tripping in this cascade.

The Italian voltage instability also affected all neighboring countries which progressively increased the reactive power transferred to Italy and reduced the voltages at the border busses.

### From Voltage Instability to Angle Instability

During Stage B the frequency recovered to nominal value but most of the voltages in north Italy were dropping. During Stage B the Milan area was apparently not affected by the incoming problems mainly seen by the northwest area and northeast edge with Slovenia which were practically lost at 3:25:32. This may have been prevented by timely local voltage based load shedding, which was not part of the Italian security plan. The voltage instability would naturally have led to the loss of synchronism of the Italian generators.

- The current on the remaining interconnection lines with France increased and the 400 kV Albertville-Rondissone line 2 tripped at 03:25:33 interrupting 682 MW import from France. This is the eighth line interruption.
- The reduced voltage value and the increased current caused the local distance relay operation with the opening of the 220 kV line from Broccarros (France) to Camporosso (Italy) at 03:25:33 interrupting 248 MW from France.
- The tenth line tripping took place at 03:25:34 between Linz (Austria) and Soverzene (Italy). The disconnecting device at Linz opened the interconnection of 220kV line, interrupting 209 MW, ahead of the fast and heavy frequency transient and low voltages at the connection with Italy.

- The loss of synchronism of the Italian system with respect to UCTE grid started Stage C, characterized by frequency decay (slope of between 3 and 1 Hz per second) and the voltages step rising at either the northwest or northeast of Italy as well as in the surrounding countries. At 03:25:34 the remaining operating interconnections between Italy and Europe were delivering a total of about 1650 MW, which was not enough to meet the required import of 6600 MW.
- The sudden lack of active power feeding the Italian loads initiated an electromechanical transient characterized by angle instability and frequency collapse. This angle instability resulted in additional cascade of lines tripping which worsened the situation. The Italian protection system began to operate based on frequency and its derivative thresholds at 03:25:33/34. This tripped the pumping load amounting to 3500 MW. This impacted on the frequency transient with a positive overshoot and a reduction of the frequency decay slope. The UCTE system reacted in the opposite direction with a fast frequency growth, therefore confirming the separation at that time.
- The persistent instability caused electromechanical oscillations between northeast Italy and eastern Europe which lasted from 03:25:34 up to the total opening of the tie with Slovenia, reached at 03:26:39 signaling the end of Stage C.
- During Stage C a number of Italian power plants and lines tripped including the remaining interconnections with UCTE. Simultaneously the load shedding protection operated on the Italian North system. At 03:25:35 two and two 220 kV lines tripped and at 03:25:42 a 220 kV line tripped due to overload.
- These trips increased the frequency of the electromechanical oscillation until, when, at 03:26:39 the tie with Slovenia was definitely open with the consequent extinction of the oscillation.
- During Stage C the persistent electromechanical oscillation significantly seen by northeast power system, caused most of the unit trips before 03:26:39 for a total amount of 3055 MW. Moreover the power plants operating at the distribution grid were lost at about 03:25:38, for a total of 1700 MW.

### From the Total Separation from UCTE at 03:26:39 up to the General Blackout at 03:28:00

This last stage follows the separation from Slovenia, at the time the angle instability disappeared either in Italy or in the surrounding countries. The voltages were high everywhere in Italy and the frequency was low but greater than 48 Hz. Stage D process is characterized by a slow decrease in frequency, accompanied by further tripping of generators out of service, load shedding determined by reaching low frequency thresholds and lines opening. At 03:26:39 all available active power contributions forced by the speed regulators were already reached and probably at that time active power control loops at some power plants were beginning to recover the extra power produced by the primary frequency regulation. Also this active power recovery has to be considered as a cause for further frequency decay. In conclusion, the total unbalance between generated power and load was:

*Loss of Generated Power*

- 6664 MW due to the separation from UCTE;
- 1700 MW due to generators on distribution grid;
- 4000 MW loss of 25 out of 50 main generators.

Loss of Loads

- 3500 MW automatic pumping shedding;
- 7000 MW load shedding by EAC;
- 1400 MW contribution of primary frequency control.

The voltages of all the Italian busses went down at 03:28:00 and all the Italian loads lost their supply as soon as the frequency reached 47.5 Hz. Thus, over the course of several minutes, the entire Italian system collapsed causing a nationwide blackout. This was the worst blackout in the history of the nation.

## *9.4 MITIGATING AND COPING*

It has been suggested that blackouts cannot be avoided and that future efforts should be directed at coping with, and minimizing the effects of any failures, and particularly cascading failures. Some people believe, however, that we currently possess the technical know-how necessary to upgrade the electric power grid to meet higher mitigation and avoidance standards than is currently available. It is important to note that the two visions complement each other because it is impractical to maintain power system reliability and security at 100%. Of course, there must always be a balance between improved system security and increased capital investment. The discussion of the present section is based on recommendations of the joint U.S. and Canadian Task Force[ii] investigating the August 14, 2003 event and the recommendations of an international panel of experts based on their experience with recent blackouts and the lessons learned[iii]. A common observation is that large disturbances often stem from a sequence of interrelated events that would otherwise be manageable if they have taken place individually. The final report of the U.S.-Canada Power System Outage Task Force cited the following as some root causes of the blackout of August 2003.

### Inadequate Understanding of the System

A significant finding was that prior to the August 2003 events, the Electric Utility did not conduct necessary long-term planning studies and did not conduct sufficient voltage stability analyses of its control area. The situation was exacerbated by employing operational voltage criteria that did not reflect actual system behavior and needs. Moreover, there was no subsequent independent review of the original analysis. In addition, some NERC planning standards were interpreted in ways that reduced the level of system operational reliability.

*Inadequate Level of Situational Awareness*

The Electric Utility failed to ensure security of its system after significant unforeseen contingencies due to the failure of various EMS system components.

*Inadequate Level of Vegetation Management (Tree Trimming)*

The Electric Utility failed to adequately manage tree growth in its transmission rights-of-way, which caused outages to take place on three 345-kV lines and one 138-kV line.

*Inadequate Level of Support from the Reliability Coordinator (RC)*

Failure of the state-estimator at the Independent System Operator and other real-time data deficiencies, caused the ISO not to become aware of the Electric Utility's system problems early enough and as a result was not in a position to offer assistance to the Electric Utility. In addition, neighboring systems did not have adequate levels of procedures and guidelines on how and when to coordinate a security limit violation observed by one of them in an external area, due to a contingency near their common boundary.

Based on the preceding observations, recommendations to mitigate blackouts were proposed in many reports. The recommendations can be categorized as follows:

- Policy level
- Planning level
- Operational level
- Data management level

## Policy Level Recommendations

The following general regulatory policy level measures can assist in reducing the risk of blackouts by improving system reliability.

National and regional reliability oversight organizations with statutory authority to enforce planning and operating reliability standards and levying effective penalties for noncompliance with the standards should be set up. These organizations must be independent from the entities (such as utilities) that they oversee. The idea of a Power System Regulatory Commission (PSRC) suggested by B. F. Wollenberg[iv] is modeled after the US Nuclear Regulatory Commission (NRC), whose mandate is to require strict adherence to the necessary safety and reliability standards in the US nuclear industry. The U.S. Energy Policy Act of 2005 designates the North American Electric Reliability Council (NERC) as a self-regulating electric power organization, subject to review by the Federal Energy Regulatory Commission, with the authority to enforce reliability rules and regulations. Mexican and Canadian authorities will back NERC's regulations with the force of law.

Industry and government funded research in power systems engineering to meet the challenges presented by the ever growing and complex power grids and their evolving loads should be promoted.

## Planning Level Recommendations

A general trend in all recent blackouts is the increased failure of aging equipment which may be driven by changing priorities for expenditures on maintenance and reinforcement of the transmission system. The following are planning level recommendations to improve the reliability of the system:

- Review and develop enforceable standards for transmission line ratings.
- Promote the requirement for investing in bulk system reliability enhancements (including the implementation of new technologies) and design transmission rates that allow investors to recover reliability-enhancement expenditures.
- Improve training and certification requirements for operators, reliability coordinators, and operator support staff by offering regular refresher training and live drills on system restoration to ensure that they remain familiar with best procedures and practices.
- Periodically review and reevaluate existing reliability standards and their development process. The standards should evolve continually to incorporate state-of-the-art practices and technological developments, taking into account experiences from major system incidents.

## Operational Level Recommendations

Taking immediate and remedial action against system events to prevent cascading requires reliable real-time data in sufficient time to allow taking decisive and appropriate remedial action by utilizing properly automated and coordinated controls. The following requirements need to be met:

- Mandate regular inspection, testing, maintenance, refurbishing, and replacement of existing substations and critical components are vital to successful mitigation and minimization of cascading events. This is motivated by the fact that cascading failures often result from equipment failure or poor coordination.

- Evaluate and adopt better real-time tools for operators and reliability coordinators.
- Maintain a relatively flat voltage profile to enhance voltage stability and hence ensure system reliability. A poor voltage profile causes generators to operate near limits with reduced reactive power reserves for contingencies, contributes to transmission lines sagging into trees due to increased line heating and increases transmission reactive power losses. Proper reactive power and voltage management require proper deployment of power plant controls to tightly regulate transmission side voltage, using under-voltage load shedding schemes and shunt reactive compensation to protect against severe disturbances.
- For generators, require the application of automatic control mechanisms such as automatic voltage regulators, and where applicable power system stabilizers. A modern excitation system automatically returns to voltage control when conditions permit. This suggests the need for prioritized replacement of old control and protection with modern digital equipment.
- Employ automatic load shedding. Operators who initiate load shedding pursuant to approved guidelines should be shielded against liability or retaliation.
- Track implementation of recommended actions to improve reliability.

The use and enhancement of special protection systems can be quite effective at times in preventing cascading outages. In the past, undesirable zone 3 relay and other generation and transmission backup relay operations have contributed to major blackouts. Many suggestions were made to deal with this issue ranging from complete elimination of third zone protection, to requiring that no third zone shall operate under "extreme" loading conditions. Horowitz and Phadke[v] note that it is essential to find alternatives to the third zone at critical system locations where restrictive loadability limits would be harmful from the point of view of cascading failures, and not a whole-sale revision of third zone settings on all protection systems. They offer technical solutions which relieve some of the loadability restrictions such that at critical locations the added expenditure of installing such newer protection systems (available in computer based relays) may well be justified.

Rapid system restoration is extremely important in order to minimize the impact of a blackout on society. Thus, means should be put into place to measure and reduce restoration times.

## Data Management Level Recommendations

It is important to incorporate lessons learned from past experience into new procedures and use such lessons to develop new and improved technologies for system control and monitoring.

- Improve the calibration of recording instruments, especially in establishing time synchronization.
- Establish predefined data reporting requirements and standardized data formats.
- Establish predefined commercial logistics, such as confidentiality agreements.
- Ensure the redundancy and reliability of remote control and telecommunication devices.

The following recommendations are intended to facilitate a smoother and faster process to collect, manage, and disseminate information following a blackout to allow postmortem analysis and information dissemination regarding probable causes and corrective actions being taken.

- Automate the means of disturbance reporting.
- Develop wide area measurement systems (WAMS)[vi].
- Establish an infrastructure to support a centralized blackout investigation.
- Extend the collection of benchmark events and dynamic signatures to determine the range of normal system behavior and perform related studies to assist proper interpretation of observed system behavior.

## New Technology Adoption Recommendations

New technologies such as Distributed Generation (DG), Flexible AC Transmission Systems (FACTS) and High Voltage DC (HVDC) can offer assistance in minimizing the occurrence and impact of widespread blackouts.

### *Distributed generation*

Distributed generation (DG) units are interfaced by power electronic converters, which can support active and reactive power locally and even provide local black-start functions, if appropriate regulatory and market conditions permit it. DG is connected to the medium-voltage and low-voltage networks providing generation support where it is most needed, in case of higher voltage network failures. Line overloading at higher voltage levels can therefore be potentially relieved during the restoration phase. However, the limitations of the energy sources (wind, sun), the link to industrial process for co-generation and the inability to control frequency and voltage (depending on the type of generation technology) may often limit the contribution of the associated forms of distributed generation in highly perturbed situations. The potential for distributed resources to provide support at lower voltage levels needs to be recognized in the development of interconnection standards for distributed generation.

### *Flexible AC Transmission Systems*

Shunt FACTS devices such as static VAr compensators (SVCs) can improve voltage control particularly in regions where old generation assets are being retired leaving large load pockets with little to no dynamic reactive support in the immediate vicinity. Another family of static compensators (STATCOM) is based on voltage source converter technology. It presents additional benefits since once at its reactive limit a STATCOM is a constant current device, while an SVC tends is a constant impedance device. Fast automatic switching of large shunt capacitor banks can also improve voltage stability.

Series devices such as thyristor controlled and conventional series capacitors improve transient stability margins on long extra-high voltage transmission corridors. More traditional devices such as phase-shifting transformers can also often be applied for controlling power flow on parallel paths.

Technologies such as Unified Power Flow Controllers (UPFC) can also control power flow on parallel transmission corridors.

### *High Voltage DC*

HVDC voltage source converters (VSC) allow independent controllability of active and reactive power and this type of HVDC system can black-start an islanded region of the system and can be applied in very weak systems. Due to the controllability of power flow, an HVDC system will not be overloaded in an emergency system condition, which significantly reduces the risk of cascading outages. Reactive power control, available with VSC, also improves voltage regulation and control and thus improving system stability by reducing the risk of voltage collapse.

---

[i] Brayley, H., Redfern, M. A., and Boo, Z. Q., "The public perception of power blackouts,", in Proc. PES Asia and Pacific Transmission and Distribution Conf., 2005, pp. 1-5

[ii] U.S.-Canada Power System Outage Task Force. (2004) Final Report on the August 14, 2003 Blackout in the United States and Canada: Causes and Recommendations. [Online]. Available: http://www.nerc.com.

[iii] Andersson, G.; Donalek, P.; Farmer, R.; Hatziargyriou, N.; Kamwa, I.; Kundur, P.; Martins, N.; Paserba, J.; Pourbeik, P.; Sanchez-Gasca, J.; Schulz, R.; Stankovic, A.; Taylor, C.; Vittal, V, "Causes of the 2003 major grid blackouts in North America and Europe, and recommended means to improve system dynamic performance," Power Systems, IEEE Transactions on, Volume 20, Issue 4, Nov. 2005 pp. 1922-1928.

[iv] Wollenberg, B. F., "From Blackout to Blackout: 1965 to 2003: how far have we come with reliability?" IEEE Power and Energy Magazine, pp. 88, January/February 2004.

[v] S.H. Horowitz and A.G. Phadke, "Third zone revisited," *IEEE Trans. Power Syst. Delivery*, pp. 23–29, Jan. 2006.

[vi] J. F. Hauer, N. B. Bhatt, K. Shah, and S. Kolluri, "Performance of WAMS East in providing dynamic information for the North East blackout of August 14, 2003," in Proc. IEEE PES General Meeting, Denver, CO, 2004, pp. 1685-1690.

# REFERENCES

Anderson, P.M. *Analysis of Faulted Power Systems*. New York: IEEE Press, 1973.

Anderson, P.M. and Fouad. *A.A Power System Control and Stability*. Ames, Iowa: The Iowa State University Press, 1977.

Andersson, G.; Donalek, P.; Farmer, R.; Hatziargyriou, N.; Kamwa, I.; Kundur, P.; Martins, N.; Paserba, J.; Pourbeik, P.; Sanchez-Gasca, J.; Schulz, R.; Stankovic, A.; Taylor, C.; Vittal, V. "Causes of the 2003 major grid blackouts in North America and Europe, and recommended means to improve system dynamic performance," *Power Systems, IEEE Transactions on*, Volume 20, Issue 4, Nov. 2005 Page(s): 1922-1928.

Arrillaga, J., Arnold, C.P., and Harker, B.J. *Computer Modeling of Electrical Power Systems*. New York: John Wiley & Sons, Inc., 1986.

Bergen, AR. *Power Systems Analysis*. Englewood Cliffs, New Jersey: Prentice-Hall, 1970.

Bergseth, F.R. and Venkata, S.S. *Introduction to Electric Energy Devices*. Englewood Cliffs, New Jersey: Prentice-Hall, 1987.

Blackburn, J.L. *Protective Relaying*. New York: Dekker, 1987.

Blackburn, J. L.. *Applied Protective Relaying*. Coral Springs, FL, Westinghouse Electric Corporation, 1979.

Blackwell, W. A and Grigsby, L. L. *Introductory Network Theory*. Boston: PWS, 1985.

Clarke, E. *Circuit Analysis of AC Power Systems*. New York: John Wiley & Sons, Inc., 1958.

Concordia, C. *Synchronous Machine - Theory and Performance*. New York: John Wiley & Sons, Inc., 1951.

Crary, S. *Power System Stability, Vol. II*. New York: John Wiley & Sons, Inc., 1955.

Elgered, O. J. *Electric Energy Systems Theory, Second Edition*. New York: McGraw-Hill, 1982.

El-Abiad, A.H. Digital Calculation of Line-to-Ground Short-Circuits by Matrix Method. *AIEE Trans*., 79, 323-332, 1960.

El-Hawary, M.E. *Electrical Power Systems Design and Analysis*. New York: IEEE Press, 1996.

EPRI. *Transmission Line Reference Book, 345 kV and Above*. Palo Alto, California: Electric Power Research Institute, 1982.

Feinberg, R. *Modern Power Transformer Practice*. New York: John Wiley & Sons, Inc., 1979.

Fink, D.G. and Beaty, H.W. *Standard Handbook for Electrical Engineers*. New York: McGraw-Hill, 1987.

Fitzgerald, A.E., Higginbotham, D.E., and Grabel, A. *Basic Electrical Engineering*. New York: McGraw-Hill, 1981.

Fitzgerald, A.E., Kingsley, C., and Umans, S. *Electric Machinery (Fourth Edition)*. New York: McGraw-Hill, 1982.

Fortescue, C.L. *Method of Symmetrical Components Applied to the Solution of Polyphase Networks*. AIEE Trans., 37,1027-1140, 1918.

Fouad, A.A. and Vijay, V. *Power System Transient Stability Analysis Using the Transient Energy Function Method*. Englewood Cliffs, New Jersey: Prentice-Hall, 1992.

Franklin, A.C. and Franklin, D.P. *The J & P Transformer Book (11th Ed.)*. London: Butterworths, 1983.

General Electric Company. *Electric Utility Systems and Practices (4th Ed)*. New York: John Wiley & Sons, Inc., 1983.

General Electric Company. *Transmission Line Reference Book - 345 kV and Above (2nd Ed)*. Palo Alto, California: Electric Power Research Institute, 1982.

Glover, J.D. *Power System Analysis and Design (Second Edition)*. Boston: PWS Publishing Company, 1994.

Gönen, T. *Electric Power Distribution Systems Engineering*. New York: McGraw-Hill, 1986.

Gross, C.A. *Power System Analysis (Second Edition)*. New York: John Wiley & Sons, 1983.

Guile, A.E. and Paterson, W. *Electrical Power Systems*. Edinburgh: Oliver & Boyd, 1969.

Gungor, B.R. *Power Systems*. New York: Harcourt, Brace, Jovanovich, Inc., 1988.

Hauer, J. F., Bhatt, N. B. Shah, K. and Kolluri, S. "Performance of WAMS East in providing dynamic information for the North East blackout of August 14, 2003," in *Proc. IEEE PES General Meeting*, Denver, CO, 2004.

Hayt, W.H. Jr. and Kemmerly, J.E. *Engineering Circuit Analysis (3rd Ed)*. New York: McGraw-Hill, 1978.

Heydt, G.T. *Computer Analysis Methods for Power Systems*. New York: MacMillan Publishing Company, 1986.

Horowitz, S.H. and A.G. Phadke, "Third zone revisited," *IEEE Trans. Power Syst. Delivery*, pp. 23–29, Jan. 2006.

Kimbark, E.W. *Power System Stability, Vol. 1: Elements of Stability Calculations*. New York: John Wiley & Sons, Inc., 1948.

Kimbark, E.W. *Power System Stability, Vol. 2: Power Circuit Breakers and Protective Relays*. New York: John Wiley & Sons, Inc., 1950.

Kimbark, E.W. *Power System Stability, Vol. 3: Synchronous Machines*. New York: John Wiley & Sons, Inc., 1956.

Kirchmayer, L.K. *Economic Operation of Power Systems*. New York: John Wiley & Sons, Inc., 1958.

Kirchmayer, L.K. *Economic Control of Interconnected Systems*. New York: John Wiley & Sons, 1959.

Knable, A. *Electrical Power Systems Engineering*. New York: McGraw-Hill, 1967.

Kundur, P. *Power System Stability and Control*. New York: McGraw-Hill, 1994.

Kusic, G.L. *Computer Aided Power System Analysis*. Englewood Cliffs, New Jersey: Prentice-Hall, 1986.

Nasar, S.A. *Electric Energy Conversion and Transmission*. New York: MacMillan Publishing Company, 1985.

Neuenswander, J.R. *Modern Power Systems*. Scranton, Pennsylvania: International Textbook Company, 1971.

Phadke, AG. and Thorpe, J.S. *Computer Relaying for Power Systems*. New York: John Wiley & Sons, Inc., 1988.

Rustebakke, H.M. *Electric Utility Systems and Practices*. New York: John Wiley & Sons, Inc., 1983.

Sarma, M.S. *Electric Machines*. Dubuque, Iowa: Brown, 1985.

Sauer, P.W. and Pai, M.A. *Power System Dynamics and Stability*. Englewood Cliffs, New Jersey: Prentice-Hall, 1998.

Singh, L.P. *Advanced Power System Analysis and Design*. New York: Halsted Press, 1983.

Stagg, G.W. and El-Abiad, A.H. *Computer Methods in Power System Analysis*. New York: McGraw-Hill Book Company, 1968.

Stevenson, W.D. Jr. *Elements of Power System Analysis, (4th Ed.)*. New York: McGraw-Hill, 1982.

Stevenson, W.D. and Grainger, 1.J. *Power System Analysis*. New York: McGraw-Hill, 1994.

Stott, B. Decoupled Newton Load Flow. *IEEE Trans. Power Apparatus and Systems*, PAS-91, 1955-1959, October 1972.

Stott, B. and Alsac, O. Fast Decoupled Load Flow. *IEEE Trans. Power Apparatus and Systems*, PAS-93, 859-869, May-June 1974.

Sullivan, R. *Power System Planning*. New York: McGraw-Hill, 1977.

Taylor, C.W. *Power System Voltage Stability*. New York: McGraw-Hill, 1977.

U.S.-Canada Power System Outage Task Force. (2004) Final Report on the August 14, 2003 Blackout in the United States and Canada: Causes and Recommendations. [Online]. Available: http://www.nerc.com.

Wadhwa, C.L. *Electrical Power Systems*. New York: John Wiley & Sons, Inc., 1983.

Wagner, C.F. and Evens, R.D. *Symmetrical Components*. New York: McGraw-Hill Book Company, 1933.

Wallach, Y. *Calculations and Programs for Power System Networks*. Englewood Cliffs, New Jersey: Prentice-Hall, 1986.

Weedy, B.M. *Electric Power Systems, (Third Edition)*. New York: John Wiley & Sons, Inc., 1979.

Weeks, W.L. *Transmission and Distribution of Electrical Energy*. New York: Harper & Row Publishers, 1981.

Westinghouse Electric Corporation. *Applied Protective Relaying*. Newark, New Jersey: Westinghouse, 1976.

Westinghouse Electric Corporation. *Electric Transmission and Distribution Reference Book*. Pittsburgh, Pennsylvania: Westinghouse, 1964.

Wollenberg, B. F., "From Blackout to Blackout: 1965 to 2003: how far have we come with reliability?" *IEEE Power and Energy Magazine*, pp. 88, January/February 2004.

Wood, AJ. and Wollenberg, B.F. *Power Generation Operation and Control*. New York: John Wiley & Sons, Inc., 1974.

Yamayee, Z.A. *Electromechanical Energy Devices and Power Systems*. New York: John Wiley & Sons, Inc., 1994.

Yu, Yao-nan. *Electric Power Systems Dynamics*. New York: Academic Press, 1983.

# INDEX

CPSIA information can be obtained
at www.ICGtesting.com
Printed in the USA
BVHW031942031022
648571BV00003B/8